Green Energy and Technology

Climate change, environmental impact and the limited natural resources urge scientific research and novel technical solutions. The monograph series Green Energy and Technology serves as a publishing platform for scientific and technological approaches to "green"—i.e. environmentally friendly and sustainable—technologies. While a focus lies on energy and power supply, it also covers "green" solutions in industrial engineering and engineering design. Green Energy and Technology addresses researchers, advanced students, technical consultants as well as decision makers in industries and politics. Hence, the level of presentation spans from instructional to highly technical.

Indexed in Scopus.

Indexed in Ei Compendex.

Allana Katiussya Silva Pereira ·
Ananias Francisco Dias Júnior
Editors

Impacts of Using Biomass as an Energy Source in Homes

Springer

Editors
Allana Katiussya Silva Pereira
"Luiz de Queiroz" College of Agriculture
University of São Paulo (ESALQ/USP)
Department of Forest Sciences
Piracicaba, São Paulo, Brazil

Ananias Francisco Dias Júnior
Federal University of Espírito Santo
(UFES), Department of Forest and Wood
Sciences
Jerônimo Monteiro, Espírito Santo, Brazil

ISSN 1865-3529 ISSN 1865-3537 (electronic)
Green Energy and Technology
ISBN 978-3-031-38823-1 ISBN 978-3-031-38824-8 (eBook)
https://doi.org/10.1007/978-3-031-38824-8

This Springer imprint is published by the registered company Springer Nature Switzerland AG
The registered company address is: Gewerbestrasse 11, 6330 Cham, Switzerland

Contents

About the Editors

Allana Katiussya Silva Pereira has been working with forest product technology, mainly in the area of wood waste, biomass and energy and technological uses and applications of products generated from the pyrolysis of forest biomass. Currently, she is PhD Student at the University of São Paulo and part of the Bioenergy and Forest-Based Bioproducts (BioEP) research group, developing research and extension projects aimed at valuing and reusing wood residues, knowledge and technology for application in biomass under the action of heat.

Ananias Francisco Dias Júnior currently works at the Forest Sciences and Wood Department, University of Espírito Santo, Brazil. He leads the research group Bioenergy and Forest-Based Bioproducts (BioEP), whose research and extension projects aim to develop knowledge and technology for the application and treatment of biomass under the action of heat, offering products and processes in an economically viable way and with attention to the social aspects. Together with his team, he develops innovative processes and products based on pyrolysis and heat, in order to value wood and charcoal as a source of energy, reduce gaseous emissions and atmospheric pollution in industrial and domestic areas, and mitigate the effects of climate change.

Energy Sources Used in Food Preparation and Impacts on Climate Change

Iara Nobre Carmona, Marina Passos de Souza, Elias Costa de Souza, Kamilla Cryslayne Alves da Silva, Allana Katiussya Silva Pereira, and Ananias Francisco Dias Júnior

Abstract One-third of the world's population does not have access to clean energy sources for food preparation. The use of inefficient technologies can increase the emission of potential greenhouse gases (GHG) and accelerate climate change. Recently, the increase in the price of Liquefied Petroleum Gas (LPG) caused by the COVID-19 pandemic made it possible to increase the use of firewood in the domestic environment, directly impacting people's health and the environment. In this context, the application of greener technologies can mitigate the harmful effects caused on people and nature. Thus, this chapter presents a description of the main primary sources of energy used in food around the world, specifically forest biomass and LPG, in addition to presenting the historical context that encompasses both energy sources. What were the main sources of energy used in the past? What has changed from the Neanderthal period to today? How was the use of Liquefied Petroleum Gas (LPG)

I. N. Carmona (✉) · E. C. de Souza · K. C. A. da Silva · A. K. S. Pereira
"Luiz de Queiroz" College of Agriculture, University of São Paulo (ESALQ/USP), Department of Forests Sciences, Av. Pádua Dias, 11, Piracicaba, São Paulo 13418-900, Brazil
e-mail: iaracarmona@usp.br

E. C. de Souza
e-mail: eliascosta@unifesspa.edu.br

K. C. A. da Silva
e-mail: kamilla.alves@usp.br

A. K. S. Pereira
e-mail: allana.florestal@gmail.com

M. P. de Souza · A. F. Dias Júnior
Federal University of Espírito Santo (UFES), Department of Forestry and Wood Sciences, Av. Governador Lindemberg, 316, Jerônimo Monteiro, Espírito Santo 29550-000, Brazil
e-mail: marinapassos58@gmail.com

A. F. Dias Júnior
e-mail: ananias.dias@ufes.br

E. C. de Souza
Federal University of South and Southeast Pará (UNIFESSPA), Institute of Xingu Studies, Subdivision Cidade nova, QD 15, sector 15, São Félix do Xingu, Pará 68380-000, Brazil

A. K. S. Pereira and A. F. Dias Júnior (eds.), *Impacts of Using Biomass as an Energy Source in Homes*, Green Energy and Technology,
https://doi.org/10.1007/978-3-031-38824-8_1

and how is it today? How was the world scenario of firewood use in domestic environments and how is it currently? And other biomasses, did they already have space? How did the COVID-19 pandemic affect the use of these energy sources (forestry biomass and LPG)? What impacts do these energy sources have on climate change? What do the COP27 discussions bring as prospects? These were the questions that guided this chapter.

Keywords Forest biomass · Energy sources · Climate changes · Fuels · Food cooking

1 Introduction

There are several sources of energy used in food preparation, from fossil fuels to renewable energy sources. Each of these sources has a different impact on climate change, either through the emission of greenhouse gases or the way they are produced. One of the most common sources of energy in food preparation is cooking gas, which is a fossil fuel derived from petroleum. The burning of cooking gas emits carbon dioxide (CO_2) into the atmosphere, contributing to the increase in the greenhouse effect and consequently to climate change.

In addition to cooking gas, electricity is also a widely used source of energy in food preparation. Most of the world's electricity is produced from fossil fuels such as coal, oil, and natural gas. The burning of these fuels emits large amounts of CO_2 into the atmosphere, increasing the greenhouse effect and contributing to climate change.

However, there are also renewable energy sources that can be used in food preparation, such as solar energy and wind energy. Solar energy can be used to heat water or to generate electricity through solar panels. Wind energy can be used to generate electricity through wind turbines. These renewable energy sources are cleaner and do not emit greenhouse gases into the atmosphere.

In summary, the energy sources used in food preparation significantly impact climate change. It is important to opt for renewable energy sources and choose sustainably produced food to reduce greenhouse gas emissions and mitigate the impacts of climate change.

2 Main Energy Sources Used in the World: Historical Context

What makes humans unique? Why and how have humans become different from other animals in how they relate to their environments and each other? The distinction of humans as a species has been the subject of study by philosophers and scientists for as long as we have historical records. Richard Wrangham, an accomplished anthropologist, has written books presenting a new theory about how and why the human lineage (a group of primates known as hominids) evolved characteristics that distinguish us from other animals. In the book "Catching Fire: How Cooking Made Us Human", [1] argues that the discovery of fire and cooking freed our human ancestors once and for all from an arboreal existence and led to a patriarchal social system and a division of work by gender. The main features that are the focus of this theory—technology, in the way fire is used and controlled—have long been heralded as features that make humans unique. His work builds on existing theories but goes a step further by incorporating new perspectives, trying to frame comprehensive and unified explanations for a set of human characteristics.

In "Catching Fire", [1] focuses on cooking as an innovation that allowed the evolution of big brains. Early members of the Neanderthal lineage that led to modern humans (species within the genus *Homo*) discovered fire's functions for cooking, warmth, and security, and eventually learned to manage and produce it [1]. With fire, humans no longer needed to sleep in trees as protection against predators and could transition to a completely terrestrial existence. With fire for cooking, many foods (such as tubers and meat) become softer, less toxic, and easier to digest. Humans have become physically and physiologically adapted to eating cooked food. Changes in the gastrointestinal tract allowed the evolution of early hominin bodies. This enabled them to allocate more resources to energy-intensive brain growth and maintenance. Reference [1] argues that adaptations found since *Homo erectus*—a species of modern human ancestor that appeared shortly after 2 million years ago—reflect this shift towards the use of cooked foods. Reference [1] also attributes to cooking the connection of male–female pairs (monogamy) and the gender division of labor in humans. He says that cooked food is a localized resource of high quality, and Neanderthals did it better. A male hominid was supposed to protect a female from aggressive attempts to steal her food. On the other hand, a male needed a female to cook for him because male hunting activities were time-consuming (Fig. 1) and, once hominids adapted to cooked food, it became impossible to efficiently process and digest raw food enough for survival [2].

Reference [1] also presents data from scientific research on nutrition and digestive processes, as well as more eclectic particulars of human survival on different diets (for example, the tribulations of raw food eaters), demonstrating that processed and cooked foods are easier to digest than raw food. Once the food is prepared and heated, energy (calories) can be more efficiently and completely extracted. Furthermore, there has been a substantial study on the anatomy of the gastrointestinal tract, linking the size and structure of certain organs to different types of [3]. Thus, the argument that consuming cooked food would allow for smaller gastrointestinal organs (which humans have relative to other animals of similar size) and provide more energy for the growth of a larger brain can be sustained. But there is still not enough evidence about fire control before the last few hundred thousand years to support Wrangham's timeline, which would allow certain hominid adaptations to be reliably credited to cooking.

But how to get energy? What has changed from the Neanderthal period to today? Over the centuries, the way to generate energy has evolved a lot. The use of food as a source of energy has always been natural to human beings, but acquiring this resource has not always been easy. And because of that, they merely served to fulfil basic survival requirements and needs. Thus, the first advance in energy generation

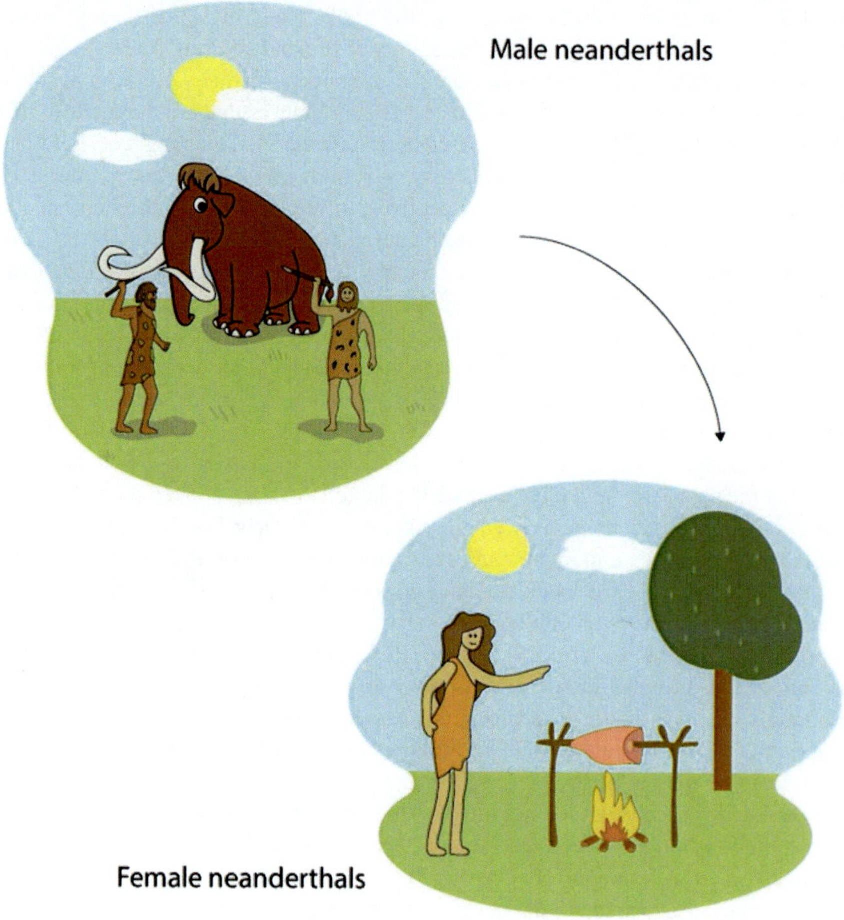

Fig. 1 Neanderthal division of labor according to sex (*Source* The authors 2023)

brought men a world of new possibilities. The discovery of fire made humanity aware of the benefits of heat, a resource that currently generally comes from the burning of fossil fuels, but which in the past, according to historical studies, happened through the use of firewood (biomass).

Around 500,000 years ago, during the Pleistocene, Europe's climate changed drastically, producing very different fauna and vegetation in glacial and interglacial periods [4]. More than 400,000 years ago, for example, in the interglacial periods, the exploitable plant and animal biomass was enormous, but it was mainly found in closed forests, dangerous and difficult to access. The Neanderthals (human groups at the time) exploited plant resources more than they did in the glacial periods, but practiced harvesting in clearings and, mainly, on the banks of rivers. In general, harvesting and hunting were enough for their survival and allowed them to grow, extend over a wider territory and even colonize a part of Asia.

Some archaeological evidence provides information about diets and landscape use. The use of fire, indicated by the presence of bonfires in many of the sites studied by archaeologists in the central and southeastern Iberian Mediterranean, was common and widespread in this period [3]. Furthermore, there is evidence of the use of fire in the Cueva Negra del Estrecho del Río Quipar spanning more than 780 km, with a rich paleontological and paleopalynological record demonstrating hot and humid environmental conditions, suggesting that the use of fire has a long history in this area.

As we have seen so far, the history of biomass and fire goes back to the roots of humanity. Biomass predates us. There is much evidence to support claims that we used biomass as an energy source between 230,000 and 1.5 million years ago (Cablevey [5]). Biomass is among the sources of renewable energy, that is, inexhaustible, whose production is part of the planet's carbon cycle, being a fuel used all over the world, and used to generate heat and cook food. From there, humans developed a fascination with what came to be known as bioenergy. Combustion was and remains the main way of converting biomass into energy. That is why biomass is still massively used in developing countries that do not have the bioenergy generation systems that more developed countries use to create alternative energy.

Humanity initially focused on using biomass for cooking and heating. Around the nineteenth century, we started to look for more modern uses of biomass materials. While fire is the oldest example of bioenergy in use, ethanol can be seen as the next big step in using carbon for energy. Ethanol has been around for a long time. Mankind discovered and used the fermentation process long before the development of civilizations. Despite this, there is no clear evidence of people distilling alcohol until twelfth century Italy [6]. Soon after people started making alcohol in the 1100s, ethanol was quickly used for cooking and lighting. People started using ethanol to create more energy. Ethanol was a very popular renewable energy source due to its simplicity and availability. It was derived from grains, which meant that the raw material was plentiful and all it took was a still to produce ethanol. It was until the substance, along with turpentine, was used to power the first engine in 1826. Ethanol fuel continued to be a popular form of fuel well into the 1890s.

As people began to explore the possibilities of ethanol as an alternative energy source, they also began to use vegetable and fish oil for heating and lighting [7]. Many civilizations used oils to generate heat and light. It is believed that even ancient cultures such as the Egyptians and Sumerians burned animal and vegetable oils. These oils were also used later in history. As populations grew, a new industry around lighting and heating emerged. People became more innovative and used the resources around them to produce light and energy. Examples include refined pine turpentine, alcohols (especially wood alcohol—methanol), and a mixture of the two, the main fuel before petroleum [8, 9].

Pine sap was a scarce renewable resource from the 1700s to the 1960s. Before oil, pine sap was a resource for which nations competed. In its raw form, pine sap was used in shipbuilding [10]. When distilled, the sap produced several chemicals that were extremely valuable at the time—the most important of which was turpentine. Turpentine has had multiple uses, but its most important use as an alternative energy source has been as lamp oil [11].

While it may seem like oil has been around forever, it was only refined and first used about 150 years ago [12]. Just like ethanol and turpentine, the energy demand has brought about a significant development in oil production. It's important to remember that oil wasn't useful until we figured out how to refine it into pieces. Various steam and internal combustion engines ran on a wide range of refined fuels. Rudolf Diesel beat them all when he created the diesel engine (Patricio Moreno [13].

An important milestone in the history of bioenergy was the beginning of the twentieth century when biomass fuel became popular again. Due to the "boom" of the automobile industry and wars, the automobile industry's lack of resources led to a return to bioenergy. Henry Ford is the most famous example of this, who switched to liquid biofuels and ethanol to power his vehicles. This became especially prevalent during World War I when fossil fuels became scarce [14]. The adversities and challenges of the time created a significant demand for ethanol.

The emerging renewable energy projects at the time had great potential. However, large-scale commercialization and mining have placed coal and oil at the forefront of the energy landscape [15]. They saturated the market and drove prices down. Furthermore, these non-renewable fuels have proven to be very efficient and practical for everyday use. This has brought a reduction in the use of bioenergy and an increase in fossil fuels. Fossil fuels became the fuel of choice in most countries and held the top position in energy consumption until the 1970s.

The geopolitical conflict that was felt in the 1970s created a fuel crisis [16]. As a result, the Organization of Petroleum Exporting Countries (OPEC) reduced oil exports. This caught the attention of governments and the academic world. Many began to look for other sources of renewable energy. This movement brought many improvements in green energy in solar panels, geothermal power plants, offshore wind farms, and hydroelectric power plants. During this period, scientists took a systematic approach to energy and coined the term biomass. Over time, the importance of bioenergy has been linked to issues such as fossil fuel pollution [17]. This period of biomass history is marked by some growing environmental concerns. Scientists have turned their attention to research on climate change and fossil fuel reduction.

Modern energy production from biomass is a vital source of renewable energy today. It has gone far beyond wind and solar energy in the quest for renewable energies. Biomass is the main source of renewable and alternative energy [18]. Biomass feedstock is processed and converted into energy in different ways. While the burning of woody biomass (forestry biomass materials, wood pellets, etc.) is an old process, innovation has brought us mass-produced energy crops converted to biofuel and biogas, and landfills that use anaerobic digestion to convert biomass into biogas for daily use.

Governments around the world have embraced the green movement and are implementing measures and protocols to raise awareness and ensure that much more green energy is produced. As we move forward, the field of biomass renewable energy technology is expected to grow. Biomass is expected to play a vital role in future energy-efficient power generation. Whether producing electricity, heat, or fuel for transport, its carbon neutrality hides many potentials. Renewable energy, whether in the form of solar, geothermal, hydroelectric, wind, or biomass, is here to stay. Biomass is an important source of renewable energy in the modern world. As biorefineries, processing plants, and businesses become more viable and popular energy solutions, they must have systems in place to ensure proper handling and production.

3 Use of Liquefied Petroleum Gas (LPG): An Overview

Liquefied Petroleum Gas (LPG), also known as autogas, is primarily composed of propane, butane, and isobutane in a variety of blends. The percentage of propane and butane in an LPG gas mixture ranges from 100% propane to 20% propane and 80% butane [19]. Small concentrations of other hydrocarbons may also be present. Depending on the source of the LPG and how it was produced, components other than hydrocarbons may also be present. It is a co-product of refining crude oil and processing natural gas. Its constituents are found in gaseous form at 20 °C and 1-atmosphere pressure (NTP) [20].

It is generally said that gaseous fuels emerged in the troubled times of World War II when gasoline shortages were common [15]. Interestingly and perhaps surprisingly, LPG was first used as a motor fuel long before the outbreak of war. The first mention of mixing propane and butane dates back to 1910 [21]. It was then that Walter O. Snelling, an American chemist who was researching the properties of gasoline, separated the gaseous fractions from the liquid ones, thus discovering the existence of propane. Two years later, in 1912, he started his first domestic propane installation, and in 1913 he patented its industrial-scale production. Later that year, the patent was purchased by Frank Philips, the founder of the oil company ConocoPhillips [22]. Even so, LPG consumption has not grown considerably.

Information on the practical use of LPG dates back to 1918 when the fuel was used for brazing lamps and metal cutting torches. However, commercial production did not begin until the 1920s. In 1928, LPG was used for the first time as an engine fuel (in a truck) and the first LPG refrigerator was manufactured [23]. In 1929, fuel sales reached 10 million gallons in the United States. LPG was rapidly gaining momentum. In the following years, the demand for Liquefied Petroleum Gas was further driven by the popularity of "airships", traveling regularly between Europe and the US [24].

The then state-of-the-art Zeppelin series dirigible balloons were powered by engines fed with the so-called Blau gas (invented by Herman Blau), very similar to butane—one of the ingredients of LPG. The use of gaseous fuel with approximately the same mass as the air was very convenient for balloons, as it did not change the total weight of a Zeppelin in the same way as liquid fuels (blimps would become considerably lighter when liquid fuels were burned, thus forcing the release of hydrogen, which was extremely dangerous) [24]. However, when the Hindenburg—the largest airship ever built—was destroyed in a disaster in 1937, killing 36 people, the era of the Zeppelin ended abruptly.

However, the LPG era did not end with the Zeppelin era. On the contrary, it flourished, because there were a large number of gas canisters left at the airfields from which airships operated. In the state of Rio de Janeiro, Brazil alone, 6,000 gas cylinders became useless, which led businessman Ernesto Igel to have the idea of buying them and promoting gas as an excellent fuel for cooking food. This is how the Brazilian company, later known as Ultragaz, emerged [25]. In 1939, the company had three distribution trucks and 166 customers [26]. Eleven years later, in 1950, there were more than 70,000 customers and today Ultragaz is one of the largest LPG operators in the world.

When World War II ended and industrial production resumed growth, LPG sales in the US surpassed 1 billion gallons [27]. Nearly 62% of all US homes had LPG installations at the time. In 1947, the first liquefied gas tanker was built and entered service. In 1950, the Chicago Transit Authority, a public transportation operator in Chicago, ordered 1000 buses powered by LPG, while in Milwaukee 270 cabs were converted in the same year [28]. In 1958, LPG sales reached 7 billion gallons, and in 1965 Chevrolet introduced 4 new LPG engines for commercial vehicles. Initial international export contracts were not made until the 1950s. However, the amount of LPG exported was still low in the 1960s—less than 1 million tons were shipped out of the US [29]. Over the next 20 years, exports grew to 17 million tons and reached 48 million tons in the year 2000.

Currently, LPG is recovered from "wet" natural gas (gas with condensable compounds of heavy petroleum) by absorption [30]. The recovered product has a low boiling point and must be distilled to remove lighter fractions. It must then be treated to remove hydrogen sulfide, carbon dioxide, and water. The finished product is transported by pipelines and by specially built sea-going tankers. LPG reaches the domestic consumer in cylinders with relatively low pressure. Most of the LPG produced is used in central heating systems (a system that provides heat for an entire building), and the second largest as raw material for chemical industries. LPG is commonly used as a fuel for gas grills and gas stoves and ovens, gas fireplaces, and portable heaters [28]. In Europe, LPG water heaters are common. It is also used as engine fuel and for backup generators. Unlike diesel, LPG can be stored almost indefinitely without degradation.

LPG is used for cooking in many countries for economic reasons, convenience, or because it is the preferred fuel source. In India, approximately 8.9 million tonnes of LPG were consumed in the domestic sector in the six months between April and September 2016, mainly for cooking [31]. The number of domestic connections is 215 million (that is, one connection for every six people) with a circulation of more than 350 million LPG cylinders [32]. Most of the need for LPG is imported. Piped gas supply in India is not yet developed on a large scale. LPG is subsidized by the Indian government for domestic users. A rise in LPG prices has been a politically sensitive issue in India as it potentially affects the voting pattern of the middle class.

Aside from electric, induction, or infrared stoves, LPG stoves are the only type of fuel available in most suburban villages and many public housing developments. In Brazil, LPG is the most common cooking fuel in urban areas, being used in virtually all households, except the cities of Rio de Janeiro and São Paulo, which have gas pipeline infrastructure [33]. Since 2001, needy families have received a government subsidy ("Vale Gás") used exclusively for the purchase of LPG [34], and, since 2003, this subsidy has been part of the government's main social welfare program ("Bolsa Família"). Also, since 2005, the national oil company Petrobras differentiates LPG destined for cooking from LPG destined for other uses, establishing a lower price for the former. This is the result of a directive from the Brazilian federal government, but its discontinuation is currently being debated.

Currently, commercially available LPG is mainly derived from fossil fuels. Burning LPG releases carbon dioxide, a greenhouse gas. The reaction also produces some carbon monoxide. LPG, however, releases less carbon dioxide (CO_2) per unit of energy than mineral coal or oil, but more than natural gas. It emits 81% of the CO2 per kWh produced by oil, 70% by mineral coal, and less than 50% by coal-generated electricity distributed by the grid [35]. Being a mixture of propane and butane, LPG emits less carbon per joule than butane, but more carbon per joule than propane.

4 Modern Biomass and Traditional Biomass: Which Ones Grew the Most Around the World?

Despite biomass being widely used as a source of energy over the years, its use has been modernized through studies that have allowed technological implementations. In addition to the various improvements in burning systems, what we call modern biomass is opposed to traditional biomass in aspects related to how it is obtained. The biomass that was traditionally used was obtained in an extractive way, that is, the natural forests were exploited without the slightest planning and indications about the management and reestablishment of the species in the place. These extractive practices still exist today, especially in more economically affected communities [36]. However, with the modernization of knowledge about silviculture and sustainable forest management, countries like Brazil have been standing out for their innovative role in the rational use of forest resources and their correct application in the area of bioenergy [37].

Energy forests, as plantations with fast-growing species, good biomass productivity per area, and selected clones to maximize energy production are called, are already known by several segments of the bioenergy sector [37]. In addition to helping to reuse unproductive or poorly productive soils, these plantations help to fix carbon dioxide (CO_2) and contribute to reducing the use of non-renewable energy sources. These fast-growing species, mainly of the *Eucalyptus* genus, stand out for the energy quality generated from the thermochemical transformation processes, being considered of high quality both for direct burning (such as firewood) and for the production of charcoal [18, 37].

These advances in the more modern use of biomass also extend to the area of sustainable forest management, where recent studies point to the high potential of using the residues of these practices to produce energy in an environmentally friendly way [38]. These residues demonstrate that, when the material classification is performed correctly, energy gains can be maximized, which improves the energy quality of the material and ensures its correct use [39]. These studies are important, especially for more remote regions or those with difficult access to other energy sources, as is the case in the North region of Brazil, where the energy source available in greater quantity is biomass [38, 40]. However, due to several exploratory activities in the area, the use of forest residues has been proven to be a sustainable and viable alternative to meet the energy needs of these populations [40].

In addition to the already mentioned advances in the quality and obtaining of biomass, some transformations are being developed, related to the modification of the biomass form, more specifically working on the process of densification of this material [41]. The densification process is related to the reduction in the volume of biomass and, consequently, an increase in its density and the amount of energy concentrated per volume, i.e., higher energy density [42]. This densification process encompasses both briquettes and pellets and allows for a significant improvement in the energy quality of the biomass, as waste with a smaller granulometry is agglutinated and transformed into larger materials, which provides benefits for its transport, storage, and further use [42, 43].

Figure 2 illustrates the process of transforming the use of biomass (i) leaving the traditional form of extractive exploration of the native forest, (ii) moving to the planting of fast-growing species for wood production, and (iii) the most recent updates of densification biomass, transformation into pellets and briquettes that can be used in burning equipment in homes. This modernization process has been taking place over the years, gradually, where, currently, there are still countries that make use of "traditional biomass", with little technology. However, there are already several countries that have innovated in their technological aspects aimed at the energy area and already make use of "modern biomass".

The number of studies evaluating these new applications of biomass, or new materials, such as agroforestry residues, for energy applications, has grown significantly in recent years [44]. The maintenance and expansion of these studies are important to guarantee a continuous supply of modern biomass to different countries of the world. From the characterization of the biomass and studies that evaluate in detail the most efficient transformation processes, from the energy point of view, it is possible to obtain a quality fuel that can be widely used in homes, reducing the consumption of wood obtained through the extractive consumption [18, 38]. The choice of biomass to be used is a factor of significant importance for energy planning in countries. Currently, with the trend of carbonization of economies and with the focus on renewable and less polluting sources, the choice for options derived from

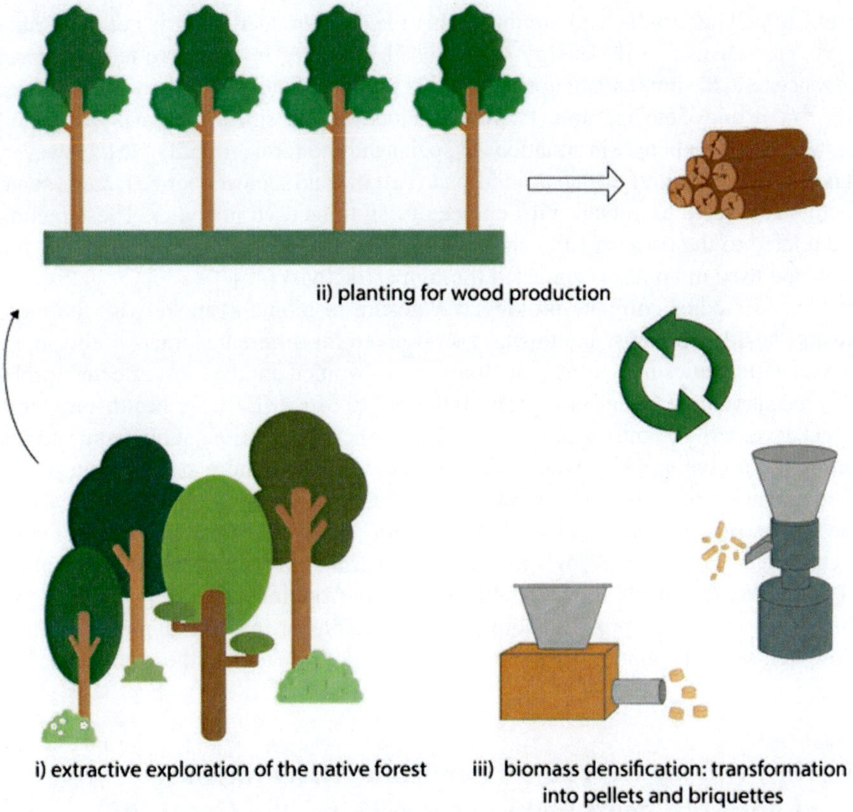

Fig. 2 The transition from "traditional" biomass to "modern" biomass (*Source* The authors 2023)

biomass becomes even more important, to the detriment of sources obtained from fossil fuels.

In addition, the insertion of these solid products produced from the densification of biomass in the market allows for an increase in the variability of renewable fuels available in markets in different countries of the world. With the correct study of lignocellulosic residues available in their respective countries or regions, technical feasibility studies can be carried out to assess the possibility of transforming these residues into solid fuels such as pellets or briquettes. Thus, two benefits are achieved directly: the first, which is the reuse of waste from a specific sector of that location; and the second is the availability of a cleaner energy source that contributes little to climate change, especially when compared to other non-renewable sources.

Even with these modernizations of biomass over the years, both in terms of fuel and in terms of burning equipment, in many parts of the world, Liquefied Petroleum Gas (LPG) has become the main energy source for cooking food in homes [45]. The impact generated on the air quality inside the homes that used firewood and started using stoves fuelled with LPG is undeniable, even knowing the environmental problems related to the exploitation of this fuel and the fact that it is not renewable [45]. This change in the energy source used in homes is related to technological advances and the implementation of different public policies in countries. In countries like Brazil, one of the important factors is socioeconomic, since, historically, firewood has been used by people in situations of social and economic fragility [45]. However, it is important to highlight that in addition to cultural and socioeconomic factors, some factors also seem to influence the choice of fuel to be used in homes. These factors are related to the price of LPG and proximity to the forest, given that much of the firewood used in homes is collected through extractivist [45].

Several studies correlate prolonged exposure to biomass smoke with the most diverse health problems, and this fact is observed for different extracts of the population, with studies indicating that children and women are the most affected, probably because they spend more time indoors [20, 46–48]. These health problems affect developing countries more significantly, where populations suffer from various economic problems [48]. Thus, it is necessary to invest more significantly in the modernization of equipment, as well as in strategies to modernize the biomass in these countries, avoiding the use of lower quality materials, and emphasizing investment in planting fast-growing species, to the detriment of unrestrained exploitation of its natural resources. Initiatives like the Clean Cooking Alliance [49] operate in different countries to provide more efficient equipment that uses biomass as fuel. However, several countries lack effective public policies to help these populations.

5 How Did the COVID-19 Pandemic Affect the Use of Forest Biomass and Liquefied Petroleum Gas (LPG) as an Energy Source?

The COVID-19 pandemic, decreed by the World Health Organization in March 2020, directly affected the lives of the entire world. Whether directly facing the health problems caused by the disease, including deaths, or the trail of economic destruction left by the impact of the paralysis of various productive sectors around the world [50]. These impacts on public health, economy, and culture of populations ended up generating several changes in habits that may, or may not, remain in these populations over the next few years. [51]. Among these changes, one of the main ones is linked to the daily lives of populations around the world: their diet.

In addition to wondering about the quantity and quality of food that will be eaten by their families, the people who are heads of these families are also concerned with issues related to the way their food is cooked, an activity present in different cultures around the world [52, 53]. The change in this traditional practice is directly related to the type of fuel used for cooking food. Due to financial needs, populations tend to migrate to cheaper and easily accessible energy sources. Avoiding, for example, dependence on the international price of oil, which can directly impact the price of LPG, and opting for available sources close to their homes, such as forest biomass.

This is a pattern that has been repeated over the years in several countries around the world, such as Brazil, a continental, diverse country with a high rate of social inequality.

Over the years, the consumption of the two main fuels used to cook food in Brazil (firewood and electricity) has varied according to the country's economic situation, which is a pattern throughout the world [54]. These economic factors are directly related to the quality, quantity and type of fuels used by communities to heat, light and, above all, cook in their homes [48]. Firewood, which was the main source of energy in homes, has been decreasing over the years, with LPG and electricity following opposite paths in this trend in Brazil.

From the 2000s onwards, with the implementation of public policies encouraging the use of electricity in homes and with the advancement of technologies, this change in the consumption behaviour of populations was even clearer, even with firewood showing a slight tendency to increase between the late 1990s to the end of the first decade of the 2000s. In recent years, with Brazil's political and financial crises, firewood consumption has grown again. This can be read as an indicator that this consumption may be related to the difficulty faced by families in purchasing LPG for cooking, even with government incentive programs, such as "Auxílio Gás" (gas aid), created by Law no. 14,237 of November 19, 2021 [55]. In addition, it is important to highlight that this data is formally collected and declared by consumers, that is, there is still much that is illegally exploited or consumed informally and not declared, given that a large part of the population that consumes firewood at home for financial reasons many times they do not buy firewood, but explore forests without quantitative control of actual consumption [45].

In addition to all the changes in the patterns of biomass consumption in homes over the years, due to different factors, the COVID-19 pandemic seems to have more significantly affected the substitution of fuels with a higher added value (LPG) for fuels with a lower added value (firewood). According to data from the National Energy Balance [54], firewood consumption in Brazilian homes increased from 7.2 to just over 7.44 million tons of oil equivalent (toe), between 2020 and 2021, an increase of 0 0.7% in the period, it is worth noting that the figures from before the pandemic (2019) showed an annual consumption of just over 7.08 million tons in households in the country. Meanwhile, the values that show the consumption of LPG, had a drop of 0.9% in the period between 2020 and 2021, going from 6.74 million toes to 6.52 million toes, with the pre-pandemic values, in 2019, in the house of 6.49 million toes.

This information is important to guide the decision-making of countries in a situation of economic weakness, as in addition to the economic factors of market fluctuations, exceptional events, such as the COVID-19 pandemic or the armed conflicts between Russia and Ukraine, end up further weakening the situation of these populations, which leads to an increase in the consumption of biomass for cooking food. Data from the National Energy Balance of Brazil demonstrate how this change in the pattern of energy consumption in homes was affected by the pandemic, however, we still do not have data on the impact of these changes in habits in the medium and long term [54]. Thus, it is important to follow up and monitor these families and carry out studies to assess the possible impact of this change on the economy and health of these people.

Often, because cooking is carried out in low-quality equipment, in addition to bringing health problems to its direct users, the smoke emitted during the combustion of biomass can significantly contribute to climate change and the decrease in air quality in different regions around the world [48]. However, it is also observed that the pandemic had a direct impact on the acceleration of the implementation of low-carbon energy sources in different countries, such as China, led mainly by the advance of solar energy and wind energy, which may have impacted the prospects of future carbon emissions in several countries [56]. Thus, the discussion of environmental issues at a global level becomes even more important, to highlight that, in addition to climate problems, health and food security problems are correlated in these situations.

6 Impacts of Forest Biomass and Liquefied Petroleum Gas (LPG) on Climate Change

Innumerable efforts are being made to use cooking fuels that are less harmful to the environment, to reduce indoor air pollution, improve air quality and reduce emissions into the atmosphere, as well as protect human health [57]. The assessment of impacts on climate change is mainly accounted for by the emission of greenhouse gases (GHG), mostly coming from fossil fuels. One of the fuel alternatives for cooking is Liquefied Petroleum Gas (LPG), which, despite being fossil, burns cleaner than other energy sources, such as mineral coal, and plays an important role in homes around the world, allowing more people to benefit from cleaner and more efficient energy, which 60% of the world's population already enjoys [30, 58].

Devices powered by LPG considerably reduce domestic air pollution when compared to rudimentary biomass burning, however, in many places there is still resistance to the use of these stoves, influenced by a combination of different socioeconomic, structural, and sociocultural factors, especially taste for food and preparation time. For these reasons, the implementation of the LPG, based stove in conjunction with the solid fuel stove can be an effective means of mitigating climate change and transacting the change in fuel type [28, 59, 60].

Another alternative to the use of GHG-emitting fuels is the use of forest biomass, which has a renewable character, in which carbon dioxide (CO_2) emissions can be reabsorbed in plant growth, through photosynthesis [61, 62]. Forest biomass has a high potential to be used in the transition from fossil fuels to bioenergy, however, its use faces some challenges related to the emission of GHG gases derived from fossil fuels. Mainly in the forest harvesting, processing, and transport processes and in the processing of biomass, which in addition to reducing the carbon sink area of forests, require inputs of materials and energy that emit GHGs [18, 62–64]. According to [65], it is necessary to adopt environmentally friendly forest management strategies to obtain forest products capable of impacting climate change mitigation. Active forest management, with high levels of harvest and efficient use of forest products, provides more climate benefits when in contrast to reducing the harvest and storing more carbon in the forest, acting in the mitigation of climate change [17].

It was generally assumed that CO_2 emissions derived from biomass had no influence on climate change, however, in the last decade, scientists have noticed that these emissions have a positive impact on climate change, with what is called the Global Warming Potential. Positive (GWPbio), during its stay in the atmosphere [2, 66, 67]. However, it is verified that the permanence of CO_2 emissions derived from biomass is shorter than those derived from fossil fuels, and is dependent on the biomass rotation time, due to the compensation effect of biomass growth to counterbalance the biogenic emission of CO_2, through photosynthesis [68–71].

If regeneration for compensation is accounted for to measure CO_2 emissions, it should be incorporated into Life Cycle Assessments (LCA—tool used to assess impacts [72]) of biomass use, as one of the positive impacts of bioenergy in mitigating climate change [73]. However, because the energy efficiency of bioenergy production based on forest biomass is low, it is necessary to have a greater amount of raw material available to produce this bioenergy, compared to the demand necessary to produce fossil fuels [62, 74]. Thus, if the GWPbio from forest biomass is included in the LCA, the impact of bioenergy on climate change can become twice as large as that of fossil fuels [75]. Despite this, studies such as the one carried out by [62], show that, although the balance of emissions generated by bioenergy (GHG derived from fossil fuels, biogenic CO_2 emissions, and regeneration for compensation) is positive and greater than the impacts of fossil fuels on climate change when integrating the negative impact generated by carbon sequestration and by stopping future carbon emissions from the equation, the final balance is reduced and much smaller than that of fossil fuels, ensuring that the use of biomass has total impacts favourable to the mitigation of climate change.

On the other hand, GHG emissions from biofuels present in their life cycle are still underestimated, since they exclude emissions due to land use or changes in land use practices. In this sense, the cultivation of energy crops and the collection of waste destined for bioenergy can increase unexpected GHG emissions, by altering the soil and, although the transition from agricultural land to energy crops has the potential to increase the content of organic matter in soil and improve its quality, this conversion faces difficult acceptance due to threats to food security [17, 76]. With this, the use of forest biomass creates a challenge for global climate policy: while fossil CO_2 emissions are partially avoided by the use of wood, other measures are needed to reduce the impacts generated by the reduction of the forest carbon sink and the changes in land use, as the goal is to reduce all GHG emissions in a few decades [65].

In addition to these obstacles, biomass burning is responsible for the release of other atmospheric pollutants, such as particulate matter (PM2.5), in which the particles that remain suspended in the atmosphere can interfere with the amount of solar radiation received by the Earth, through the dispersion and absorption processes, causing heating or cooling of the atmosphere, depending on the physical–chemical and optical properties of the particles [77, 78]. However, there are big differences in the emissions generated by the combustion of biomass, depending on which stove is used and how it is used. New, clean-burning stoves emit significantly lower amounts of local air pollutants than traditional stoves; and emissions resulting from incomplete combustion are strongly related to the way biomass is burned, with the frequency with which wood is placed in the stove and how much air is available for combustion, to which new stoves have greater control of the combustion process and the airflow [79].

7 Prospects for Fuels Arising from Discussions at COP27

During the 26th Conference of the Parties (COP), held in Glasgow—Scotland in 2021, world leaders agreed to accelerate their actions due to the alarming increase in global temperatures released by the Intergovernmental Panel on Climate Change (IPCC). However, arriving at the 27th COP, held in Sharm el Sheikh—Egypt, it was possible to see that this did not materialize, and the world went through countless consequences of climate change, such as floods in Pakistan, the dry season in the United States, hunger in Africa and heat waves in Europe [80].

COP27 brought some guidelines with the Adaptation Agenda of Sharm-El-Sheikh, seeking to be established as "the COP of implementation". The Agenda goes through some important points, aiming to protect more than 4 billion people from the growing climate consequences. These include (i) transitioning the world to sustainable and more climate-resilient agriculture, increasing production by 17% and reducing emissions by 21%, without expanding agricultural frontiers and improving livelihoods; (ii) the protection and restoration of 400 million ha of areas in a critical situation, both in terrestrial and freshwater ecosystems, supporting indigenous peoples and local communities and also, together with them, transforming 2 billion ha into a management area sustainable; (iii) the investment of 4 billion dollars in the protection of 15 million ha of mangroves, with a view to actions to protect all of them in the future; (iv) expanding access to clean cooking to 2,4 billion people and mobilizing up to USD 300 billion to adapt and encourage 2000 of the world's largest companies to integrate climate risk and develop climate change adaptation and mitigation action plans [82].

At least 230 million dollars were donated to the Adaptation Fund to help less developed countries to adapt to the new targets, as well as the creation of a specific fund for developing countries to deal with Loss and Damage caused by climate change. Based on the FAO report on the state of the world's forests, a Partnership between Forestry and Climate Leaders was launched, to implement the commitments established at COP26 and halt forest loss, biodiversity loss, as well as halt land degradation until 2030, including mobilization of financial support for indigenous peoples and local communities. COP27 also launched the Food and Agriculture for Sustainable Transformation (FAST) initiative to address climate impacts on both agriculture and food security.

COP27 also reaffirmed the Paris Agreement target of limiting global warming to 1.5 °C, but with data from the IPCC report presented at the event, commitments made by countries still increase GHG emissions by 10.6% by 2030, which could cause warming of up to 2.5 °C by the end of the century. To meet the Paris Agreement target, emissions should decrease by 43% by 2030, a scenario that is far from becoming a reality. With that, a work plan was launched, aiming to accelerate efforts and gradually reduce the use of energy from mineral coal and eliminate subsidies to fossil fuels [81]. In this context, the countries participating in COP27 will have a lot to change in their policies, from energy bases to human rights policies, as well as the distribution of subsidies. It will be a long road, which requires commitment, so there is still a habitable world to fight for.

8 Conclusions

Historically, for both humans and the animal world in general, the fundamental question about food and energy has been that, to survive, an individual must acquire at least as much food energy as is expended in basal metabolism, reproduction, and acquisition of food. In recent years, the main point of interaction between food, energy, and health has changed radically. Access to unprecedented levels of usable energy and non-renewable energy transformation practices are responsible for most of the man-made greenhouse gas emissions that are causing climate change. This change, in turn, poses major risks to the health of the population, including affecting food production and nutrition.

The use of forest-based biomass has the potential to contribute to the mitigation of climate change. However, although biomass can supply a significant part of primary energy demand with low or negative GHG emissions, these parallel targets are sensitive to uncertain economic, technical, and political futures. In addition to being an abundantly available and renewable source, forest biomass is used by different crops over the years in different countries around the world. Research that defines strategies aimed at mitigating gaseous emissions from this energy source is important to ensure compliance with climate goals in developed and developing countries, without the need to resort to more fossil fuels.

References

1. Wrangham R (2009) Catching fire: how cooking made us human—Richard Wrangham—Google Livros. Basic books. https://books.google.com.br/books?hl=pt-BR&lr=&id=ebEOupKz-rMC&oi=fnd&pg=PP10&dq=Catching+Fire.+How+Cooking+Made+Us+Human&ots=sXVs4_8MDW&sig=5oNICKZBgs-M4MNk9VnsDm9PWAw#v=onepage&q=CatchingFire. How Cooking Made Us Human&f=false. Accessed 4 Mar 2023

2. Yan Y (2018) Integrate carbon dynamic models in analyzing carbon sequestration impact of forest biomass harvest. Sci Total Environ 615:581–587. https://doi.org/10.1016/J.SCITOTENV.2017.09.326
3. Salazar-García DC, Power RC, Sanchis Serra A, Villaverde V, Walker MJ, Henry AG (2013) Neanderthal diets in central and southeastern Mediterranean Iberia. Quat Int 318:3–18. https://doi.org/10.1016/j.quaint.2013.06.007
4. Sørensen B (2009) Energy use by Eem Neanderthals. J Archaeol Sci 36(10):2201–2205. https://doi.org/10.1016/j.jas.2009.06.003
5. Cablevey News (2003) The history of biomass as a renewable energy source—Cablevey® Conveyors. https://cablevey.com/the-history-of-biomass-as-a-renewable-energy-source/. Accessed 19 Feb 2023
6. Jákl J (2021) Distilled beverages. Brill
7. Phuah E-T, Yap JW-L, Lau C-W, Lee Y-Y, Tang T-K (2022) Vegetable oils and animal fats: sources, properties and recovery 1–26. https://doi.org/10.1007/978-981-16-5113-7_1
8. Mansoori GA, Agyarko LB, Estevez LA, Fallahi B, Gladyshev G, Santos RG dos, Niaki S, Perišić O, Sillanpää M, Tumba K, Yen J (2021) Fuels of the future for renewable energy sources (Ammonia, Biofuels, Hydrogen). https://doi.org/10.48550/arxiv.2102.00439
9. Stafford W, De Lange W, Nahman A, Chunilall V, Lekha P, Andrew J, Johakimu J, Sithole B, Trotter D (2020) Forestry biorefineries. Renew. Energy 154:461–475. https://doi.org/10.1016/J.RENENE.2020.02.002
10. Carvalho C, Fonseca N, Vieira De Castro F (2008) Notas sobre a Tecnologia de Construção Naval nos Estaleiros Navais Portugueses do Século XVI
11. Hudaya T, Widjaja O, Rionardi A, Soerawidjaja TH (2016) Synthesis of biokerosene through electrochemical hydrogenation of terpene hydrocarbons from turpentine oil. J Eng Technol Sci 48(6):655. https://doi.org/10.5614/j.eng.technol.sci.2016.48.6.2
12. Gupta R, Wohar M (2017) Forecasting oil and stock returns with a Qual VAR using over 150 years off data. Energy Econ 62:181–186. https://doi.org/10.1016/J.ENECO.2017.01.001
13. Patricio Moreno Montalvo HI, Patricio Pineda Maigua DI, Alberto Santos Correa III L (2022) Evolución e historia de los motores diesel. Polo del Conoc 7(10):744–760. https://doi.org/10.23857/PC.V7I10.4754
14. Johnstone P, McLeish C (2020) World wars and the age of oil: exploring directionality in deep energy transitions. Energy Res Soc Sci 69:101732. https://doi.org/10.1016/J.ERSS.2020.101732
15. Fischer-Kowalski M, Rovenskaya E, Krausmann F, Pallua I, Mc Neill JR (2019) Energy transitions and social revolutions. Technol Forecast Soc Change 138:69–77. https://doi.org/10.1016/J.TECHFORE.2018.08.010
16. De Freitas H, Gil C, Da Amazônia G, Brasileiro M, Empreendimentos G (2021) O PENSAMENTO GEOPOLÍTICO DE GOLBERY DO COUTO E SILVA E OS POVOS TRADICIONAIS NA AMAZÔNIA: UMA RELAÇÃO TENSA. Rev Geopolítica Transfront 1(1):120–140
17. Gustavsson L, Haus S, Lundblad M, Lundström A, Ortiz CA, Sathre R, Le TN, Wikberg PE (2017) Climate change effects of forestry and substitution of carbon-intensive materials and fossil fuels. Renew Sustain Energy Rev 67:612–624. https://doi.org/10.1016/J.RSER.2016.09.056
18. Reid WV, Ali MK, ChristopherI, Field B, Correspondence WV, Reid D, Packard L (2020) The future of bioenergy. Glob Chang Biol 26(1):274–286. https://doi.org/10.1111/GCB.14883
19. Lee I, Bae DJ, Lee WK, Yang CM, Cho SW, Nam J, Lee DY, Jang AR, Shin HS, Hwang JY, Hong S, Kim KS (2019) Rapid synthesis of graphene by chemical vapor deposition using liquefied petroleum gas as precursor. Carbon N Y 145:462–469. https://doi.org/10.1016/J.CARBON.2019.01.004
20. Mazumder S, Lee A, Dube B, Mehra D, Khaing P, Taneja S, Yan B, Chillrud SN, Bhandari N, D'Armiento JM (2019) A clean fuel cookstove is associated with improved lung function: effect modification by age and secondhand tobacco smoke exposure. Sci Rep 9(1):1–8. https://doi.org/10.1038/s41598-018-37887-8

21. Raslavičius L, Keršys A, Mockus S, Keršiene N, Starevičius M (2014) Liquefied petroleum gas (LPG) as a medium-term option in the transition to sustainable fels and transport. Renew Sustain Energy Rev 32:513–525. https://doi.org/10.1016/J.RSER.2014.01.052
22. Minadeo R (2022) FUSÕES E AQUISIÇÕES (F&A's) NA INDÚSTRIA PETROLÍFERA: UMA VISÃO PANORÂMICA. Rev Estud Debate 29(2):28–60. https://doi.org/10.22410/ISSN.1983-036X.V29I2A2022.2970
23. James RW, Missenden JF (1992) The use of propane in domestic refrigerators. Int J Refrig 15(2):95–100. https://doi.org/10.1016/0140-7007(92)90033-Q
24. Hunt JD, Byers E, Balogun AL, Leal Filho W, Colling AV, Nascimento A, Wada Y (2019) Using the jet stream for sustainable airship and balloon transportation of cargo and hydrogen. Energy Convers Manag X 3:100016. https://doi.org/10.1016/J.ECMX.2019.100016
25. Colomer M, Lyra M, Pires-Alves C, Delorme Prado LC (2020) The market of LPG (Liquefied petroleum gas) in Brazil: a brief structural, institutional-historical perspective. Entrep Hist 99(2):21–34. https://doi.org/10.3917/EH.099.0021
26. Baryshevaa A, Jevgenij E, Markova A, Sergejs S (2023) Liquefied petroleum gas: market overview and pricing models. Econophysica. https://econophysica.ru/researches/liquefied-petroleum-gas-market-overview-and-pricing-models/. Accessed 5 Mar 2023
27. Taneja S, Singh P, Sharma A, Singh G (2021) Use of alcohols and biofuels as automotive engine fuel 161–183. https://doi.org/10.1007/978-981-16-1256-5_10
28. Nuño Martínez N, Mäusezahl D, Hartinger SM (2020) A cultural perspective on cooking patterns, energy transfer programmes and determinants of liquefied petroleum gas use in the Andean Peru. Energy Sustain Dev 57:160–167. https://doi.org/10.1016/J.ESD.2020.06.007
29. Ou S, Lin Z, Manente V, Bouchard J, He X, Lu Z, Gan Y, Zhou Y, Przesmitzki S, De Castro Gomez DJ, Aburas N, Lilley W, Calendini PO (2022) Light-duty vehicle transportation policy and implication on greenhouse gas emissions. ACS Symp Ser 1412:21–81. https://doi.org/10.1021/BK-2022-1412.CH002
30. de Mello SB (2022) GLP: potencial de sobra para uma matriz energética mais limpa – Sindigás. Sindigás
31. Patnaik S, Jha S (2020) Caste, class and gender in determining access to energy: a critical review of LPG adoption in India. Energy Res Soc Sci 67:101530. https://doi.org/10.1016/J.ERSS.2020.101530
32. Mariselvam V, Dharshini MS (2021) IoT based level detection of gas for booking management using integrated sensor. Mater Today Proc 37(Part 2):789–792. https://doi.org/10.1016/J.MATPR.2020.05.825
33. Lucon O, Coelho ST, Goldemberg J (2004) LPG in Brazil: lessons and challenges. Energy Sustain Dev 8(3):82–90. https://doi.org/10.1016/S0973-0826(08)60470-6
34. Gioda A (2019) Residential fuelwood consumption in Brazil: environmental and social implications. Biomass Bioenerg 120:367–375. https://doi.org/10.1016/J.BIOMBIOE.2018.11.014
35. Gürbüz H, Şöhret Y, Akçay H (2019) Environmental and enviroeconomic assessment of an LPG fueled SI engine at partial load. J Environ Manage 241:631–636. https://doi.org/10.1016/J.JENVMAN.2019.02.113
36. WHO WHO (2018) Household air pollution and health
37. de Paula Protásio, Roque Lima MD, Scatolino MV, Silva AB, Rodrigues de Figueiredo IC, Gherardi Hein PR, Trugilho PF (2021) Charcoal productivity and quality parameters for reliable classification of Eucalyptus clones from Brazilian energy forests. Renew Energy 164:34–45. https://doi.org/10.1016/j.renene.2020.09.057
38. Lima MDR, Patrício EPS, Barros Junior U de O, Silva R de CC, Bufalino L, Numazawa S, Hein PRG, Protásio T de P (2021) Colorimetry as a criterion for segregation of logging wastes from sustainable forest management in the Brazilian Amazon for bioenergy. Renew Energy 163:792–806. https://doi.org/10.1016/j.renene.2020.08.078
39. de Paula Protásio, da Costa JS, Scatolino MV, Lima MDR, de Assis MR, da Silva MG, Bufalino L, Dias Junior AF, Trugilho PF (2022) Revealing the influence of chemical compounds on the pyrolysis of lignocellulosic wastes from the Amazonian production chains. Int J Environ Sci Technol 19(5):4491–4508. https://doi.org/10.1007/s13762-021-03416-w

40. Lima MDR, Simetti R, de Assis MR, Trugilho PF, Carneiro ACO, Bufalino L, Hein PRG, de Paula Protásio T (2020b) Charcoal of logging wastes from sustainable forest management for industrial and domestic uses in the Brazilian Amazonia. Biomass Bioenergy 142:105804. https://doi.org/10.1016/j.biombioe.2020.105804
41. Oliveira PRS, Trugilho PF, Oliveira TJP (2022) Briquettes of acai seeds: characterization of the biomass and influence of the parameters of production temperature and pressure in the physical-mechanical and energy quality. Environ Sci Pollut Res 29(6):8549–8558. https://doi.org/10.1007/s11356-021-15847-6
42. de Souza EC, Gomes JPS, Pimenta AS, de Azevedo TKB, Pereira AKS, Gomes RM, Brito JO, Dias Júnior AF (2022) Briquette production as a sustainable alternative for waste management in the tannin extraction industry. Environ Sci Pollut Res. https://doi.org/10.1007/s11356-022-23490-y
43. Ferreira G, Brito TM, da Silva JGM, Minini D, Dias Júnior AF, Arantes MDC, Batista DC (2022) Wood waste pellets as an alternative for energy generation in the Amazon region. BioEnergy Res. https://doi.org/10.1007/s12155-022-10446-w
44. Azevedo SG, Santos M, Antón JR (2019) Supply chain of renewable energy: a bibliometric review approach. Biomass Bioenerg 126:70–83. https://doi.org/10.1016/j.biombioe.2019.04.022
45. Gioda A, Tonietto GB, de Leon AP (2019) Exposure to the use of firewood for cooking in Brazil and its relation with the health problems of the population. Cien Saude Colet 24(8):3079–3088. https://doi.org/10.1590/1413-81232018248.23492017
46. Gordon SB, Bruce NG, Grigg J, Hibberd PL, Kurmi OP, Lam K bong H, Mortimer K, Asante KP, Balakrishnan K, Balmes J, Bar-Zeev N, Bates MN, Breysse PN, Buist S, Chen Z, Havens D, Jack D, Jindal S, Kan H, Mehta S, Moschovis P, Naeher L, Patel A, Perez-Padilla R, Pope D, Rylance J, Semple S, Martin WJ (2014) Respiratory risks from household air pollution in low and middle income countries. Lancet Respir Med 2:823–860
47. Kaur-Sidhu M, Ravindra K, Mor S, John S, Aggarwal AN (2019) Respiratory Health status of rural women exposed to liquefied petroleum gas and solid biomass fuel emissions. Air, Soil Water Res 12:117862211987431. https://doi.org/10.1177/1178622119874314
48. WHO WHO (2017) Burning opportunity: clean household energy for health, sustainable development, and wellbeing of women and children. World Health Organization
49. Clean Cooking Alliance (2022) Clean Cooking Alliance
50. Mogaji E, Adekunle I, Aririguzoh S, Oginni A (2022) Dealing with impact of COVID-19 on transportation in a developing country: Insights and policy recommendations. Transp Policy 116:304–314. https://doi.org/10.1016/j.tranpol.2021.12.002
51. Wassler P, Talarico C (2021) Sociocultural impacts of COVID-19: a social representations perspective. Tour Manag Perspect 38:100813. https://doi.org/10.1016/j.tmp.2021.100813
52. Li H, Leng X, Hu J, Cao A, Guo L (2023) When cooking meets confucianism: Exploring the role of traditional culture in cooking energy poverty. Energy Res Soc Sci 97:102956. https://doi.org/10.1016/j.erss.2023.102956
53. Lindgren SA (2020) Clean cooking for all? A critical review of behavior, stakeholder engagement, and adoption for the global diffusion of improved cookstoves. Energy Res Soc Sci 68:101539. https://doi.org/10.1016/j.erss.2020.101539
54. Empresa de Pesquisa Energética (2022) Brazilian Energy Balance. Rio de Janeiro
55. Brasil (2021) LEI Nº 14.237, DE 19 DE NOVEMBRO DE 2021. Brazil
56. Li K, Qi S, Shi X (2022) The COVID-19 pandemic and energy transitions: evidence from low-carbon power generation in China. J Clean Prod 368:132994. https://doi.org/10.1016/j.jclepro.2022.132994
57. Oke DO, Fakinle BS, Sonibare JA, Akeredolu FA (2020) Evaluation of emission indices and air quality implications of liquefied petroleum gas burners. Heliyon 6(8):e04755. https://doi.org/10.1016/J.HELIYON.2020.E04755
58. Isihak S, Akpan U, Adeleye M (2012) Interventions for mitigating indoor-air pollution in Nigeria: a cost-benefit analysis. Int J Energy Sect Manag 6(3):417–429. https://doi.org/10.1108/17506221211259655/FULL/PDF

59. Hartinger SM, Lanata CF, Hattendorf J, Verastegui H, Gil AI, Wolf J, Mäusezahl D (2016) Improving household air, drinking water and hygiene in rural Peru: a community-randomized–controlled trial of an integrated environmental home-based intervention package to improve child health. Int J Epidemiol 45(6):2089–2099. https://doi.org/10.1093/IJE/DYW242
60. Jetter JJ, Kariher P (2009) Solid-fuel household cook stoves: characterization of performance and emissions. Biomass Bioenerg 33(2):294–305. https://doi.org/10.1016/J.BIOMBIOE.2008.05.014
61. Darda S, Papalas T, Zabaniotou A (2019) Biofuels journey in Europe: currently the way to low carbon economy sustainability is still a challenge. J Clean Prod 208:575–588. https://doi.org/10.1016/J.JCLEPRO.2018.10.147
62. Hao H, Dai L, Wang K, Xu J, Liu W (2021) An updated framework for climate change impact assessment of bioenergy and an application in poplar biomass. Appl Energy 299:117323. https://doi.org/10.1016/J.APENERGY.2021.117323
63. Guo J, Gong P, Brännlund R (2019) Impacts of increasing bioenergy production on timber harvest and carbon emissions. J For Econ 34(3–4):311–335. https://doi.org/10.1561/112.00000500
64. Jåstad EO, Bolkesjø TF, Trømborg E, Rørstad PK (2020) The role of woody biomass for reduction of fossil GHG emissions in the future North European energy sector. Appl Energy 274:115360. https://doi.org/10.1016/J.APENERGY.2020.115360
65. Soimakallio S, Saikku L, Valsta L, Pingoud K (2016) Climate change mitigation challenge for wood utilization—the case of Finland. https://doi.org/10.1021/ACS.EST.6B00122
66. Liu W, Zhang Z, Xie X, Yu Z, Von Gadow K, Xu J, Zhao S, Yang Y (2017) Analysis of the global warming potential of biogenic CO_2 emission in life cycle assessments. Sci Reports 71 7(1):1–8. https://doi.org/10.1038/srep39857
67. Patel M, Zhang X, Kumar A (2016) Techno-economic and life cycle assessment on lignocellulosic biomass thermochemical conversion technologies: a review. Renew Sustain Energy Rev 53:1486–1499. https://doi.org/10.1016/J.RSER.2015.09.070
68. Cherubini F, Strømman AH (2011) Life cycle assessment of bioenergy systems: state of the art and future challenges. Bioresour Technol 102(2):437–451. https://doi.org/10.1016/J.BIORTECH.2010.08.010
69. Guest G, Cherubini F, Strømman AH (2013) The role of forest residues in the accounting for the global warming potential of bioenergy. GCB Bioenergy 5(4):459–466. https://doi.org/10.1111/GCBB.12014
70. Liu W, Wang J, Debangsu Bhattacharyya C, Cafferty K, Shawn Grushecky M, Schuler J, Singh K, Spatari S (2015) Economic and environmental analyses of biomass utilization for bioenergy products in the Northeastern United States. ProQuest
71. Pelletier C, Rogaume Y, Dieckhoff L, Bardeau G, Pons MN, Dufour A (2019) Effect of combustion technology and biogenic CO_2 impact factor on global warming potential of wood-to-heat chains. Appl Energy 235:1381–1388. https://doi.org/10.1016/J.APENERGY.2018.11.060
72. Hunt RG, Franklin WE (1996) LCA—how it came about—personal reflections on the origin and the development of LCA in the USA. Int J Life Cycle Assess 1(1):4–7. https://doi.org/10.1007/BF02978624/METRICS
73. Liu W, Yu Z, Zhu Q, Zhou X, Peng C (2020) Assessment of biomass utilization potential of *Caragana korshinskii* and its effect on carbon sequestration on the Northern Shaanxi Loess Plateau China. L Degrad Dev 31(1):53–64. https://doi.org/10.1002/LDR.3425
74. Chaves AMB, do Vale AT, Melido RCN, Zoch VP (2013) CARACTERÍSTICAS ENERGÉTICAS DA MADEIRA E CARVÃO VEGETAL DE CLONES DE Eucalyptus spp. Biosf Enciclopédia Científico Conhecer-Goiânia, Cent 9
75. Liu W, Xu J, Xie X, Yan Y, Zhou X, Peng C (2020) A new integrated framework to estimate the climate change impacts of biomass utilization for biofuel in life cycle assessment. J Clean Prod 267:122061. https://doi.org/10.1016/J.JCLEPRO.2020.122061
76. Njakou Djomo S, Witters N, Van Dael M, Gabrielle B, Ceulemans R (2015) Impact of feedstock, land use change, and soil organic carbon on energy and greenhouse gas performance of biomass cogeneration technologies. Appl Energy 154:122–130. https://doi.org/10.1016/J.APENERGY.2015.04.097

77. Artaxo P, Rizzo L V, Paixão M, De Lucca S, Oliveira PH, Lara LL, Wiedemann KR, Andreae MO, Holben B, Schafer J, Correia AL, Pauliquevis TM (2009) Partículas de Aerossóis na Amazônia: Composição, Papel no Balanço de Radiação, Formação de Nuvem e Ciclos de Nutrientes. Glob Chang Geophys Monogr Ser 186. https://doi.org/10.1029/2008GM000778
78. Lima FDM, Pérez-Martínez PJ, de Fatima AM, Kumar P, de Miranda RM (2020) Characterization of particles emitted by pizzerias burning wood and briquettes: a case study at Sao Paulo, Brazil. Environ Sci Pollut Res 27(29):35875–35888. https://doi.org/10.1007/S11356-019-07508-6/METRICS
79. Solli C, Reenaas M, Strømman AH, Hertwich EG (2009) Life cycle assessment of wood-based heating in Norway. Int J Life Cycle Assess 14(6):517–528. https://doi.org/10.1007/S11367-009-0086-4/METRICS
80. Friedman L (2022) What is COP27 and what's at stake? What to know about the UN's climate summit. The New York Times. New York Times
81. Gurgel e Silva JP (2023) COP 27 e a participação brasileiralPolitize! Politize
82. COP27 (2022) COP27—COP27 Presidency launches Sharm El-Sheikh Adaptation Agenda

Biomass as a Biofuel Used in Food Preparation: Qualitative Variables that Contribute to People's Quality of Air and Life

Gabriela Fontes Mayrinck Cupertino, Fernanda Aparecida Nazário de Carvalho, Fabíola Martins Delatorre, Kamilla Cryslayne Alves da Silva, Daniel Saloni, Allana Katiussya Silva Pereira, and Ananias Francisco Dias Júnior

Abstract The use of biomass for cooking food is a daily activity for many families and can cause an increase in morbidity and mortality in different parts of the world. Improper burning of this material can generate various polluting compounds that negatively impact the environment and people's health. Wood-burning stoves and fireplaces emit significant amounts of harmful pollutants, including several carcinogenic compounds, with CO and NO_x being two of the primary gaseous pollutants from wood smoke. Thus, the objective of this chapter was to discuss the main variables that contribute to people's air quality and life. For this, we debated biomass as a

G. F. M. Cupertino (✉) · F. A. N. de Carvalho · F. M. Delatorre · A. F. Dias Júnior
Federal University of Espírito Santo (UFES), Department of Forestry and Wood Sciences, Av. Governador Lindemberg, 316, Jerônimo Monteiro, Espírito Santo 29550-000, Brazil
e-mail: gabriela.mayrinck01@gmail.com

F. A. N. de Carvalho
e-mail: fernandacarvalhonaz@gmail.com

F. M. Delatorre
e-mail: fabiolamdelatorre@gmail.com

A. F. Dias Júnior
e-mail: ananias.dias@ufes.br

K. C. A. da Silva · A. K. S. Pereira
"Luiz de Queiroz" College of Agriculture, University of São Paulo (ESALQ/USP), Department of Forests Sciences, Av. Pádua Dias, 11, Piracicaba, São Paulo 13418-900, Brazil
e-mail: kamilla.alves@usp.br

A. K. S. Pereira
e-mail: allana.florestal@gmail.com

D. Saloni
North Carolina State University (NCSU), College of Natural Resources, Department of Forest Biomaterials, Raleigh, NC 27695, USA
e-mail: desaloni@ncsu.edu

A. K. S. Pereira and A. F. Dias Júnior (eds.), *Impacts of Using Biomass as an Energy Source in Homes*, Green Energy and Technology,
https://doi.org/10.1007/978-3-031-38824-8_2

biofuel in food cooking, the combustion of biomass in air quality and people's health, the evolution of cooking methods, the use and effect of biomass use in domestic environments, and how to tackle the challenge of clean cooking. It was observed that the methods of preparing food with forest biomass are often carried out incorrectly, often with outdated preparation methods. This occurs mainly in developing countries by families without prior knowledge of the risks, or when there is knowledge, there need to be more economic subsidies.

Keywords Air pollution · Respiratory diseases · Biomass burning

1 Introduction

The dependence on biomass has had its origins since ancient times. Defined as any organic material derived from plants, animals, and waste that can be used as a source of energy, this material has an important place in the history of humanity [55]. Biomass was the first energy source that humans were able to control and use its energy [10]. When burned, this material releases energy through heat, which can be used for cooking or heating. Until two centuries ago, biomass was the world's primary source of energy [20]. Over the years, new sources of energy generation have developed, but biomass, along with kerosene, remains the main source of energy for cooking food for about 3 billion people worldwide. The proportion of the population that depends on biomass for cooking is higher in sub-Saharan Africa and India [74]. That's right, one-third of humanity primarily cooks with biomass [32]. For centuries, this material has been a part of society's daily life and continues to be used today. In various regions of the planet, biomass, such as wood, harvest waste, wet cake, or charcoal, is used for cooking food and heating and lighting residences.

Biomass is a traditional and widely used fuel for cooking food in many countries worldwide. Often related to cultural practices, biomass is considered a sustainable and eco-friendly option for cooking food, especially in rural areas. In countries such as India, China, Africa, Brazil, and Indonesia, biomass is considered the primary raw material for preparing food in rural areas [7, 49, 55, 56]). Cooking with biomass has traditionally been practiced in developing countries where other fuel sources are scarce [45]. However, we emphasize that there is a small portion of the population that uses biomass in a more "noble" way, in preparing food for people with higher purchasing power, such as in barbecues, pizzerias, and other dishes considered valuable by the fuel, such as the "dirty steak" made with food directly on charcoal embers.

One of the advantages of cooking food with biomass is that this raw material is considered a renewable energy source. Unlike fossil fuels, which are finite and non-renewable, biomass can be grown and harvested continuously [23]. This makes this cooking practice an eco-friendly option for cooking food. There is a range of methods for cooking food from biomass, from direct contact with the heat source to open and closed ovens that use pots and other utensils for boiling, cooking, and baking based on indirect contact with the heat source generated by the fuel [22, 63]. We emphasize that there are various possibilities for using biomass for cooking food. However, cooking with biomass has its challenges.

One of the main challenges is reducing smoke and other pollutants emitted during the cooking process, which can have adverse health effects [25, 46]. W To address this issue, several studies have been conducted worldwide, aiming to evaluate the impact of indoor air pollution caused by biomass burning on human health and seeking solutions to this situation. Effective solutions to reduce indoor air pollution require the development and implementation of policies and programs that promote the adoption of cleaner cooking technologies and fuels. Additionally, education and training to communities on the importance of indoor air quality and the risks associated with indoor air pollution are vital to achieve significant results. In this context, it is essential to consider qualitative variables of biomass that can influence air quality and, consequently, human health. This chapter aims to provide insights into the factors affecting biomass combustion and the resulting air pollution. By understanding the key factors influencing biomass combustion, policymakers, researchers, and communities can work together to develop and implement effective solutions to reduce indoor air pollution and improve human health. Overall, addressing the issue of indoor air pollution caused by burning biomass for cooking and heating purposes is crucial to achieving the Sustainable Development Goals and improving the health and well-being of people around the world.

2 Biomass Combustion X Air Quality

Despite biomass's numerous benefits, its use has become a concerning factor [98]. One of the main challenges is that biomass as fuel is subject to the release of smoke and other hazardous pollutants to human health and the environment, mainly related to the equipment used in combustion. When biomass is burned in traditional stoves or fires, it releases various harmful pollutant gases into the air. These combustion gases can contain particulate matter, nitrogen oxides (NO_x), sulfur oxides (SO_x), carbon monoxide (CO), volatile organic compounds (VOCs), and other pollutants (such as particulate matter) (Fig. 1) [36, 99].

NO_x can contribute to the formation of ground-level ozone and cause respiratory problems, while SO_x can cause respiratory and cardiovascular issues and contribute to acid rain [93, 100]. CO is considered a toxic gas that can reduce oxygen delivery to the human body and cause various health problems [57]. VOCs can contribute to the formation of ground-level ozone and other secondary pollutants [73]. Particulate matter (PM) is a mixture of solid and liquid particles suspended in the air that can negatively impact air quality and human health [18]. But why do these compounds form during biomass burning? The quantity and type of emissions produced by biomass combustion depend on several factors, such as the type and quality of biomass, the combustion technology used, and the operating conditions [2, 33].

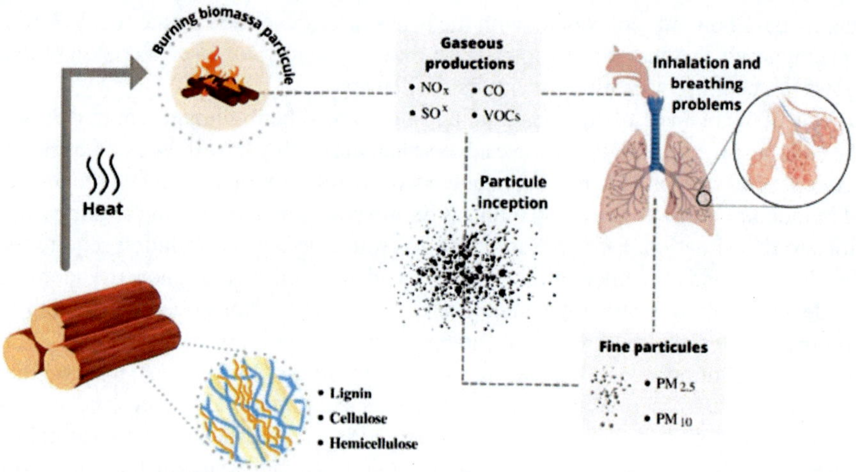

Fig. 1 Diagram of biomass conversion and gas release (*Source* The authors 2023)

A factor that can influence gas emissions during the combustion process of a solid fuel is its chemical composition. In general, the chemical composition of biomass fuels is less complex than solid fossil fuels [48]. Biomass is mainly composed of carbohydrates, such as cellulose, hemicellulose, and lignin, which are compounds of carbon, hydrogen, and oxygen [29, 83]. Although it has a chemical composition rich in carbon, oxygen, and hydrogen, many elements in its composition can be problematic during combustion, especially nitrogen (N) and sulfur (S), which can directly influence gas emissions [48]. Studies suggest that NO_x emissions from biomass combustion are directly related to the nitrogen content of the fuel [47, 61].

Regarding sulfur, the behavior of emissions is like that of nitrogen. The amount of SO_x emissions produced during combustion is related to the amount of sulfur in the biomass [13]. For example, some biomasses, such as agricultural residues, may contain higher levels of sulfur-containing compounds than others. Additionally, the presence of sulfur may depend on soil conditions and fertilizers used during biomass growth.

Another factor inherent to the characteristics of biomass that favors a greater amount of gas emissions during the combustion process is moisture. High moisture content limits thermal performance and increases pollutant emissions after slow or incomplete combustion [18, 54]. In addition to the intrinsic characteristics of biomass, how the material is combusted plays an essential role in gas emissions during its burning. For example, well-designed and operated combustion systems can produce more emissions than well-designed and operated systems [74]. When biomass is burned, incomplete combustion can occur due to a lack of oxygen, leading to the formation of carbon monoxide (CO) emissions [86]. In addition to the gases mentioned, volatile organic compounds (VOCs) are also a threat to air quality and may be related to both the biomass combustion process and the chemical characteristics of the material [65]. These emissions go far beyond air quality. Issues related to climate change and human health are among the main concerns that need to be considered regarding gas emissions from biomass burning.

3 Is Biomass a Problem for Human Health?

Cooking food is a necessary daily activity for human survival, but it can also pose health risks, especially when done with traditional equipment that relies on biomass fuels. Using biomass fuels, such as wood, crop residues, and animal dung, is common in many developing countries, particularly in rural areas where access to modern energy sources is limited. Although biomass fuel is an important energy source for millions of people, it also presents significant health risks, especially for women and children. The smoke and vapors released while cooking with biomass fuels contain harmful pollutants such as fine particulate matter, carbon monoxide, nitrogen oxides, and volatile organic compounds, which can have serious health consequences [4, 24]. Human health and gas emissions from biomass burning are closely interconnected, as exposure to these emissions can significantly impact the health of individuals, particularly those living nearby or in areas where biomass burning is common. Biomass burning is the combustion of organic matter, such as wood, crop residues, and animal waste, and is a significant source of air pollution in rural and urban areas [79]. Respiratory diseases are one of the primary health impacts of gas emissions from biomass burning [17, 39, 69].

Exposure to particulate matter, carbon monoxide, nitrogen oxides, sulfur dioxide, and other pollutants released during biomass burning can cause or exacerbate respiratory diseases such as asthma, chronic obstructive pulmonary disease (COPD), and bronchitis [27]. These pollutants can also increase the risk of lung cancer and other respiratory infections. In addition to respiratory diseases, biomass burning emissions can lead to cardiovascular diseases [34]. O Fine particulate matter released during biomass burning can penetrate deep into the lungs (Fig. 2) and enter the bloodstream, causing inflammation and increasing the risk of heart attacks and strokes [52, 87]. This particulate matter can also cause vasoconstriction, narrowing blood vessels and reducing blood flow to vital organs [52]. The most significant risk factors for these diseases are exposure to fine particles, carbon monoxide, and other harmful pollutants released during biomass burning. However, given a series of problems related to biomass use, the following question arises: is biomass really the villain?

4 Is Biomass Really the Villain?

Considering that the proper use of energy ensures the continuity of life and the sustainable development of communities, energy consumption management is one of the main topics on the agenda of all countries [1]. One of the energy sources in developing countries is biomass-based fuel [41]. This creates a conflicting factor:

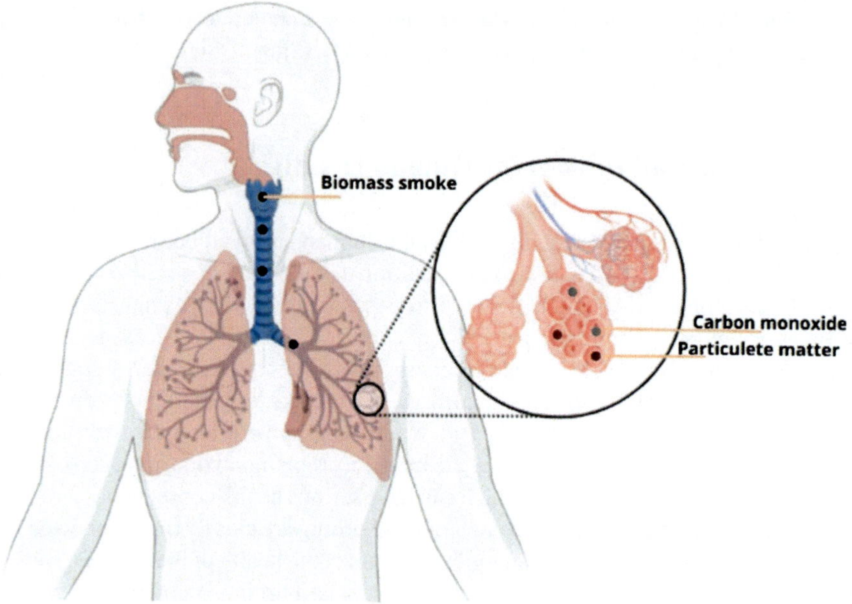

Fig. 2 Demonstration of the occurrence of cardiovascular diseases due to long-term exposure to smoke from burning biomass (*Source* The authors 2023)

the incidence of diseases due to high exposure to combustion compounds. Using biomass, especially wood, and charcoal, for cooking food is one of humanity's oldest and most widespread methods. Over time, the use of equipment to handle fire has become a common practice that has been perpetuated for generations. Even with technological advances in recent times, humans still face a series of challenges in the best way to use fire, using various types of stoves to facilitate food preparation [40]. One of the types of stoves used in developing countries is a wood stove. This type is used for cooking food and residential heating. However, these stoves need to present an ideal standard. Some lack chimneys; others have inefficient or poorly constructed chimneys, often subject to leaks in their operating system and ventilation methods [21].

In biomass-based stoves (popularly known as wood-burning stoves), the combustion of fuels often occurs incompletely, contributing to the generation of gases, fine particles, and compounds such as benzene, formaldehyde, and benzo[a]pyrene, which are highly carcinogenic to humans [21]. These components are also responsible for the growing incidence of respiratory infections, such as pneumonia, tuberculosis, chronic obstructive pulmonary disease, newborn malnutrition, cataracts, and even cardiovascular problems [3]. This is due to the need for more access to cooking and food preparation technologies, resulting from a confluence of factors ranging from the stove model to cooking in enclosed spaces. Almeida [5] analyzed the inefficiency of traditional wood-burning stoves and observed that their low-efficiency results in excessive and rapid firewood consumption due to incomplete energy utilization and faster fuel burning. In addition, the lack of insulation in the combustion chamber results in poor air circulation, leading to increased production and emission of toxic volatiles and soot particle [30]. These factors result in air pollution, leading to the health problems mentioned earlier.

The World Health Organization (WHO) has stated that approximately 2.4 billion people worldwide use wood and dung fires or stoves to prepare food (OMS 2022). This creates a problematic situation due to a combination of socio-environmental and economic factors, ranging from the stove model, the location of its use (in an open or closed area), and who operates it [59]. It is essential that biomass stoves meet the WHO Guidelines emission targets, as this is a determining factor in contributing to the health of those exposed to them daily.

5 Combustion with an Excess of Oxygen and Its Emissions

The biomass stove is a physical arrangement in which fuel is added in the presence of air, leading to the release of heat [28]. However, it is known that these devices are also characterized by low energy efficiency and high gas emissions. Air entry into the combustion chamber, either in the form of primary or secondary air, sustains the combustion process [74]. However, an excess of air can cause incomplete combustion and inadequate heat transfer due to reduced temperatures in the combustion chamber [88]. During this process of incomplete combustion of biomass, some toxic components are emitted, among them one of the main classes of organic pollutants, polycyclic aromatic hydrocarbons (PAHs) [64], black carbon (BC) [85], organic carbon (OC), carbon monoxide, nitrogen oxides (NO_x), sulfur dioxide (SO_2), ammonia (NH_3), benzene ($C_6 H_6$), formaldehyde (CH_2O) and others. Due to the low efficiency of combustion of these traditional stoves, large quantities of harmful effluents are produced to the environment and human health due to the incomplete burning of biomass fuel.

Biomass burning is a phenomenon capable of playing a fundamental role in terrestrial and atmospheric dynamics [60]. When a fuel particle is burned, a balance of ash fraction occurs due to simultaneous physical and chemical transformations, which are influenced by high temperatures [9]. The process of modifying the raw material and its constituent elements via heat in the presence of oxygen is called combustion [60]. Wood-burning stoves mostly do not have a mechanism for controlling the combustion process and require constant supervision by the operator. In many cases, they are operated improperly, contributing to incomplete combustion. This incomplete combustion results in the formation of gases such as carbon monoxide, breathable particulate matter, and a variety of greenhouse gases responsible for affecting human health and even contributing to climate change [70]. Another factor that affects the combustion process is the physical conditions of the biomass. In general, firewood for domestic burning does not follow a quality standard. Factors such as high moisture content, shape, size, density, and ash content can influence the combustion process. It is known that various uncontrollable factors are influential in the combustion process. Due to the harmful effects on the environment and human health, the development of new technologies to control the combustion process in traditional stoves and the possible reduction in the emission of these effluents has become relevant.

6 The Evolution of Ways of Cooking

The mastery of the stove begins with the discovery of the usability of fire and archaeological excavations in Chou Kutien, China [37]. Historically, it is known that using fire was responsible for our development as humans, mainly when related to meal preparation. Fire improved the ways of better utilizing the nutrients in food. This resulted in the evolution of the human organism, which stopped eating raw meals and began cooking them. Around half a million years ago, the mastery of fire popularized a scenario of differentiated nutritional habits, allowing for increased nutritional gain through processes that gradually improved throughout human evolution, such as baking, drying, and even storing. These techniques were responsible for allowing settlement and exploring the potential of plants. They even led to the improvement of different cooking methods, such as boiling stones in leather or wooden containers [12].

The cooking methodologies of prehistory evolved with the use of fire. Subsequently, they improved to other traditional and widely used methods, such as the "three-stone" "wood stove type," which operates based on wood burning. In its arrangement, three large, undefined stones support pots and vessels over an open bottom (Fig. 3) [53]. The designs of three-stone or "trempe" stoves are as their name implies. Compared to open fires, these stoves are considered inefficient. They are remarkably similar in their functioning methodology, allowing a large part of the heat to dissipate and gradually waste its fuel (wood) [95]. Over time, these stoves have steadily improved. Using a mixture of mud, water, and animal waste and some changes in shape, the traditional oven passed from "three stones" to the "Chulha." This improvement occurred exponentially, with the goal of better utilizing fuel-burning efficiency and indirectly minimizing emissions produced. Thanks to these adjustments, "modern" ovens could escape the smoke produced during biomass combustion outside of the kitchen in some stoves [50]. For example, the Chulha (fireplace in Hindi), traditionally used in India and other countries but not named, generally consists of a clay stove that operates based on biomass or humus-based fuels and is used for cooking meals (Fig. 3a) [43].

Fixed Chulhas with smoke removal apparatus is known as smokeless Chulhas and are modified versions of traditional Chulhas. As an example, the "Mid Chulha" does not have a smoke director and has a "U" shape. Reference [58] reiterated that Chulhas are designed for better fuel performance, promoting slightly more efficient combustion due to the presence of the chimney. Fuel consumption is reduced, and exhaust gases are removed through the chimney. The authors also considered that this cooking equipment has a 25–35% thermal efficiency and consumes less fuel. Another

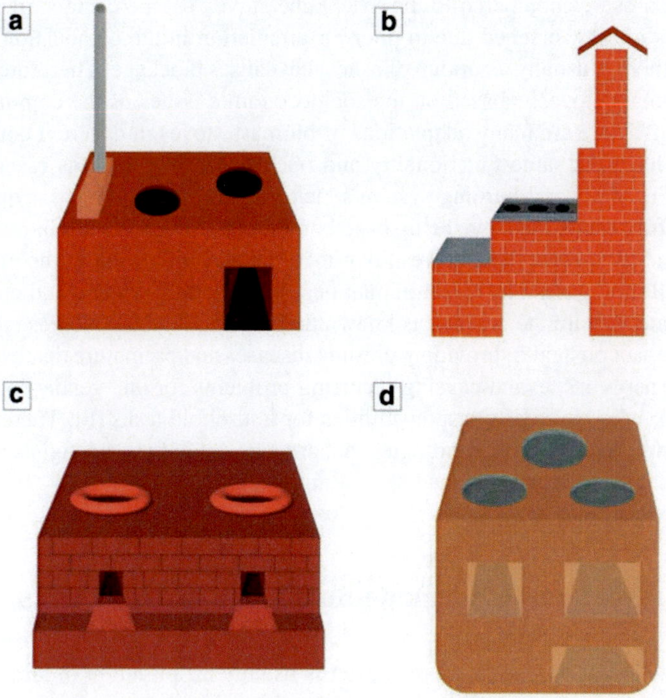

Fig. 3 Representation of stoves from different countries. Where: **a** Chulha stove; **b** traditional masonry stove; **c** Popular kiln in Kenya: Chepkube; **d** Traditional stove in Kenya: Cheproketes (*Source* The authors 2023)

traditional wood-burning stove model in Brazilian communities is made of masonry, cement, clay, and perforated metal plate for up to 5 pots (which may vary) (Fig. 3b). Some models of this oven already have a location for the exhaust gases to escape outside the residence (chimneys), but many still need this mechanism. Moreover, these wood-burning stoves have a low energy yield, emitting a considerable amount of gases and particulates inside the residences, harming the operator's health [40].

In other countries such as Kenya, besides the three-stone stove, which has become popular, different types of stoves are also used for cooking, such as the Chepkube stove (Fig. 3c) with a 30.8% adoption rate among the population and then the Cheprocket stove with 8.9% usability [42]. When compared to other stoves, it can be observed that the Chepkube stove has a more comprehensive air intake. The Chepkube stove is an improved stove that typically features a clay lining. This stove is considered improved due to its fuel efficiency and, thus, lower emissions [84]. However, Cheproketes stoves (Fig. 3d) have lower energy efficiency and promote higher CO exposures compared to the Chepkube stove. The percentage of incomplete combustion is worsened due to poor air circulation in the combustion chambers, where there is usually a buildup of ashes that causes blockage. Therefore, using different biomass stoves is based on the socioeconomic issues of the communities that use them. There are many adaptations of biomass stoves in different countries, but they all have the same functionality and operating principle. It is essential to understand that the wood-burning system's lack of standardization and acquisition can be harmful despite the stove being based on biomass. Even though biomass is a renewable energy source, these stoves favor incomplete combustion of the material, releasing polluting gases that affect human health, especially women and children, and exacerbate the climate crisis. It is known that traditional technologies can have a negative impact on health, providing chronic diseases and premature deaths, going beyond the energy issue, and causing recurring problems for one gender in particular: Women, who have daily responsibilities for household tasks [8]. Therefore, it is necessary to employ new technologies that are accessible for cooking, accessible to all, and enable clean and sustainable cooking.

7 Traditional Fuel: Inadequate Burning in Environments

Regarding air pollution, the first association is usually air pollution in large urban centers, with images of pollutants being emitted by vehicles or even companies. However, this scenario is not limited to these areas only; a considerable portion of the world's population lives with another source of pollution: burning forest biomass for domestic use (wood, charcoal, etc.). Approximately three billion people depend on traditional biomass for cooking and heating, causing various socio-environmental problems. About 35% of the population still used firewood for cooking in 2019, with 37% depending on Liquefied Petroleum Gas (LPG), natural gas, or biogas, and 10% on electricity [19]. The impact on health and the environment from using traditional fuels is significant—and affects women disproportionately, as they are primarily responsible for household chores. At least about 2.6 million people were without access to clean cooking in 2019, and slow progress predicts that by 2030—the target year to achieve the Sustainable Development Goal (SDG)—2.4 billion people will continue to use traditional fuel [19].

As seen in the data presented above, a large part of the population still uses traditional stoves (such as wood stoves) for cooking, using inefficient raw materials that often result in "inadequate burning," causing various health and environmental damages. The health risks associated with biomass burning in household environments have been linked to acute respiratory infections, chronic obstructive pulmonary disease (COPD), pneumoconiosis, pulmonary tuberculosis, and adverse effects during pregnancy. Such effects are more likely to occur in developing countries, where women and children are particularly vulnerable to the health impacts of indoor air pollution from cooking with biomass fuels [16, 26]. Women, in particular, often spend many hours a day cooking over open fires or traditional stoves, exposing them to high levels of smoke and other pollutants. Children are also at risk, as they spend a significant amount of time indoors and are more susceptible to respiratory infections due to the development of their immune system. Domestic air pollution is the fourth most important risk factor for the global disease burden and, according to reports, causes approximately 3.5 million premature deaths annually [33, 35]. Of these deaths, a large proportion is attributed to respiratory diseases, such as COPD, pneumonia, and lung cancer [31].

This type of domestic pollution affects the health of women who cook and children, who are especially susceptible to declining lung function, chronic bronchitis, and respiratory infections, and is the leading cause of death in children in developing countries [91, 92]. Reference [38] state that exposure to domestic pollution from biomass burning during the first trimester of pregnancy can contribute to lower fetal weight gain. The elements analyzed in the present research were ozone (O_3), nitrogen dioxide (NO_2), sulfur dioxide (SO_2), particulate matter with a diameter of less than 10 μm (PM10), and carbon monoxide (CO). The research demonstrated that 4.6% of newborns had a birth weight of less than 2500 g, which was statistically significant. Inadequate burning results in the formation of potentially toxic substances, such as carbon monoxide (CO), ammonia (NH_3), and methane (CH_4), among others, with fine particles containing particles smaller or equal to 10 μm (PM10), i.e., inhalable particles, being the most toxic pollutants that have been studied. Constituting the largest percentage (94%) of fine and ultrafine particles, these particles reach the deepest parts of the respiratory system, causing severe diseases [96, 99, 100].

8 Effect of Biomass Burning on Health

As the ability to control fire is often considered the distinguishing feature of pre-human evolution and wood is the oldest fuel, it is true that exposure to wood smoke is as old as humanity. Since domestic use dominates the total fuel demand in many developing countries, especially in rural areas, biomass will likely continue to be the primary energy source for a considerable portion of humanity. Per capita use is highly dependent on local conditions. Countries with abundant wood supply, such as Finland, Sweden, and Canada, burn more biomass fuels per capita than most other countries, while those with low supply, such as South Korea and Singapore, burn less [80–82].

The notion that wood smoke, being a natural substance, must be benign to human beings is sometimes still heard. It is well established, however, that wood stoves, fire-places, and forest and agricultural fires emit significant amounts of health-harmful pollutants, including several carcinogenic compounds (e.g., polycyclic aromatic hydrocarbons, benzene, aldehydes, and respirable particles). Some of these toxic pollutants during biomass burning, according to researchers Naesher et al. (2007) and [6] were represented in Fig. 4.

CO and NO, two of the primary pollutants in wood smoke, add to atmospheric levels of these gases emitted by other combustion sources. The health effects of

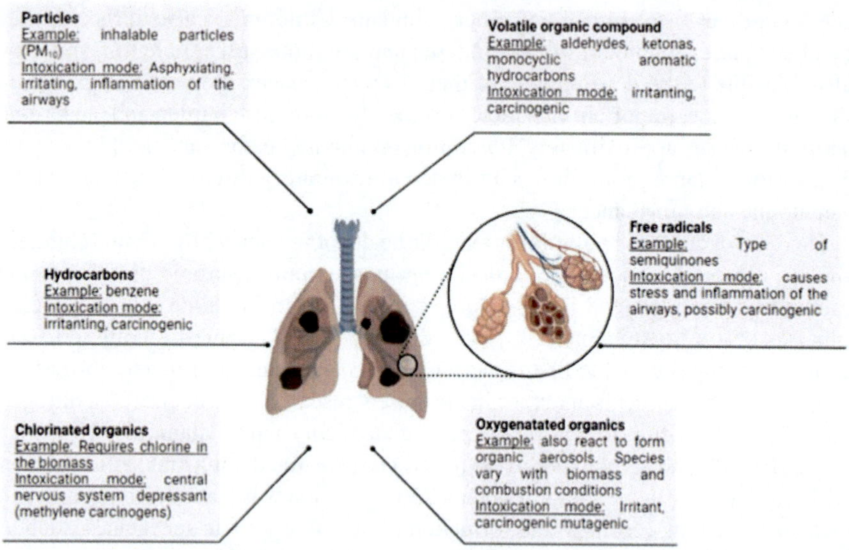

Fig. 4 Primary pollutants harmful to health from biomass combustion (*Source* The authors 2023)

exposure to these gases and other components of wood smoke (e.g., benzene) are well-known in the scientific community. Emissions from biomass burning contain a wide variety of solid, liquid, and gaseous constituents that can change rapidly over time, temperature, sunlight, and interaction with other pollutants [77, 78]. Several constituents are known to be hazardous to human health but are not specifically regulated or even thoroughly evaluated for their health effects. Wood-burning particles are typically smaller than 1 μm, with a peak in the size distribution between 0.15 and 0.4 μm. This is the case with other combustion mixtures, such as diesel fuel and tobacco smoke. In fact, most of the particle mass in old wood smoke is formed by such condensation processes. Fine particles in this size range (0.15 and 0.4 μm) efficiently escape the mucociliary defense system and are deposited in the peripheral airways, where they can exert toxic effects. Particles in this size range are not easily removed by gravitational settling and thus can be transported long distances [72, 75, 76].

The transport of biomass combustion particles over hundreds of kilometers has been extensively documented [22, 44]. Studies conducted by [44] show that modern stoves emit 95% fewer PAH16 and 13 fewer total suspended particles (TSP) than wood stoves. Many gaseous species are converted into other gases or particles during transport. The "black carbon" from biomass emissions are believed to contribute to regional and global climate change and adverse health effects in some parts of the world [66–68]. Methoxyphenols are a class of chemicals derived from the pyrolysis of lignin wood polymer. This class of chemicals covers a range of volatilities, from relatively volatile (e.g., guaiacol) to exclusively particle-associated. These chemicals are relatively abundant in wood smoke, although the most abundant compounds are predominantly in the vapor phase [96, 97, 99]. Accurate chemical analysis of methoxyphenols has proved to be an analytical challenge. Many methoxyphenols have been considered chemically reactive—a property that would impair their suitability as tracers for biomass smoke [15, 66]. Methoxyphenols have been used as tracers of wood smoke in multivariate source apportionment models to determine the proportion of urban fine PM derived from wood burning [11, 14].

9 Meeting the Challenge of Clean Cooking

Due to incomplete combustion and the lack of adequate air pollution control devices, emissions from burning biomass fuels contribute substantially to air pollution [90, 92, 94]. Various cultural, economic, and social factors directly influence the adoption of clean cooking solutions. However, some options are available to rapidly expand adoption in an environmentally sustainable manner and bring various co-benefits. Renewable alternatives such as improved biomass stoves, biogas, ethanol, and solar cookers already contribute to expanded access. Renewable energy-enabled electric cooking has also begun to play an important role [71, 85, 89]. Biogas-based solutions have been deployed for a long time to expand access to clean cooking and improve agricultural waste management.

However, inadequate policies banning traditional fuels and the high price of clean and renewable energy have weakened the motivation of households to use renewable energy. Meanwhile, low income and a lack of clear understanding of pollutants from burning biomass fuels have led families to adopt biomass fuels as a substitute for clean energy, which can further aggravate regional air pollution [51, 62]. Due to the current energy dilemma, a national cessation of biomass fuel burning and promoting an energy transition are incredibly challenging. Some measures that government agencies can take include reviewing and implementing national air quality standards in accordance with the latest WHO guidelines, monitoring air quality and identifying pollutant sources, and supporting and incentivizing the transition to exclusive use of clean domestic energy for cooking, lighting, and heating, and including air pollution themes for both domestic and urban centers in basic cycle disciplines, and providing tools for engagement in the health sector for citizens who do not have access to this knowledge.

10 Conclusions

The use of biomass as a fuel source can also generate harmful gas emissions, including carbon monoxide (CO), nitrogen oxides (NO_x), and volatile organic compounds (VOCs). These emissions can have detrimental effects on human health, such as respiratory and cardiovascular diseases, and contribute to air pollution, which is a significant public health concern worldwide. The amount and composition of gaseous emissions from biomass combustion depend on several factors, including the type of biomass used and characteristics of the biomass involved in the process, such as humidity, density, chemical composition, and even the presence of resin. Furthermore, inefficient combustion processes increase gaseous emissions. Various measures can be taken to mitigate the negative impacts of gas emissions from biomass combustion. These include using sustainable and low-emission biomass sources, improving combustion efficiency, implementing emission controls and monitoring systems, and promoting cleaner technologies such as gas stoves and ovens. Developing and adopting these measures can help minimize the negative impacts of biomass combustion and facilitate the sustainable and low-carbon use of this renewable energy source.

References

1. Abbaspour M, Karbassi A, Asadi MK, Moharamnejad N, Khadivi S, Moradi MA (2013) Energy demand model of the household sector and its application in developing metropolitan cities (case study: Tehran). Polish J Environ Stud 22(2):319–329
2. Adeniran JA, Yusuf RO, Abdulkadir MO, Yusuf M-NO, Abdulraheem KA, Adeoye BK, Sonibare JA, Du M (2020) Evaporation rates and pollutants emission from heated cooking oils and influencing factors. Environ Pollut 266:115169. https://doi.org/10.1016/j.envpol.2020.115169
3. Agrawal S, Yamamoto S (2015) Effect of Indoor air pollution from biomass and solid fuel combustion on symptoms of preeclampsia/eclampsia in Indian women. Indoor Air 25(3):341–352. https://doi.org/10.1111/ina.12144
4. Akteruzzaman M, Rahman MA, Rabbi FM, Asharof S, Rofi MM, Hasan MK, Muktadir Islam MA, Khan MAR, Rahman MM, Rahaman MH (2023) The impacts of cooking and indoor air quality assessment in the southwestern region of Bangladesh. Heliyon 9(1):e12852. https://doi.org/10.1016/j.heliyon.2023.e12852
5. Almeida MM (2017) Protótipo de um fogão à lenha como alternativa aos modelos tradicionais menos eficientes. Universidade Estadual do Centro-Oeste
6. Arbex AA, Cançado JED, Pereira LAA, Braga ALF, Saldiva PHN (2004) Queima de biomassa e efeitos sobre a saúde. J Bras Pneumol 30(2):158–175
7. Bensch G, Jeuland M, Peters J (2021) Efficient biomass cooking in Africa for climate change mitigation and development. One Earth 4(6):879–890. https://doi.org/10.1016/j.oneear.2021.05.015
8. Boudewijns EA, Trucchi M, van der Kleij RMJJ, Vermond D, Hoffman CM, Chavannes NH, van Schayck OCP, Kirenga B, Brakema EA (2022) Facilitators and barriers to the implementation of improved solid fuel cookstoves and clean fuels in low-income and middle-income countries: an umbrella review. Lancet Planet Health 6(7):e601–e612. https://doi.org/10.1016/S2542-5196(22)00094-8
9. Brand MA, Henne RA, Schein VAS, Pereira ER (2021) Mapeamento dos problemas associados à geração e tratamento das cinzas na combustão da biomassa florestal em caldeira. Problem mapping of the generation and treatment of forest biomass ashes in boiler, pp 1167–1192
10. Brito JO (2007) O uso energético da madeira. Estudos Avançados 21(59):185–193. https://doi.org/10.1590/S0103-40142007000100015
11. Browning KG, Koenig JQ, Checkoway H, Larson TV, Pierson WE (1990) A questionnaire study of respiratory health in areas of high and low ambient wood smoke pollution. Pediatr Asthma Allergy Immunol 4(3):183–191. https://doi.org/10.1089/pai.1990.4.183

12. Carneiro H (2003) Comida e sociedade: Uma história da alimentação. Elsevier, Rio de Janeiro
13. Chen J, Li C, Ristovski Z, Milic A, Gu Y, Islam MS, Wang S, Hao J, Zhang H, He C, Guo H, Fu H, Miljevic B, Morawska L, Thai P, LAM YF, Pereira G, Ding A, Huang X, Dumka UC (2017) A review of biomass burning: emissions and impacts on air quality, health and climate in China. Sci Total Environ 579:1000–1034. https://doi.org/10.1016/j.scitotenv.2016.11.025
14. Chen L, Verrall K, Tong S (2006) Air particulate pollution due to bushfires and respiratory hospital admissions in Brisbane, Australia. Int J Environ Health Res 16(3):181–191. https://doi.org/10.1080/09603120600641334
15. Chew FT, Ooi BC, Hui JKS, Saharom R, Goh DYT, Lee BW (1995) Singapore's haze and acute asthma in children. The Lancet 346(8987):1427. https://doi.org/10.1016/S0140-6736(95)92443-4
16. Dutta A, Ray MR, Banerjee A (2012) Systemic inflammatory changes and increased oxidative stress in rural Indian women cooking with biomass fuels. Toxicol Appl Pharmacol 261(3):255–262. https://doi.org/10.1016/j.taap.2012.04.004
17. Enyew HD, Mereta ST, Hailu AB (2021) Biomass fuel use and acute respiratory infection among children younger than 5 years in Ethiopia: a systematic review and meta-analysis. Public Health 193:29–40. https://doi.org/10.1016/j.puhe.2020.12.016
18. Fachinger F, Drewnick F, Borrmann S (2021) How villages contribute to their local air quality—the influence of traffic- and biomass combustion-related emissions assessed by mobile mappings of PM and its components. Atmos Environ 263:118648. https://doi.org/10.1016/j.atmosenv.2021.118648
19. FAO F and AO of the UN (2022) Renewable energy for agri-food systems
20. Ghorashi SA, Khandelwal B (2023) Toward the ultra-clean and highly efficient biomass-fired heaters. A review. Renew Energy 205:631–647. https://doi.org/10.1016/j.renene.2023.01.109
21. Gioda A, Tonietto GB, de Leon AP (2019) Exposição ao uso da lenha para cocção no Brasil e sua relação com os agravos à saúde da população. Cien Saude Colet 24(8):3079–3088. https://doi.org/10.1590/1413-81232018248.23492017
22. Gordon SB, Bruce NG, Grigg J, Hibberd PL, Kurmi OP, Lam KH, Mortimer K, Asante KP, Balakrishnan K, Balmes J, Bar-Zeev N, Bates MN, Breysse PN, Buist S, Chen Z, Havens D, Jack D, Jindal S, Kan H, Mehta S, Moschovis P, Naeher L, Patel A, Perez-Padilla R, Pope D, Rylance J, Semple S, Martin WJ (2014) Respiratory risks from household air pollution in low and middle income countries. Lancet Respir Med 2(10):823–860. https://doi.org/10.1016/S2213-2600(14)70168-7
23. Hoang AT, Nizetic S, Ong HC, Chong CT, Atabani AE, Pham VV (2021) Acid-based lignocellulosic biomass biorefinery for bioenergy production: Advantages, application constraints, and perspectives. J Environ Manage 296:113194. https://doi.org/10.1016/j.jenvman.2021.113194
24. Jewitt S, Smallman-Raynor M, Binaya KC, Robinson B, Adhikari P, Evans C, Karmacharya BM, Bolton CE, Hall IP (2022) Domesticating cleaner cookstoves for improved respiratory health: using approaches from the sanitation sector to explore the adoption and sustained use of improved cooking technologies in Nepal. Soc Sci Med 308:115201. https://doi.org/10.1016/j.socscimed.2022.115201
25. Kajumba PK, Okello D, Nyeinga K, Nydal OJ (2022) Assessment of the energy needs for cooking local food in Uganda: a strategy for sizing thermal energy storage with cooker system. Energy Sustain Dev 67:67–80. https://doi.org/10.1016/j.esd.2022.01.005
26. Kim Y, Knowles S, Manley J, Radoias V (2017) Long-run health consequences of air pollution: evidence from Indonesia's forest fires of 1997. Econ Hum Biol 26:186–198. https://doi.org/10.1016/j.ehb.2017.03.006
27. Krecl P, Targino AC, Lara C, Oukawa GY, Soares J, Mollinedo EM (2023) Detecting local and regional air pollution from biomass burning at a suburban site. Atmos Environ 297:119591. https://doi.org/10.1016/j.atmosenv.2023.119591
28. Kshirsagar MP, Kalamkar VR (2014) A comprehensive review on biomass cookstoves and a systematic approach for modern cookstove design. Renew Sustain Energy Rev 30:580–603. https://doi.org/10.1016/j.rser.2013.10.039

29. Lage S, Gentili FG (2023) Chemical composition and species identification of microalgal biomass grown at pilot-scale with municipal wastewater and CO_2 from flue gases. Chemosphere 313:137344. https://doi.org/10.1016/j.chemosphere.2022.137344
30. Lupepsa BF, Machado G (2019) Fogão à lenha portátil metálico de tecnologia melhorada para o menor consumo de lenha. ENCICLOPÉDIA Biosf 16:2337–2343. https://doi.org/10.18677/encibio
31. Laumbach RJ, Kipen HM (2012) Respiratory health effects of air pollution: update on biomass smoke and traffic pollution. J Allergy Clin Immunol 129(1):3–11. https://doi.org/10.1016/j.jaci.2011.11.021
32. Lenz L, Bensch G, Chartier R, Kane M, Ankel-Peters J, Jeuland M (2023) Releasing the killer from the kitchen? Ventilation and air pollution from biomass cooking. Dev Eng 8:100108. https://doi.org/10.1016/j.deveng.2023.100108
33. Li L, Cheng Y, Dai Q, Liu B, Wu J, Bi X, Choe T-H, Feng Y (2022) Chemical characterization and health risk assessment of VOCs and PM2.5-bound PAHs emitted from typical Chinese residential cooking. Atmos Environ 291:119392. https://doi.org/10.1016/j.atmosenv.2022.119392
34. Liao W, Liu X, Kang N, Song Y, Wang L, Yuchi Y, Huo W, Mao Z, Hou J, Wang C (2023) Associations of cooking fuel types and daily cooking duration with sleep quality in rural adults: effect modification of kitchen ventilation. Sci Total Environ 854:158827. https://doi.org/10.1016/j.scitotenv.2022.158827
35. Lim SS, Vos T, Flaxman AD, Danaei G, Shibuya K, Adair-Rohani H, AlMazroa MA, Amann M, Anderson HR, Andrews KG, Aryee M, Atkinson C, Bacchus LJ, Bahalim AN, Balakrishnan K, Balmes J, Barker-Collo S, Baxter A, Bell ML, Blore JD, Blyth F, Bonner C, Borges G, Bourne R, Boussinesq M, Brauer M, Brooks P, Bruce NG, Brunekreef B, Bryan-Hancock C, Bucello C, Buchbinder R, Bull F, Burnett RT, Byers TE, Calabria B, Carapetis J, Carnahan E, Chafe Z, Charlson F, Chen H, Chen JS, Cheng AT-A, Child JC, Cohen A, Colson KE, Cowie BC, Darby S, Darling S, Davis A, Degenhardt L, Dentener F, Des Jarlais DC, Devries K, Dherani M, Ding EL, Dorsey ER, Driscoll T, Edmond K, Ali SE, Engell RE, Erwin PJ, Fahimi S, Falder G, Farzadfar F, Ferrari A, Finucane MM, Flaxman S, Fowkes FGR, Freedman G, Freeman MK, Gakidou E, Ghosh S, Giovannucci E, Gmel G, Graham K, Grainger R, Grant B, Gunnell D, Gutierrez HR, Hall W, Hoek HW, Hogan A, Hosgood HD, Hoy D, Hu H, Hubbell BJ, Hutchings SJ, Ibeanusi SE, Jacklyn GL, Jasrasaria R, Jonas JB, Kan H, Kanis JA, Kassebaum N, Kawakami N, Khang Y-H, Khatibzadeh S, Khoo J-P, Kok C, Laden F, Lalloo R, Lan Q, Lathlean T, Leasher JL, Leigh J, Li Y, Lin JK, Lipshultz SE, London S, Lozano R, Lu Y, Mak J, Malekzadeh R, Mallinger L, Marcenes W, March L, Marks R, Martin R, McGale P, McGrath J, Mehta S, Memish ZA, Mensah GA, Merriman TR, Micha R, Michaud C, Mishra V, Hanafiah KM, Mokdad AA, Morawska L, Mozaffarian D, Murphy T, Naghavi M, Neal B, Nelson PK, Nolla JM, Norman R, Olives C, Omer SB, Orchard J, Osborne R, Ostro B, Page A, Pandey KD, Parry CD, Passmore E, Patra J, Pearce N, Pelizzari PM, Petzold M, Phillips MR, Pope D, Pope CA, Powles J, Rao M, Razavi H, Rehfuess EA, Rehm JT, Ritz B, Rivara FP, Roberts T, Robinson C, Rodriguez-Portales JA, Romieu I, Room R, Rosenfeld LC, Roy A, Rushton L, Salomon JA, Sampson U, Sanchez-Riera L, Sanman E, Sapkota A, Seedat S, Shi P, Shield K, Shivakoti R, Singh GM, Sleet DA, Smith E, Smith KR, Stapelberg NJ, Steenland K, Stöckl H, Stovner LJ, Straif K, Straney L, Thurston GD, Tran JH, Van Dingenen R, van Donkelaar A, Veerman JL, Vijayakumar L, Weintraub R, Weissman MM, White RA, Whiteford H, Wiersma ST, Wilkinson JD, Williams HC, Williams W, Wilson N, Woolf AD, Yip P, Zielinski JM, Lopez AD, Murray CJ, Ezzati M (2012) A comparative risk assessment of burden of disease and injury attributable to 67 risk factors and risk factor clusters in 21 regions, 1990–2010: a systematic analysis for the Global Burden of Disease Study 2010. The Lancet 380(9859):2224–2260. https://doi.org/10.1016/S0140-6736(12)61766-8
36. Logue JM, Klepeis NE, Lobscheid AB, Singer BC (2014) Pollutant exposures from natural gas cooking burners: a simulation-based assessment for Southern California. Environ Health Perspect 122(1):43–50. https://doi.org/10.1289/ehp.1306673

37. Manoj Kumar, Sachin Kumar, Tyagi SK (2013) Design, development and technological advancement in the biomass cookstoves: a review. Renew Sustain Energy Rev 26:265–285. https://doi.org/10.1016/j.rser.2013.05.010
38. Medeiros A, Gouveia N (2005) Relação entre baixo peso ao nascer e a poluição do ar no Município de São Paulo. Rev Saude Publica 39(6):965–972. https://doi.org/10.1590/S0034-89102005000600015
39. Mitra P, Chakraborty D, Nayek S, Kundu S, Mishra D, Dan U, Mondal NK (2023) Biomass using tribal women exhibited respiratory symptoms, hypertensive risks and abnormal pulmonary function. Chemosphere 311:136995. https://doi.org/10.1016/j.chemosphere.2022.136995
40. Moraes AM De, Trigoso F, Martins G (2008) O fogão a lenha no semiárido do Piauí : métodos tradicionais e alternativas tecnológicas eficientes
41. Moreira MAC, Barbosa MA, Jardim JR, Queiroz MCC, Inácio LU (2013) Doença pulmonar obstrutiva crônica em mulheres expostas à fumaça de fogão à lenha. Rev Assoc Med Bras 59(6):607–613. https://doi.org/10.1016/j.ramb.2013.09.001
42. Munyao CM, Kiptoo KKG, Odinga C, Simiyu GM (2022) Assessment of improved biomass cookstove technologies and kitchen characteristics on indoor air quality and fuel consumption in rural settings of western, Kenya. Eur J Health Sci 7(5):58–83. https://doi.org/10.47672/ejhs.1228
43. Muralidharan V, Sussan T, Limaye S, Koehler K, Williams D, Rule A, Juvekar S, Breysse P, Salvi S, Biswal S (2015) Field testing of alternative cookstove performance in a rural setting of western India. Int J Environ Res Public Health 12(2):1773–1787. https://doi.org/10.3390/ijerph120201773
44. Naeher LP, Brauer M, Lipsett M, Zelikoff JT, Simpson CD, Koenig JQ, Smith KR (2007) Woodsmoke health effects: a review. Inhal Toxicol 19(1):67–106. https://doi.org/10.1080/08958370600985875
45. Nix E, Betang E, Baame M, Abbott M, Saligari S, Shupler M, Čukić I, Puzzolo E, Pope D, Mbatchou B, Anderson de Cuevas R (2022) Complex dynamics in sustaining clean cooking and food access through a pandemic: A COVID-19 impact study in peri-urban Cameroon. Energy Sustain Dev 71:167–175. https://doi.org/10.1016/j.esd.2022.09.017
46. Nuño N, Mäusezahl D, Hartinger SM, Riley-Powell AR, Verastegui H, Wolf J, Muela J, Paz-Soldán VA (2023) Acceptance and uptake of improved biomass cookstoves in Peru—learning from system level approaches to transform large-scale cooking interventions. Energy Res Soc Sci 97:102958. https://doi.org/10.1016/j.erss.2023.102958
47. Obernberger I, Biedermann F, Widmann W, Riedl R (1997) Concentrations of inorganic elements in biomass fuels and recovery in the different ash fractions. Biomass Bioenergy 12(3):211–224. https://doi.org/10.1016/S0961-9534(96)00051-7
48. Olave RJ, Forbes EGA, Johnston CR, Relf J (2017) Particulate and gaseous emissions from different wood fuels during combustion in a small-scale biomass heating system. Atmos Environ 157:49–58. https://doi.org/10.1016/j.atmosenv.2017.03.003
49. Padhi A, Bansal M, Habib G, Samiksha S, Raman RS (2022) Physical, chemical and optical properties of PM2.5 and gaseous emissions from cooking with biomass fuel in the Indo-Gangetic Plain. Sci Total Environ 841:156730. https://doi.org/10.1016/j.scitotenv.2022.156730
50. Panwar NL (2010) Performance evaluation of developed domestic cook stove with Jatropha Shell. Waste Biomass Valorization 1(3):309–314. https://doi.org/10.1007/s12649-010-9040-8
51. Park D, Barabad ML, Lee G, Kwon S-B, Cho Y, Lee D, Cho K, Lee K (2013) Emission characteristics of particulate matter and volatile organic compounds in cow dung combustion. Environ Sci Technol 47(22):12952–12957. https://doi.org/10.1021/es402822e
52. Pratali L, Marinoni A, Cogo A, Ujka K, Gilardoni S, Bernardi E, Bonasoni P, Bruno RM, Bastiani L, Vuillermoz E, Sdringola P, Fuzzi S (2019) Indoor air pollution exposure effects on lung and cardiovascular health in the High Himalayas, Nepal: an observational study. Eur J Intern Med 61:81–87. https://doi.org/10.1016/j.ejim.2018.10.023

53. Preble CV, Hadley OL, Gadgil AJ, Kirchstetter TW (2014) Emissions and climate-relevant optical properties of pollutants emitted from a three-stone fire and the Berkeley-Darfur stove tested under laboratory conditions. Environ Sci Technol 48(11):6484–6491. https://doi.org/10.1021/es5002715
54. Price-Allison A, Lea-Langton AR, Mitchell EJS, Gudka B, Jones JM, Mason PE, Williams A (2019) Emissions performance of high moisture wood fuels burned in a residential stove. Fuel 239:1038–1045. https://doi.org/10.1016/j.fuel.2018.11.090
55. Qin Z, Zhuang Q, Cai X, He Y, Huang Y, Jiang D, Lin E, Liu Y, Tang Y, Wang MQ (2018) Biomass and biofuels in China: toward bioenergy resource potentials and their impacts on the environment. Renew Sustain Energy Rev 82:2387–2400. https://doi.org/10.1016/j.rser.2017.08.073
56. Rhofita EI, Rachmat R, Meyer M, Montastruc L (2022) Mapping analysis of biomass residue valorization as the future green energy generation in Indonesia. J Clean Prod 354:131667. https://doi.org/10.1016/j.jclepro.2022.131667
57. Ribeiro IO, do Santos EO, Batista CE, Fernandes KS, Ye J, Medeiros AS, e Oliveira RL, de Sá SS, de Sousa TR, Kayano MT, Andreoli RV, Machado CMD, Surratt JD, Junior SD, Martin ST, de Souza RAF (2020) Impact of biomass burning on a metropolitan area in the Amazon during the 2015 El Niño: the enhancement of carbon monoxide and levoglucosan concentrations. Environ Pollut 260:114029. https://doi.org/10.1016/j.envpol.2020.114029
58. Saravanna JY, Kantamnen R, Fasil N, Sivamani S, Hariram V, Micha Premkumar T, Mohan T (2017) Modelling and analysis of water heating using recovered waste heat from hot flue gases of Chulha. ARPN J Eng Appl Sci 12(21):6164–6171
59. Santos GHF, do Nascimento RS, Alves GM (2017) Biomassa como energia renoáel no Brasil. UNINGÀ Rev 29:6–13
60. Santos PR, Pereira G, da Silva Cardozo F, Mataveli GAV, Moraes EC (2021) Desenvolvimento e implementação do ciclo diurno da queima de biomassa no PREP-CHEM-SRC. Geography Department University of Sao Paulo 41:e174236. https://doi.org/10.11606/eISSN.2236-2878.rdg.2021.174236
61. Sartor K, Restivo Y, Ngendakumana P, Dewallef P (2014) Prediction of SO_x and NO_x emissions from a medium size biomass boiler. Biomass Bioenergy 65:91–100. https://doi.org/10.1016/j.biombioe.2014.04.013
62. Sha Q, Lu M, Huang Z, Yuan Z, Jia G, Xiao X, Wu Y, Zhang Z, Li C, Zhong Z, Zheng J (2019) Anthropogenic atmospheric toxic metals emission inventory and its spatial characteristics in Guangdong province, China. Sci Total Environ 670:1146–1158. https://doi.org/10.1016/j.scitotenv.2019.03.206
63. Shen G, Lin W, Chen Y, Yue D, Liu Z, Yang C (2015) Factors influencing the adoption and sustainable use of clean fuels and cookstoves in China—a Chinese literature review. Renew Sustain Energy Rev 51:741–750. https://doi.org/10.1016/j.rser.2015.06.049
64. Shen G, Tao S, Chen Y, Zhang Y, Wei S, Xue M, Wang B, Wang R, Lu Y, Li W, Shen H, Huang Y, Chen H (2013) Emission characteristics for polycyclic aromatic hydrocarbons from solid fuels burned in domestic stoves in rural China. Environ Sci Technol 47(24):14485–14494. https://doi.org/10.1021/es403110b
65. Shi Y, Xi Z, Simayi M, Li J, Xie S (2020) Scattered coal is the largest source of ambient volatile organic compounds during the heating season in Beijing. Atmos Chem Phys 20(15):9351–9369. https://doi.org/10.5194/acp-20-9351-2020
66. Smith KR, Apte MG, Yuqing M, Wongsekiarttirat W, Kulkarni A (1994) Air pollution and the energy ladder in Asian cities. Energy 19(5):587–600. https://doi.org/10.1016/0360-5442(94)90054-X
67. Smith KR, Ezzati M (2005) How environmental health risks change with development: the epidemiologic and environmental risk transitions revisited. Annu Rev Environ Resour 30(1):291–333. https://doi.org/10.1146/annurev.energy.30.050504.144424
68. Smith MA, Jalaludin B, Byles JE, Lim L, Leeder SR (1996) Asthma presentations to emergency departments in Western Sydney during the January 1994 Bushfires. Int J Epidemiol 25(6):1227–1236. https://doi.org/10.1093/ije/25.6.1227

69. Sousa AC, Pastorinho MR, Masjedi MR, Urrutia-Pereira M, Arrais M, Nunes E, To T, Ferreira AJ, Robalo-Cordeiro C, Borrego C, Teixeira JP, Taborda-Barata L (2022) Issue 1—"Update on adverse respiratory effects of outdoor air pollution" Part 2): outdoor air pollution and respiratory diseases: perspectives from Angola, Brazil, Canada, Iran, Mozambique and Portugal. Pulmonology 28(5):376–395. https://doi.org/10.1016/j.pulmoe.2021.12.007
70. Still D, Bentson S, Li H (2015) Results of laboratory testing of 15 cookstove designs in accordance with the ISO/IWA tiers of performance. EcoHealth 12(1):12–24. https://doi.org/10.1007/s10393-014-0955-6
71. Streets D, Hao J, Wu Y, Jiang J, Chan M, Tian H, Feng X (2005) Anthropogenic mercury emissions in China. Atmos Environ 39(40):7789–7806. https://doi.org/10.1016/j.atmosenv.2005.08.029
72. Sullivan J, Sheppard L, Schreuder A, Ishikawa N, Siscovick D, Kaufman J (2005) Relation between short-term fine-particulate matter exposure and onset of myocardial infarction. Epidemiology 16(1):41–48. https://doi.org/10.1097/01.ede.0000147116.34813.56
73. Sun Y, Wang S, Yang Q, Li J, Wang L, Zhang S, Yang H, Chen H (2023) Environmental impact assessment of VOC emissions from biomass gasification power generation system based on life cycle analysis. Fuel 335:126905. https://doi.org/10.1016/j.fuel.2022.126905
74. Sutar KB, Kohli S, Ravi MR, Ray A (2015) Biomass cookstoves: a review of technical aspects. Renew Sustain Energy Rev 41:1128–1166. https://doi.org/10.1016/j.rser.2014.09.003
75. Sutherland ER, Make BJ, Vedal S, Zhang L, Dutton SJ, Murphy JR, Silkoff PE (2005) Wildfire smoke and respiratory symptoms in patients with chronic obstructive pulmonary disease. J Allergy Clin Immunol 115(2):420–422. https://doi.org/10.1016/j.jaci.2004.11.030
76. Tan WC, Qiu D, Liam BL, Ng TP, Lee SH, van Eeden SF, D'Yachkova Y, Hogg JC (2000) The human bone marrow response to acute air pollution caused by forest fires. Am J Respir Crit Care Med 161(4):1213–1217. https://doi.org/10.1164/ajrccm.161.4.9904084
77. Tepper A, Comstock GW, Levine M (1991) A longitudinal study of pulmonary function in fire fighters. Am J Ind Med 20(3):307–316. https://doi.org/10.1002/ajim.4700200304
78. Tesfaigzi Y (2002) Health effects of subchronic exposure to low levels of wood smoke in rats. Toxicol Sci 65(1):115–125. https://doi.org/10.1093/toxsci/65.1.115
79. Tesfaldet YT, Chanpiwat P (2023) The effects of meteorology and biomass burning on urban air quality: the case of Bangkok. Urban Clim 49:101441. https://doi.org/10.1016/j.uclim.2023.101441
80. Torigoe K, Hasegawa S, Numata O, Yazaki S, Matsunaga M, Boku N, Hiura M, Ino H (2000) Influence of emission from rice straw burning on bronchial asthma in children. Pediatr Int 42(2):143–150. https://doi.org/10.1046/j.1442-200x.2000.01196.x
81. Traynor GW, Apte MG, Carruthers AR, Dillworth JF, Grimsrud DT, Gundel LA (1987) Indoor air pollution due to emissions from wood-burning stoves. Environ Sci Technol 21(7):691–697. https://doi.org/10.1021/es00161a010
82. Triche EW, Belanger K, Bracken MB, Beckett WS, Holford TR, Gent JF, McSharry J-E, Leaderer BP (2005) Indoor heating sources and respiratory symptoms in nonsmoking women. Epidemiology 16(3):377–384. https://doi.org/10.1097/01.ede.0000158225.44414.85
83. Vassilev SV, Baxter D, Andersen LK, Vassileva CG (2010) An overview of the chemical composition of biomass. Fuel 89(5):913–933. https://doi.org/10.1016/j.fuel.2009.10.022
84. Wamalwa P, Okoti M, Mutembei H, Mandila B, Kisiangani B (2022) Adoption of improved biomass cook stoves: case study of Baringo and West Pokot Counties in Kenya. J Sustain Bioenergy Syst 12(02):21–36. https://doi.org/10.4236/jsbs.2022.122003
85. Wang L, Wang S, Zhang L, Wang Y, Zhang Y, Nielsen C, McElroy MB, Hao J (2014) Source apportionment of atmospheric mercury pollution in China using the GEOS-Chem model. Environ Pollut 190:166–175. https://doi.org/10.1016/j.envpol.2014.03.011
86. Wei W, Zhang W, Hu D, Ou L, Tong Y, Shen G, Shen H, Wang X (2012) Emissions of carbon monoxide and carbon dioxide from uncompressed and pelletized biomass fuel burning in typical household stoves in China. Atmos Environ 56:136–142. https://doi.org/10.1016/j.atmosenv.2012.03.060
87. WHO (2014) Guidelines for indoor air quality: household fuel combustion

88. Witt MB (2005) An improved wood cookstove: Harnessing fan driven forced draft for cleaner combustion. Hartford CT Trinity Coll
89. Wu J, Kong S, Wu F, Cheng Y, Zheng S, Yan Q, Zheng H, Yang G, Zheng M, Liu D, Zhao D, Qi S (2018) Estimating the open biomass burning emissions in central and eastern China from 2003 to 2015 based on satellite observation. Atmos Chem Phys 18(16):11623–11646. https://doi.org/10.5194/acp-18-11623-2018
90. Xing X, Zhou Y, Lang J, Chen D, Cheng S, Han L, Huang D, Zhang Y (2018) Spatiotemporal variation of domestic biomass burning emissions in rural China based on a new estimation of fuel consumption. Sci Total Environ 626:274–286. https://doi.org/10.1016/j.scitotenv.2018.01.048
91. Xu HM, Cao JJ, Ho KF, Ding H, Han YM, Wang GH, Chow JC, Watson JG, Khol SD, Qiang J, Li WT (2012) Lead concentrations in fine particulate matter after the phasing out of leaded gasoline in Xi'an, China. Atmos Environ 46:217–224. https://doi.org/10.1016/j.atmosenv.2011.09.078
92. Xu J-W, Martin RV, Henderson BH, Meng J, Öztaner YB, Hand JL, Hakami A, Strum M, Phillips SB (2019) Simulation of airborne trace metals in fine particulate matter over North America. Atmos Environ 214:116883. https://doi.org/10.1016/j.atmosenv.2019.116883
93. Yadav IC, Linthoingambi Devi N, Li J, Syed JH, Zhang G, Watanabe H (2017) Biomass burning in Indo-China peninsula and its impacts on regional air quality and global climate change-a review. Environ Pollut 227:414–427. https://doi.org/10.1016/j.envpol.2017.04.085
94. Yan Y, Lin J, Chen J, Hu L (2016) Improved simulation of tropospheric ozone by a global-multi-regional two-way coupling model system. Atmos Chem Phys 16(4):2381–2400. https://doi.org/10.5194/acp-16-2381-2016
95. Yang N (2010) Fuel-efficient stoves: a comparison of Berkeley-Darfur stoves to three-stone fires 1–17
96. Yu O, Sheppard L, Lumley T, Koenig JQ, Shapiro GG (2000) Effects of ambient air pollution on asthma symptoms in Seattle-area children enrolled in the CAMP study. Environ Health Perspect 108(12):1209–1214. https://doi.org/10.1289/ehp.001081209
97. Zelikoff JT, Chen LC, Cohen MD, Schlesinger RB (2002) The toxicology of inhaled woodsmoke. J Toxicol Environ Health Part B 5(3):269–282. https://doi.org/10.1080/10937400290070062
98. Zhang R, Xu G, Li B, Wang Z, Gao J, Li J, Sun Y, Xu G (2023) Analysis of the pollution emission system of large-scale combustion of biomass briquette fuel in China. Process Saf Environ Prot 169:928–936. https://doi.org/10.1016/j.psep.2022.11.088
99. Zheng S, Shen H, Shen G, Chen Y, Ma J, Cheng H, Tao S (2022) Vertically-resolved indoor measurements of air pollution during Chinese cooking. Environ Sci Ecotechnol 12:100200. https://doi.org/10.1016/j.ese.2022.100200
100. Zong Z, Shi X, Sun Z, Tian C, Li J, Fang Y, Gao H, Zhang G (2022) Nitrogen isotopic composition of NO_x from residential biomass burning and coal combustion in North China. Environ Pollut 304:119238. https://doi.org/10.1016/j.envpol.2022.119238

Combustion Equipment Used in Food Preparation Around the World: What Is Its Influence on Air Pollution and How to Mitigate These Harmful Effects?

Álison Moreira da Silva, João Gilberto Meza Ucella Filho, Kamilla Cryslayne Alves da Silva, Tayná Rebonato Oliveira, Allana Katiussya Silva Pereira, and Ananias Francisco Dias Júnior

Abstract The use of wood or biomass-burning stoves for cooking is a common practice in many countries. Approximately 3 billion people cook and heat their homes with fires and stoves that burn biomass (wood, animal dung or agricultural waste) and charcoal. However, this practice can lead to serious environmental and public health problems, as the inefficient burning of biomass in these stoves produces large amounts of polluting gases and fine particles that are harmful both to those using the equipment and to the environment. To mitigate these impacts, there are several strategies that can be adopted. One of them is the use of more efficient stoves that consume less fuel and produce less smoke. These stoves can be of various types, from the simplest ones that use only a small amount of wood, to the most advanced ones that have combustion systems and filters to reduce the emissions of polluting gases. In addition, there are also other alternatives, such as the use of solar ovens, which use the energy of the sun to cook food, or the use of oven filters, which can

Á. M. da Silva (✉) · K. C. A. da Silva · A. K. S. Pereira
"Luiz de Queiroz" College of Agriculture, University of São Paulo (ESALQ/USP), Department of Forests Sciences, Av. Pádua Dias, 11, Piracicaba, São Paulo 13418-900, Brazil
e-mail: alison_vni@hotmail.com

K. C. A. da Silva
e-mail: kamilla.alves@usp.br

A. K. S. Pereira
e-mail: allana.florestal@gmail.com

J. G. M. Ucella Filho · T. R. Oliveira · A. F. Dias Júnior
Federal University of Espírito Santo (UFES), Department of Forestry and Wood Sciences, Av. Governador Lindemberg, 316, Jerônimo Monteiro, Espírito Santo 29550-000, Brazil
e-mail: 16joaoucella@gmail.com

T. R. Oliveira
e-mail: taynarebonato1@gmail.com

A. F. Dias Júnior
e-mail: ananias.dias@ufes.br

A. K. S. Pereira and A. F. Dias Júnior (eds.), *Impacts of Using Biomass as an Energy Source in Homes*, Green Energy and Technology,
https://doi.org/10.1007/978-3-031-38824-8_3

be installed in existing equipment to reduce the emission of polluting gases. Despite these alternatives, there are still many gaps in scientific knowledge about the use of wood or biomass-burning stoves for cooking. In this context, the objective of this chapter is to address how the equipment used in food preparation leads to a significant increase in gas emissions. We will also address alternatives to overcome the reality of using rudimentary technologies, such as solar ovens and oven filters. Finally, we will show the existing gaps within this niche of study and what research should focus on in future studies.

Keywords Food safety · Biomass · Household energy · Environmental health

1 Introduction

The use of wood or biomass-burning stoves for cooking is a common practice in various regions of the world, especially in rural and low-income communities. These stoves can range from the most rudimentary, such as clay stoves and open fires, to more modern equipment, such as wood-fired ovens and stoves with more efficient burners. However, despite being considered a more accessible and economical alternative, these stoves can generate serious environmental and public health impacts.

Since ancient times, wood or biomass-burning stoves have been used worldwide. In Africa, for example, the "jiko" is a type of clay stove used for cooking food [64]. In Asia, the use of "tandoors" or cylindrical clay ovens to cook bread and other foods is common [33]. In Latin America, wood-burning stoves are often used in rural areas. In recent years, more modern equipment has emerged, such as stoves with more efficient burners and wood-fired ovens with thermal insulation, which reduce the amount of fuel needed to cook food.

However, despite the economic and cultural benefits, the use of wood or biomass equipment can generate serious environmental and public health impacts. The combustion of biomass produces large amounts of polluting gases that contribute to the greenhouse effect and climate change [51]. In addition, inhalation of smoke generated by these equipments can cause serious respiratory problems, such as lung diseases and lung cancer [103]. Among the main gases emitted by wood or biomass equipment are carbon dioxide, methane, and carbon monoxide [51]. Carbon dioxide is one of the main greenhouse gases, while methane is even more potent in terms of greenhouse effect. Carbon monoxide is a toxic gas that can cause poisoning and even death in high concentrations.

There are several strategies to mitigate gas emissions from wood or biomass cooking equipment. One of them is the adoption of more efficient stoves, which consume less fuel and produce less smoke. Another strategy is the use of renewable energy sources, such as solar energy or biogas, for food preparation [100]. Additionally, community awareness about the environmental and public health impacts of using wood or biomass equipment is crucial for the adoption of more sustainable practices. In this sense, the objective of this chapter is to address the main biomass-based equipment used in food preparation, from the oldest to the most modern, and their influence on gas emissions and climate change. We will also present the latest technologies for wood or biomass equipment in food preparation, with the aim of identifying innovative and sustainable solutions to reduce emissions and improve the energy efficiency of these devices. Moreover, we will present strategies to mitigate gas emissions from these cooking devices, taking into account technical, economic, and social aspects. Finally, we will also present scientific advances in this research field to determine existing gaps and what needs to be the focus of study in this area in the coming years.

2 Equipment for Food Preparation: History and Technological Advances

The history of stoves begins with the discovery of fire by our ancestors. After archaeological excavations in the village of Zhoukoudian (Chou Kutien), China, it was found that *Homo erectus pekingnensis*, or Peking Man, used biomass fuel inside caves to keep warm about 400,000 years ago [14]. Research conducted at the site later enabled the discovery of burned items such as charcoal, stones, and bones. These findings were widely accepted as the oldest reliable evidence of hominids using fire in the world [12, 78]. During the Paleolithic period, open fires (bonfires) (Fig. 1a) were used for warmth and protection against animal attacks, and only in the Middle Paleolithic period did the use of fire become known for cooking food [66, 102]. One of the earliest cooking methods was to roast meat in a type of oven that consisted of holes in the ground. The fire is made inside the hole with biomass fuels, and only when the flames are extinguished are the meats that are arranged in alternating layers between stones inserted into the pits and finished by covering them with soil (Fig. 1a). These methods are still used in some places today [102].

The initial period, prior to the seventeenth century, was marked by low-energy efficiency stoves with high smoke emissions due to the use of weak materials in their construction, known as "traditional stoves" [66]. The stove took on its basic familiar form about 12,000 years ago with the development of its design, which progressed from an open fire to a closed firebox with the goal of balancing pots on the fire. The

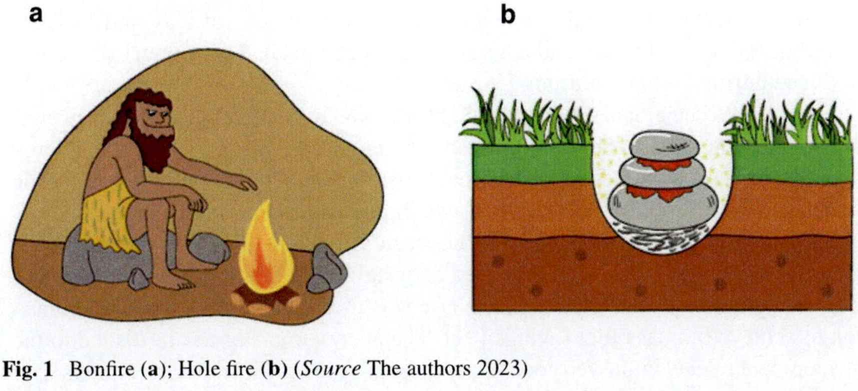

Fig. 1 Bonfire (**a**); Hole fire (**b**) (*Source* The authors 2023)

Fig. 2 Three-stone stove (*Source* The authors 2023)

three-stone wood stove (trempé), the first model of a closed stove and still used today in underdeveloped countries, consists of an arrangement of three similarly shaped and sized stones, which are placed on a flat surface to maintain the stability of the pots that will be placed on top, with biomass arranged in the center of the stones (Fig. 2). Its dimensions are defined relative to the size of the utensil that will be used [66].

With the purpose of obtaining a more efficient stove, models were increasingly improved over time. As time went by, stoves evolved and took on various shapes and sizes, made with different types of materials and adapted to different cultures and needs [59]. With the development of the enclosed fire, the three-stone model became a clay stove in the shape of a "U". In this model, the stove is made with a mixture of clay and straw, fixed to the ground in a "U" shape, modeling a semi-open oven, where the front has an opening for the biomass feed and there may be protrusions on its surface to improve the support of cooking utensils and facilitate air intake (Fig. 3) [66]. The mechanism for improving this model is the induction of secondary air for efficient combustion, as the air prevents the slip of unburned biomass, thus promoting complete burning of the fuel that was used. The first improved clay stove reported in the literature was created in India in 1947 and was called the "Magan Chulha". Aimed at solving problems such as smoke production for health improvement and forest conservation, the model is now produced with blocks of clay and includes curved ducts, a metal grate, and a ceramic chimney attached to conduct the smoke produced during biomass burning [5, 58].

In the 1970s, the global oil crisis brought attention to energy issues [87]. However, during this time, researcher [29] published a report on "Wood: The Other Energy Crisis," which highlighted the unsustainable use of wood in rural areas for domestic activities such as cooking and heating, which he deemed subtle but equally devastating. According to [102] the period between 1970 and 1980 was known as the "first phase" or "first wave" of improved cookstove (IC) development. The primary goal of professionals during this first phase was to achieve more efficient combustion to achieve biomass fuel savings [91]. The energy crisis of the 1970s led to the development of new improved cookstoves through research, introducing models with

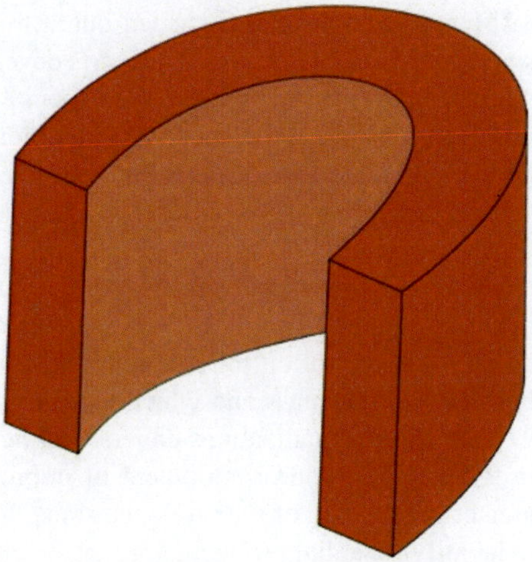

Fig. 3 "U" type stove (*Source* The authors 2023)

low biomass consumption in locations that heavily relied on this fuel for cooking, as attention was focused on environmental and energy conservation issues [30, 39, 66]. It was in this context that the first projects were implemented. After a major earthquake hit Guatemala in 1976, the country received international aid to rebuild destroyed homes, and improvements were made to kitchens. As a result, they developed a stove model called the "Lorena stove," made with clay blocks filled with a sand-clay mixture (Fig. 4). This stove model soon became popular in Guatemala and other Central American countries due to its efficiency, product promotion, and further research by the Aprovecho Research Center (ARC) [102]. The Lorena stove model has been modified over the years by other IC programs in South Asian countries, where [5] showcases several programs and their respective modifications based on the Guatemala model.

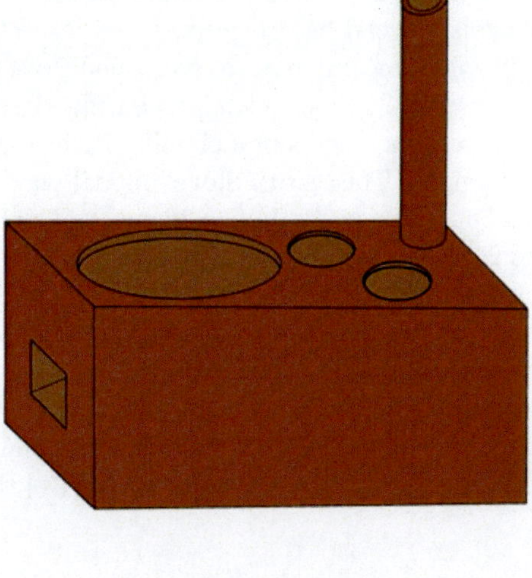

Fig. 4 Lorena stove (*Source* Adapted from [37]

Many of the programs carried out at this time were not accepted by the public since the creation of stoves by experts developed rapidly and extremely technically, without even considering the issue of social context, given that the majority of stove users in the world are women who were not involved in the development process as their empirical knowledge was considered inferior to the technical knowledge of the experts [19, 88].

The beginning of the second phase was marked by two scenarios that changed the course of improved cookstove organizations. Firstly, the understanding that the deficit of wood was not linked to subsistence consumption of rural communities, but rather to deforestation for subsequent cultivation or grazing of livestock, practices linked to agribusiness and which became a threat to energy resources [19, 36, 65, 97]. Secondly, organizations understood the importance of community participation in the planning and development of improved cookstoves, which would result in greater acceptance of the products, since their involvement would make it possible to identify potential problems for that specific community [10, 102].

At this moment, several stove improvement programs were created in various parts of the world, but mainly in underdeveloped countries that had biomass as their main source of energy for domestic activities. The stoves were adapted over the years according to culinary habits and traditions. Improved stoves were developed in various countries around the world, such as Ethiopia, Eritrea, Ghana, Ivory Coast, Kenya, Madagascar, Malawi, Mozambique, Niger, Bangladesh, Indonesia, Nepal, Pakistan, Thailand, Bolivia, Argentina, Cuba, Peru, and Mexico [3]. However, India and China are countries that have developed more widespread programs and are therefore chosen to be explored throughout the chapter.

The Indian National Program on Improved Chulha (NPIC) was established in 1983 and only launched on a full-scale basis in mid-1985 [42]. As previously discussed, the main purpose was to improve the management of energy resources, and the NPIC included several objectives in its project to achieve this goal: (i) conservation of biomass, (ii) removal of kitchen smoke; (iii) updating of deforestation data; (iv) reduction of women's labor-intensive work; and (v) providing opportunities for extra income for local communities [5]. Despite the program having 80 different models, all can be categorized into 6 different groups: (a) fixed mud stove, with and without a chimney, (b) fixed mud stove coated with ceramics, with or without a chimney; (c) portable metal stove without a chimney; (d) portable metal stove coated with ceramics, without a chimney; and (e) portable stove with a separate chimney system [3]. The "Astra Stove" model was designed by the Centre for Sustainable Technologies at the Indian Institute of Science in Bangalore. The Astra stove is considered one of the best stoves available in the country, offering more than 40% efficiency [53]. These results were achieved through improvements such as controlled entry of primary and secondary air, facilitating proper combustion, inclusion of a chimney with appropriate measurements, and the overall design [5]. In addition to improvements in efficient combustion, the NPIC program demonstrated other progress, such as female empowerment, and some states promoted improvements in cooking through these popular programs [3].

As discussed earlier, each stove model corresponds to the local reality of the community. The most popular ICs in Sri Lanka, a country located south of India, is the "Anagi Stove" introduced by the Ceylon Electricity Board (CEB) in 1986. This stove is built with two pieces of clay pot and can be fueled by wood, leaves, coconut shells, and other biomass residues. Its unique design was developed to meet the needs and culinary habits of the local culture in Sri Lanka [5].

Another program with a high rate of dissemination of stoves was the Chinese National Improved Stove Program (NISP), created by the government in the early 1980s and considered the world's largest public financing initiative for stove improvements. With more than 100 million units, the NISP is one of the most successful programs [106]. The FL model is a stove developed based on local cuisine and built with brick, cement, and cast iron. This stove has two design models: the FL Partial Composite Stove (FL-PCS) and the FL Complete Composite Stove (FL-CCS). The difference between the models is in the material they are constructed with, the FL-PCS has the combustion chamber and cast iron grate while the other parts are self-constructed, whereas the FL-CCS is entirely made of cast iron and includes a ready-to-install chimney [105].

As previously discussed, the second phase was mainly characterized by the need to save wood and reduce hard work, but reducing smoke was a secondary issue [91]. After research, it was found that the use of chimneys only shifted the smoke problem outside the houses but did not eliminate it [94]. As a consequence, in the early 1990s, it was observed that the use of wood for cooking food could pose health risks due to the smoke produced during combustion, and in this problem, agents realized that the use of ICs could bring, once again, a solution to this impasse [34, 92]. At the World Summit on Sustainable Development held in Johannesburg in 2002, the U.S. Environmental Protection Agency (EPA) announced the "Partnership for Clean Indoor Air" project aimed at mitigating the health risks of people using biomass fuels in their homes [13]. In this third phase, improved stove programs were improved with an emphasis on health issues [7]. Single-pot metal stoves without chimneys were the most launched models during this period [59].

As a result, the Indian government, after the country's Ministry of New and Renewable Energy (MNRE) considered the NPIC a failure, relaunched a "National Biomass Stove Initiative" project in December 2009 with the main objective of increasing the use of improved biomass stoves, but with more efficient, longer-lasting, effective, and easy-to-use products. Other objectives of the program are to develop and implement biomass stoves that provide cleaner energy, reduce the hard work of women and children who use traditional Chulhas, and reduce climate change by reducing black carbon emissions during biomass burning. The project has more than 40 different models divided into fixed and portable types (natural and forced draft) and can be made of metal, ceramic, and terracotta/ceramic [70].

Despite technological advancements in improving wood stoves, researchers around the world have been seeking more sustainable and low-cost alternatives for the development of cooking equipment. This interest is due to three important factors: (i) improving the quality of life of the population; (ii) reducing the emission of polluting gases released during biomass burning that influence climate change; and (iii) seeking renewable energy sources in order to reduce the use of petroleum and natural gas. These points are directly related to the Sustainable Development Goals (SDGs) established by the WHO (World Health Organization), specifically: 3—Good health and well-being; 7—Affordable and clean energy; 10—Reduced inequalities; 12—Responsible consumption and production; and 13—Climate action. Based on this, solar ovens stand out as a recent, accessible, and efficient technology for cooking that can reduce the use of wood stoves and consequently their social and environmental impacts [27, 74].

Solar stoves, also known as solar ovens, are devices that use solar radiation as the primary source of energy, mainly for cooking food, but also have potential for important processes such as pasteurization and sterilization [44, 46] Solar stoves are based on the principle of concentrating solar energy and using this heat for cooking food, without the need for any other fuel besides sunlight to operate these devices [77]. It is reported that the first solar ovens were developed in the eighteenth century, with physicist Horace de Saussure being the first researcher to use this equipment to cook fruits [20, 84]. However, according to [84], the World Wars and the energy crisis intensified the interest in using these types of stoves.

Considering that around 2.4 billion people, mainly located in poor or developing countries, cook using polluting fuels [22] world organizations such as the Solar Cookers International (SCI) have intensified efforts to introduce the use of solar stoves in the daily lives of these populations for sustainable and environmentally friendly cooking, with the main mission of improving human health, economic well-being, and the environment [4, 86].

Solar cookers can be found in different shapes, varying according to the manufacturers and the improvements made by researchers focused on developing more efficient devices. However, solar cookers fall into two categories based on the heat transfer method, namely the direct method, which includes box-type cookers, and the indirect method, which includes solar panel and parabolic cookers. Box-type solar cookers are characterized as an isolated device with a multiple or single glass lid, which has reflectors that direct the sun's rays into the box. The interior of the box is coated with black paint to improve the absorption capabilities of the solar light [71]. According to [20], the first containers created for cooking food using solar radiation were of the box type. However, it was only in the 1970s that an efficient and viable box-type solar oven model for domestic use was developed by Barbara Kerr, a resident of Arizona in the United States [84]. Generally, this cooker model can reach temperatures of up to 100 °C, a characteristic that enables boiling food [107]. However, the main limitation of this device is the lack of energy storage, which makes it difficult to use when it is not receiving solar radiation. Within this context, research is being developed mainly to improve the characteristics of these cookers in order to make them more efficient. Reference [21], when evaluating solar cookers with and without thermal energy storage, concluded that devices made up of Bayburt stone are an excellent material to complement box-type cookers due to their constant, efficient, and continuous cooking ability, as well as their capacity for thermal energy storage for different periods of the day. The advantages of box-type cookers are linked to the simplicity of the equipment's construction, ease of use, no risk of fires or burns to the operator, and excellent energy capacity, whether reinforced by mirrors or not [60, 71, 77, 84].

Among the indirect heat transfer stoves, the solar panel and parabolic stoves stand out. Solar panel stoves are attractive equipment for travelers or people living alone, and they are easy to construct and do not require expensive materials for assembly. The basic principle of this type of device involves heat transfer to a container intended for cooking food through reflective plates [71]. This device is capable of providing a cooking temperature of around 140 °C, but it is entirely dependent on solar radiation, so it becomes unfeasible for use on cloudy days or at night [55, 77]. Unlike solar panel stoves, parabolic solar stoves stand out as one of the best models of solar stoves for their ability to reach high temperatures. The first models of this device were developed in the 1950s by researcher Ghai in India, who evaluated different ways of obtaining the most efficient parabolic stove. This type of equipment is a concentrator stove, meaning that it is a solar stove in which the pot is placed at the focus of a concentration mirror, so the cooking temperature can reach from 300 to 450 °C [52, 71]. The main disadvantage of this type of equipment is the constant need for sun, the operator's need to monitor the food preparation process to prevent burning, and the risk of fire during the activity [61]. To improve the efficiency of using this type of device, recent research is focused on developing mechanisms that can store energy in parabolic solar ovens to make them viable for use on days with low solar radiation, at night, and for indoor use [6, 31, 67, 69].

The main advantage of using solar cookers is the absence of toxic gas emissions during food preparation. Additionally, these devices can be constructed using a variety of low-cost and highly efficient materials. In a study conducted by [82], different types of waste materials from humanitarian supply packaging were used to develop solar cookers. The authors concluded that the prototypes developed were suitable for domestic use, with those made of cardboard being able to heat food and boil water in relatively short times. Therefore, solar cookers are considered promising technological innovations for replacing biomass-fueled stoves, thereby reducing the emission of gases that affect human health and intensify climate change.

3 Cooking Equipment and Gaseous Emissions: Implications for Climate Change and Human Health

3.1 The Influence of Equipment on Gas Emissions

Around the world, there are approximately 3 billion people who depend on traditional stoves to cook their food, primarily using solid biomass fuels [9]. Given their increasing use, the global research community has begun to shift its focus towards the health impacts associated with emissions during the combustion of fuels in kitchen stoves [9, 93]. Concerns over the type of biomass used are closely related to concerns over exposure to toxic gases that are generated during combustion. However, it is equally important to evaluate the efficiency of the stoves to which this biomass will be used, as more efficient stoves tend to minimize the propagation of smoke.

Both the type of fuel used during cooking and the inefficiency of the stoves used can be harmful to human health, depending on the quantity and types of gases present in the ambient air. A new approach, where the main focus is to advance towards improving these devices to have high combustion efficiency and low emissions, and not to produce any significant pollution, should be given top priority. Stoves containing well-operated chimneys tend to avoid indoor air pollution, however, they will still transfer it to the outside, which could result in substantial human exposure. To truly achieve high performance, more reliable materials should be used in their construction. Advanced ceramics or metal alloys, as well as other components, should be made under good quality control by factories, complemented by other modern production techniques [99].

The search for more advanced and efficient stove designs has been going on for a long time, and it is possible to observe an evolution over the years. To face the challenges related to traditional biomass stoves, scientists have created the concept of improved stoves [8]. However, it was only after the 1980s that parameters such as heat transfer, fluid mechanics, standard testing, etc. were finally taken into consideration to improve stove efficiency [11].

Forced-air biomass stoves, such as those used for cooking and heating, are considered clean technology, containing fans that assist in providing air for the combustion of fuel [17]. They can potentially reduce carbon monoxide (CO) emissions by more than 50% compared to traditional stoves [89, 90]. However, the airflow, distribution, and velocity can directly influence pollutant emissions from incomplete combustion of the biomass used when the O_2 content is not comparable to the combustion rate [24, 25]. For a wood stove, it is recommended that there be two limits and two main zones, the air flow and the mass flow. The air flow can usually be modeled as a limit with known velocity, or the mass flow [57].

Higher temperatures followed by good air conditions can make stoves more efficient, generating lower concentrations of small aerodynamic particles (PM) and carbon monoxide (CO) [47]. This makes the presence of an excess air control system important to avoid excessive unsatisfactory air circulation that can result in higher particle mass concentrations [54].

The open burning of these fuels, on the other hand, can have catastrophic consequences for the environment and human health. This includes the destruction of various ecosystems, increased air pollution, changes in soil properties and consequent increase in erosion, water pollution, and contribution to climate change [63, 80].

3.2 *Main Gases Emitted During Food Preparation*

The inefficient combustion of wood is a major source of atmospheric pollutants that generate gases such as carbon monoxide (CO), sulfur dioxide (SO_2), nitrogen dioxide (NO_x), and particulate matter less than 2.5 μm (PM2.5), which cause numerous lung diseases [50, 72, 75, 96]. PM is one of the pollutants that poses the greatest risk to human health, in addition to contributing to climate change. PM is a group of solid and liquid particles that are suspended in the atmosphere and have a diameter of less than 2.5 μm [63].

Carbon monoxide is a colorless, odorless, and non-irritating gas that can be fatal in high concentrations, as it binds to hemoglobin in the blood, reducing the amount of oxygen available. It is a product of incomplete combustion of carbon-containing substances [68]. SO_2 is also a colorless gas emitted from fuel combustion. Its formation results from particles containing sulfur and can cause a range of health problems [15]. Nitrogen oxide (NO_x) can act alone or in combination with PM, directly affecting cardiovascular disease. Typically, NO_x is more attributed to motor vehicle emissions [16], but its presence cannot be ruled out when gases from biomass fuel combustion are released.

3.3 Effects of Polluted Gases in Food Preparation on Human Health

Domestic air pollution is generated through the use of inefficient biomass technologies used in cooking or heating. It is estimated that around 3.8 million deaths worldwide are due to indoor air pollution from these technologies, which generate a series of harmful pollutants to human health, including small particles that penetrate the lungs and bloodstream [49]. Intoxication begins with the incomplete burning of the biomass used, emitting pollutants such as carbon monoxide, nitrogen dioxide, sulfur dioxide, among others, as well as volatile organic compounds such as aromatic hydrocarbons and benzene [76]. The particles emitted during burning are captured by the nasal cavity, airways, and thoracic cavity, with smaller particles tending to lodge in areas such as lung alveoli [22].

Organic compounds such as benzene and polycyclic aromatic hydrocarbons are carcinogenic and mutagenic compounds. Prolonged exposure to these pollutants tends to cause cellular alterations in exposed cells, such as those in the nasopharynx, oral cavity, and airways and lungs [28].

Several research studies report that diseases generated by inhaling this smoke occur more frequently in females. There is a disproportionate prevalence of chronic pulmonary diseases and lung cancer, even in those without a history of smoking [56]. This is because in most cases, women are responsible for household tasks, including cooking. Therefore, during the repetitive process of cooking for their family, women would be exposed to emissions generated by biomass fuels for hours, accumulating higher concentrations of these particulates in their bodies. Children who spend more time with their mothers during the cooking process are also diagnosed with these diseases [40].

Intense exposure to household air pollution tends to increase the prevalence of lung diseases, including asthma, tuberculosis, chronic obstructive pulmonary disease

(COPD), interstitial lung disease, acute infections, lung cancer, as well as non-lung diseases such as nasopharyngeal cancer and ischemic heart disease [23, 26, 32, 83]. COPD is a common chronic inflammatory disease characterized by restriction of airflow in the affected person, as well as symptoms associated with the respiratory system. In general, COPD is considered preventable and treatable. On the other hand, its progression can cause respiratory failure, hypertension, pulmonary heart disease, which will seriously affect the patient's quality and life expectancy [104].

In some cases, diseases resulting from excessive exposure to this smoke may come silently, and patients may be asymptomatic or present nonspecific signs such as cough, sputum production, and dyspnea [56]. In these cases, it is recommended that the exposed person undergo periodic examinations to evaluate their condition.

4 Reducing Gas Emissions in Cooking Equipment: Innovative Solutions for a Healthy and Sustainable Future

In recent years, the evaluation of emission reduction strategies for residential biomass combustion has been investigated with growing interest. However, the evaluation of emission factors from biomass combustion is complex due to the dynamics of combustion being influenced by many parameters, such as the technology employed in the stoves, the type of biomass used, and operating conditions. Therefore, strategies can be oriented towards optimizing the combustion process and/or improving the fuel (primary measures), or they can focus on flue gas cleaning technologies (secondary measures).

When the focus is directed towards the use of primary measures in the combustion concept, studies have addressed projects that contemplate the supply of air through the implementation of "staged combustion with air" [18, 43, 62, 108]. In practice, staged combustion with air prioritizes effective mixing of pyrolysis gases and air with a satisfactory residence time. Typically, air supply occurs in two or three stages, depending on the project, with devolatilization of the fuel (primary) and oxidation of the combustion products (secondary and tertiary). Thus, the introduction of heated secondary air at the top of the primary combustion chamber further increases the ignition of the combustion gases.

Regarding energy crops, it is possible to reduce particulate emissions by improving the fuel used. This involves the use of additives such as aluminum silicates (sewage sludge), calcium-based additives (limestone), and sulfur (lignosulfonate) to capture alkaline gas compounds and thus reduce particulate formation [38]. Furthermore, lower emissions can be achieved simply through good stove operating practices. Reference [85] conducted a study on the impact of operator behavior on emissions from manually operated systems. They modified the amount of fuel loaded into the stove and the air flow configuration and found that the operator can have up to six times the influence on particulate emissions from wood-burning stoves. Reference

[101] conducted a study on how operating conditions affect emissions from a wood-burning stove. Two ignition techniques (bottom-up versus top-down), hot start versus cold start, and different fuel loads (low, medium, and high load) were examined. The results indicated that these variables can have a significant impact on emissions. For example, the top-down ignition technique was able to reduce PM10 by more than 50% compared to the traditional bottom-up ignition technique.

Regarding secondary measures, there are several studies on devices that can be applied to combustion gases, controlling pollution in a certain way. Examples include electrostatic precipitators [73, 79, 109], condensation purifiers [41, 79] and catalytic converters [35, 45, 100].

Small-scale electrostatic precipitators (ESPs) are designed to remove inorganic particles present in combustion gases [79]. However, studies report a series of problems, mainly related to the increased toxicity of gases after combustion [48, 100]. This happens because the depollution device can alter the composition of particulate emissions and, as a result, some compounds may become enriched in the combustion gas after the ESP, altering the toxicity of the released particles. In practice, the use of electrostatic precipitators to control PM emissions from small-scale combustion appliances is not economically viable. In addition, their effectiveness is affected by the resistivity of particles, which constantly varies according to the composition of emitted particles [100].

Widely used, including in the residential heating sector, catalytic combustors have a surprising efficiency when it comes to converting polluting gases into non-harmful gases to the environment and human health. Catalytic converters are equipment whose interior has honeycomb-shaped holes composed of precious metals such as platinum, palladium, and rhodium [45]. In the first stage, the gases undergo a reduction, losing oxygen or electrons. In the second stage, the gases undergo an oxidation process. In these processes, gases resulting from incomplete biomass combustion are converted into CO_2, H_2O, and N_2 [45]. A major problem resulting from the use in combustors is related to the low temperature of the combustion gases when using these devices in inefficient combustion appliances. When installed at the beginning of the chimney, temperatures above 500 °C are hardly reached. This is especially problematic during ignition, when emissions can be very high, but the temperature is very low. Another disadvantage is related to the presence of ash, soot, and creosote that encrust on the surface of the catalyst, decreasing its efficiency and requiring constant cleaning of catalytic converters [100].

Condensation purifiers are used in furnaces to reduce emissions of polluting gases into the atmosphere by condensing them, thus removing solid and liquid particles present. In this way, the condensed gases are filtered before being released into the atmosphere. Condensation purifiers are often used in centralized heating systems that use solid fuels such as wood, pellets, or charcoal. Installing a condensation purifier in a furnace requires some modifications to the exhaust system. It is necessary to add a condensation chamber to the system to allow the gases to be cooled and condensed. It is also important to ensure that the condensation purifier has an adequate filtering system to remove all solid and liquid particles present in the combustion gases [35]. Condensation purifiers are a relatively new technology and may be a costly option for many people. However, they are highly efficient in reducing emissions of pollutants and can help improve local air quality and reduce negative impacts on public health. It is important to remember that installing a condensation purifier is not a complete solution to air pollution but rather a part of a larger effort to reduce emissions of pollutants.

Understanding the particulate matter emission factors is crucial for developing effective pollution control strategies. Several studies have shown that detailed analysis of emission factors, considering the technology and fuel type used, can result in large differences in emissions, especially when comparing older residential devices to modern energy conversion technologies, such as automated wood and pellet stoves and highly efficient boilers. While new biomass burning technologies perform much better than traditional systems, their adoption has been gradual. These systems offer significant advantages, such as automatic control, as well as reducing fuel consumption and pollutant emissions. In inefficient manual combustion devices, the main source of particulate matter (PM) is incomplete combustion, which is mainly composed of carbonaceous material, including soot and condensable organic compounds, which have high toxicity and carcinogenic properties [79, 100]. In contrast, properly operated automatic biomass combustion systems mainly emit inorganic salts [100]. Under non-ideal conditions, such as frequently observed in many small-scale combustion devices, the concentration of condensable organic compounds in the combustion gas can be up to 10 times higher than the mass concentration of solid particles collected directly on hot filters in the chimney.

Although there are several options to reduce pollutant formation, secondary strategies for emissions reduction are not well developed for household-scale applications. This is because their installation would increase costs and maintenance, as well as raise safety concerns. Before any implementation in the market, several operational and failure risks still need to be addressed, and test methods for determining removal efficiency need to be standardized and improved. Additionally, recent studies have shown that these reduction technologies can affect the physicochemical characteristics of particulate matter, which can lead to unwanted changes in toxicological responses [48, 100]. For these reasons, it is necessary to study and establish relationships between these alterations and toxicological effects in future research. Under real-life operating conditions, automatic heating appliances have a significantly lower impact on air quality than manually operated stoves. Therefore, it is expected that replacing obsolete technology with modern, high-standard stoves will lead to a substantial reduction in emissions from residential biomass combustion.

5 Advances in Studies on the Use of Combustion Equipment in Food Preparation and Its Environmental Impacts

In order to understand the progress of research over the years on the use of equipment and the environmental and human health impacts of gases released during food preparation around the world, we conducted a bibliometric analysis on the subject and investigated: (i) the number of publications over the years; (ii) countries and institutions that research the most on the topic under analysis; and (iii) keywords as indicative of the most studied topics. The Scopus reference database was used as a data source for bibliometric analysis because it is one of the largest international and multidisciplinary databases of scientific publications [95]. To carry out the survey comprehensively, we used keywords considering the central theme of the present study. Therefore, the following command was employed: (TITLE ("food preparation" OR "food equipment" OR "stove" OR "electric stove" OR "wood burning stove" OR "gas oven" OR "solar ovens") AND TITLE-ABS-KEY ("food safety" OR "food contamination" OR "HPA" OR "climate changes" OR "air pollution" OR "environmental impacts") AND TITLE-ABS-KEY ("biomass" OR "firewood" OR "coal" OR "charcoal" OR "fire" OR "cooking gas")) AND (LIMIT-TO (DOCTYPE, "ar") OR LIMIT-TO (DOCTYPE, "cp") OR LIMIT-TO (DOCTYPE, "ch")). Data collection was carried out in February 2022, and then processed using the "biblioshiny" function in the "bibliometrix" package (ARIA; CUCCURULLO, 2017) of the R Core Team software [81].

307 research studies were found, comprising 268 articles published in scientific journals, 36 conference proceedings, and 3 book chapters, with a total of 1344 authors, including 12 single-authored documents and 1332 multi-authored ones. The first three research articles on the topic analyzed in this survey were published in 1980, and the periods of highest publication intensity were concentrated between the years 2000 and 2022, with 26 documents published in 2017 and 2021, respectively (Fig. 5). This result can be justified by the worldwide interest in seeking measures aimed at reducing environmental impacts caused by gas emissions, including those released during food preparation, which intensify the process of climate change, highlighting the Kyoto Protocol and the Paris Agreement [1].

A total of 57 countries were identified as having conducted research related to the theme under study (Fig. 6). The United States, China, and the United Kingdom are the countries that conduct the most research on gas emissions from food production equipment, with a scientific production frequency of 149, 82, and 27, respectively. These regions are also where the main institutes that conduct research on this theme are located, with the University of California (49 published articles), Tsinghua University (24 published articles), Colorado State University (19 published articles), and Peking University (14 published articles) standing out. The interest of these countries in conducting research on this topic is entirely linked to the fact that both countries are among the world's largest economic powers and, consequently, the largest greenhouse gas (GHG) emitters, accelerating the process of climate change

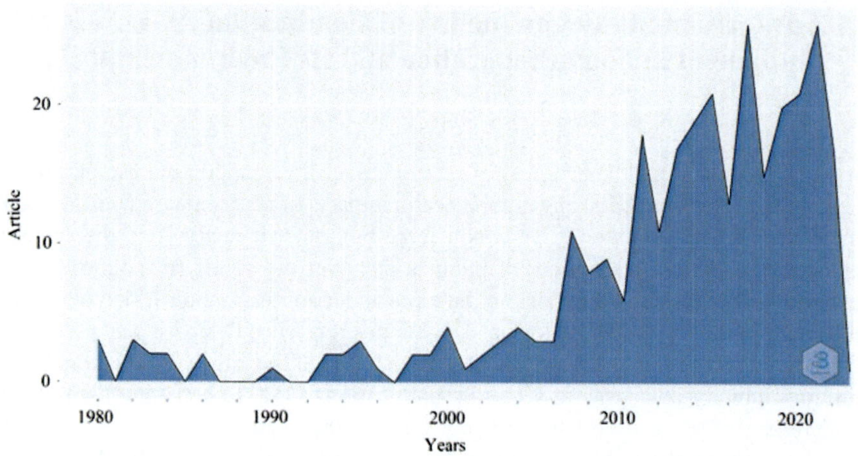

Fig. 5 Scientific production over the years (*Source* The authors 2023)

[2]. However, the countries responsible for the largest release of gases from the burning of biomass in food preparation are classified as poor or developing countries [22]. Nonetheless, they are the ones that least seek measures to minimize the environmental and human health impacts resulting from this practice, as evidenced in this research survey.

A total of 754 keywords were identified by the authors for indexing the articles in databases, with the most frequent terms being indoor air pollution (n = 42), biomass (n = 35), household air pollution (n = 33), particulate matter (n = 28), carbon monoxide (n = 16), cookstove (n = 12), and China (n = 11). Keywords serve to indicate the thematic studies within a research niche [98]. Figure 7 shows the most commonly used keywords over a temporal scale, with recent studies focused

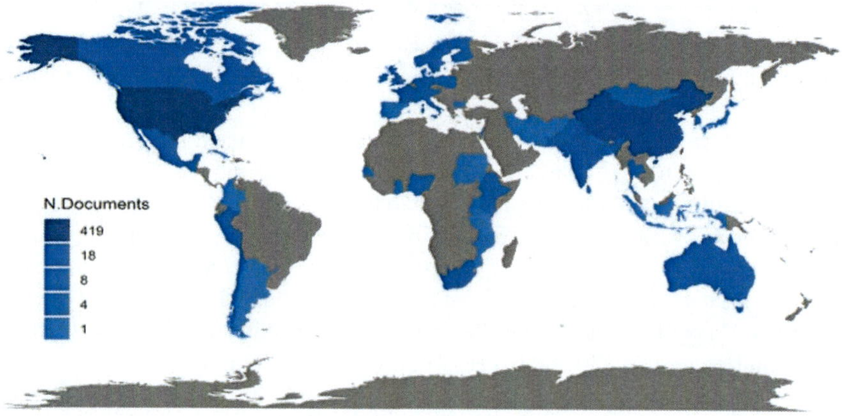

Fig. 6 Scientific production of the countries (*Source* The authors 2023)

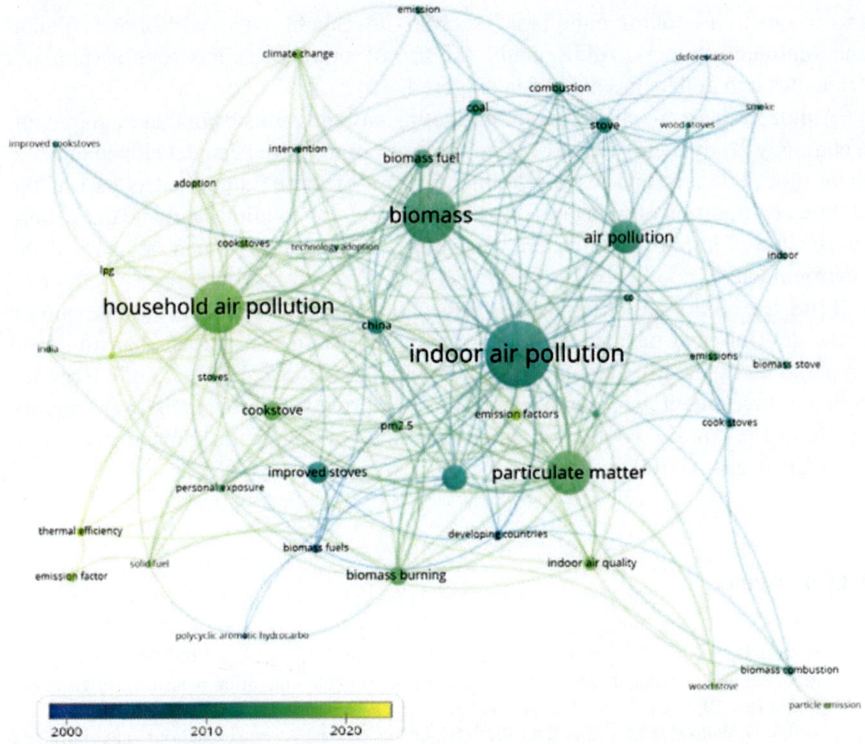

Fig. 7 Network of interactions between keywords over time (*Source* The authors 2023)

on researching the impacts and seeking solutions to air pollution in domestic environments and the particulate matter released during biomass burning for food preparation. This result highlights the global concern to find new cooking equipment that is less harmful to the environment and human health.

6 Conclusions

Throughout history, biomass-fueled cooking equipment such as ovens, stoves, grills, and many others have been widely used worldwide, from the oldest to the most modern ones. These devices are essential for food preparation in many cultures. However, the use of these devices has a significant impact on gas emissions and climate change because the gases emitted during the burning of wood and biomass contribute to air pollution and global warming. The main gases emitted are carbon dioxide (CO_2), carbon monoxide (CO), and particulate matter (PM), which can cause respiratory and cardiovascular problems in humans. To mitigate these gas emissions, many efforts have been made to develop more efficient technologies for biomass and wood-fired cooking equipment, such as stoves and ovens with more efficient and ventilated burners. Additionally, the use of alternative fuels such as biomass briquettes and pellets has also been explored.

Although there are challenges in mitigating gas emissions from these equipment, technology is advancing rapidly and many innovations have been developed to make them more efficient and less polluting. The use of these equipment is part of the culture and tradition of many people, and the search for solutions to make them more sustainable is important to ensure the safety and health of human beings and the environment.

From the bibliometric review, it is also evident the need for further studies aimed at the development of new equipment for more sustainable and less harmful food preparation, especially in countries where biomass is the main source of energy for cooking. Investment in research in this area will provide improvement in the quality of life of the world's population and reduce the environmental impacts caused by greenhouse gas (GHG) emissions.

References

1. Abbass K, Qasim MZ, Song H, Murshed M, Mahmood H, Younis I (2022) A review of the global climate change impacts, adaptation, and sustainable mitigation measures. Environ Sci Pollut Res 29(28):42539–42559. https://doi.org/10.1007/s11356-022-19718-6
2. Althor G, Watson JEM, Fuller RA (2016) Global mismatch between greenhouse gas emissions and the burden of climate change. Sci Rep 6(1):20281. https://doi.org/10.1038/srep20281
3. Anhalt J, Holanda S (2009) Implementation of a dissemination strategy for efficient cook stoves in Northeast Brazil (10604030):29
4. Arunachala UC, Kundapur A (2020) Cost-effective solar cookers: a global review. Sol Energy 207:903–916. https://doi.org/10.1016/j.solener.2020.07.026
5. Rahman ML (2010) Improved Cooking Stoves in South Asia. SAARC Energy Centre, Islamabad, Pakistan
6. Asmelash H, Bayray M, Kimambo C, Gebray P, Sebbit A (2014) Performance test of parabolic trough solar cooker for indoor cooking. Momona Ethiopian J Sci 6(2):39. https://doi.org/10.4314/mejs.v6i2.109621
7. Bailis R, Cowan A, Berrueta V, Masera O (2009) Arresting the killer in the kitchen: the promises and pitfalls of commercializing improved cookstoves. World Dev 37(10):1694–1705
8. Bank W (2011) Problems and opportunities. Cogeneration Distrib Gener J 26(1):5–6. https://doi.org/10.4324/9780203360835_chapter_11
9. Bargman JM, Breborowicz A, Rodela H, Abraham G, Oreopoulos DG (1988) Absorption of recombinant human erythropoietin (EPO) from the peritoneal-cavity in rabbits. In: Kidney international. Blackwell Science Inc., 350 Main St., Malden, MA 02148, p 243
10. Barnes DF, Openshaw K, Smith KR, van der Plas R (1994) What makes people cook with improved biomass stoves. World Bank technical paper 242:2004
11. Berry J, Anstey N, Spillane N, Birzer C (2019) Design for mass production and dissemination of clean cookstoves in developing countries. World Engineers Convention (October):628–646
12. Black D (1932) Evidences of the use of fire by Sinanthropus. Bull Geol Soc China 11:107–108
13. Blunck M, Griebenow C, Rammelt M, Zimm C (2011) Carbon markets for improved cooking stoves: a GIZ guide for project operators. GIZHERA—poverty-oriented Basic Energy Services, Eschborn, Germany
14. Bronowski J (1973) The ascent of man little brown and company

15. Brook RD, Franklin B, Cascio W, Hong Y, Howard G, Lipsett M, Luepker R, Mittleman M, Samet J, Smith SC, Tager I (2004) Air pollution and cardiovascular disease: a statement for healthcare professionals from the expert panel on population and prevention science of the American Heart Association. Circulation 109(21):2655–2671. https://doi.org/10.1161/01.CIR.0000128587.30041.C8
16. Brook RD, Rajagopalan S, Pope CA, Brook JR, Bhatnagar A, Diez-Roux AV, Holguin F, Hong Y, Luepker RV, Mittleman MA, Peters A, Siscovick D, Smith SC, Whitsel L, Kaufman JD (2010) Particulate matter air pollution and cardiovascular disease: an update to the scientific statement from the American heart association. Circulation 121(21):2331–2378. https://doi.org/10.1161/CIR.0b013e3181dbece1
17. Carter EM, Shan M, Yang X, Li J, Baumgartner J (2014) Pollutant emissions and energy efficiency of Chinese gasifier cooking stoves and implications for future intervention studies. Environ Sci Technol 48(11):6461–6467. https://doi.org/10.1021/es405723w
18. Choi M, Kim K, Li X, Deng K, Park Y, Seo M, Sung Y, Choi G (2020) Strategic combustion technology with exhaust tube vortex flame: combined effect of biomass co-firing and air-staged combustion on combustion characteristics and ash deposition. Energy 203:117839. https://doi.org/10.1016/J.ENERGY.2020.117839
19. Crewe E (1997) The silent tradition of development cooks. In: Discourses of development: anthropological perspective, pp 59–80
20. Cuce E, Cuce PM (2013) A comprehensive review on solar cookers. Appl Energy 102:1399–1421. https://doi.org/10.1016/j.apenergy.2012.09.002
21. Cuce PM (2018) Box type solar cookers with sensible thermal energy storage medium: a comparative experimental investigation and thermodynamic analysis. Sol Energy 166:432–440. https://doi.org/10.1016/j.solener.2018.03.077
22. Cupertino GFM, Souza EC de, Souza Júnior AD de, Delatorre FM, Santos Júnior AJ dos, da Silva ÁM, Pereira AKS, Mesquita JG, Ucella Filho JGM, Oliveira TR, Cezário LFC, de Oliveira MP, Curty TA, Dias Júnior AF (2022) Biomass use and exposure to residential air pollution increase the risk of servere COVID-19. In: Biomassa: recursos, aplicações e tecnologias em pesquisas. Editora Científica Digital, pp 33–47
23. Dean Hosgood H, Boffetta P, Greenland S, Lee YCA, Mclaughlin J, Seow A, Duell EJ, Andrew AS, Zaridze D, Szeszenia-Dabrowska N, Rudnai P, Lissowska J, Fabiánová E, Mates D, Bencko V, Foretova L, Janout V, Morgenstern H, Rothman N, Hung RJ, Brennan P, Lan Q (2010) In-home coal and wood use and lung cancer risk: a pooled analysis of the international lung cancer consortium. Environ Health Perspect 118(12):1743–1747. https://doi.org/10.1289/ehp.1002217
24. Deng M, Li P, Shan M, Yang X (2020) Optimizing supply airflow and its distribution between primary and secondary air in a forced-draft biomass pellet stove. Environ Res 184(July 2019):109301. https://doi.org/10.1016/j.envres.2020.109301
25. Deng M, Nie Y, Yuan Y, Ma R, Shan M, Yang X (2022) The impact of oxygen content in the primary air supply on fuel burning rate and pollutant emissions in a forced-draft biomass stove. Fuel 321(x):124129. https://doi.org/10.1016/j.fuel.2022.124129
26. Dennis RJ, Maldonado D, Norman S, Baena E, Martinez G (1996) Woodsmoke exposure and risk for obstructive airways disease among women. Chest 109(1):115–119. https://doi.org/10.1378/chest.109.1.115
27. Devan PK, Bibin C, Gowtham S, Hariharan G, Hariharan R (2020) A comprehensive review on solar cooker with sun tracking system. Mater Today Proc 33:771–777. https://doi.org/10.1016/j.matpr.2020.06.124
28. Dutta A, Ray MR (2014) Hypertension and respiratory health in biomass smoke-exposed premenopausal Indian women. Air Qual Atmos Health 7(2):229–238. https://doi.org/10.1007/s11869-013-0228-5
29. Eckholm E (1975) The other energy crisis: firewood. Worldwatch paper
30. Eckholm E, Foley G, Barnard G, Timberlake L (1984) Fuelwood: the energy crisis that won't go away. Earthscan

31. El Moussaoui N, Talbi S, Atmane I, Kassmi K, Schwarzer K, Chayeb H, Bachiri N (2020) Feasibility of a new design of a Parabolic Trough Solar Thermal Cooker (PSTC). Sol Energy 201:866–871. https://doi.org/10.1016/j.solener.2020.03.079
32. Epton MJ, Dawson RD, Brooks WM, Kingham S, Aberkane T, Cavanagh JAE, Frampton CM, Hewitt T, Cook JM, McLeod S, McCartin F, Trought K, Brown L (2008) The effect of ambient air pollution on respiratory health of school children: a panel study. Environ Health 7:1–11. https://doi.org/10.1186/1476-069X-7-16
33. Erarslan A (2021) A traditional wooden corbelled dome construction technique from Anatolia. The Eastern Anatolian Tandoor house with its wooden "swallow-dome" type of roof 21(4):1275–1303. https://doi.org/10.1080/1346758120211929243, https://doi.org/10.1080/13467581.2021.1929243
34. Ezzati M, Bailis R, Kammen DM, Holloway T, Price L, Cifuentes LA, Barnes B, Chaurey A, Dhanapala KN (2004) Energy management and global health* 29:383–419. https://doi.org/10.1146/annurev.energy29062103121246, https://doi.org/10.1146/ANNUREV.ENERGY.29.062103.121246
35. Feng Y, Li Y, Cui L, Yan L, Zhao C, Dong Y (2019) Cold condensing scrubbing method for fine particle reduction from saturated flue gas. Energy 171:1193–1205. https://doi.org/10.1016/J.ENERGY.2019.01.065
36. Foley G, Moss P, Timberlake L (1984) Stoves and trees: how much wood would a woodstove save if a woodstove could save wood? Earthscan
37. Foster C (2007) Rocket Lorena Stove—Appropedia. License CC-BY-SA-4.0. https://www.appropedia.org/Rocket_Lorena_Stove. Accessed 3 May 2023
38. Fournel S, Palacios JH, Godbout S, Heitz M (2015) Effect of additives and fuel blending on emissions and ash-related problems from small-scale combustion of reed canary grass. Agriculture 5(3):561–576. https://doi.org/10.3390/AGRICULTURE5030561
39. Gill J (1987) Improved stoves in developing countries: a critique. Energy Policy 15(2):135–144
40. Gogoi D, Sazid A, Bora J, Deka P, Balachandran S, Hoque RR (2021) Particulate matter exposure in biomass-burning homes of different communities of Brahmaputra Valley. Environ Monit Assess 193(12). https://doi.org/10.1007/s10661-021-09624-8
41. Grigonyte J, Nuutinen I, Koponen T, Lamberg H, Tissari J, Jokiniemi J, Sippula O (2014) Evaluation of a heat exchanger designed for efficient fine particle precipitation in small-scale wood combustion. Energy Fuels 28(9):6058–6065. https://doi.org/10.1021/EF500958X/ASSET/IMAGES/LARGE/EF-2014-00958X_0006.JPEG
42. Hanbar RD, Karve P (2002) National Programme on Improved Chulha (NPIC) of the Government of India: an overview. Energy Sustain Dev 6(2):49–55. https://doi.org/10.1016/S0973-0826(08)60313-0
43. He Z, Liu S, Wang S, Liu W, Li Y, Feng X (2022) Reduced pollutant emissions and slagging rate of biomass pellet combustion by optimizing the multilayer distribution of secondary air. ACS Omega 7(33):28962–28973. https://doi.org/10.1021/ACSOMEGA.2C02587/ASSET/IMAGES/MEDIUM/AO2C02587_M011.GIF
44. Herez A, Ramadan M, Khaled M (2018) Review on solar cooker systems: economic and environmental study for different Lebanese scenarios. Renew Sustain Energy Rev 81:421–432. https://doi.org/10.1016/j.rser.2017.08.021
45. Hukkanen A, Kaivosoja T, Sippula O, Nuutinen K, Jokiniemi J, Tissari J (2012) Reduction of gaseous and particulate emissions from small-scale wood combustion with a catalytic combustor. Atmos Environ 50:16–23. https://doi.org/10.1016/J.ATMOSENV.2012.01.016
46. Jassim MM, Abbood MH, Rashid FL (2022) Design and construction solar oven sterilizer. Int J Heat Technol 40(2):641–645. https://doi.org/10.18280/ijht.400235
47. Johansson LS, Leckner B, Gustavsson L, Cooper D, Tullin C, Potter A (2004) Emission characteristics of modern and old-type residential boilers fired with wood logs and wood pellets. Atmos Environ 38(25):4183–4195. https://doi.org/10.1016/j.atmosenv.2004.04.020
48. Kaivosoja T, Jalava PI, Lamberg H, Virén A, Tapanainen M, Torvela T, Tapper U, Sippula O, Tissari J, Hillamo R, Hirvonen MR, Jokiniemi J (2013) Comparison of emissions and

toxicological properties of fine particles from wood and oil boilers in small (20–25 kW) and medium (5–10 MW) scale. Atmos Environ 77:193–201. https://doi.org/10.1016/J.ATMOSENV.2013.05.014

49. Kansiime WK, Mugambe RK, Atusingwize E, Wafula ST, Nsereko V, Ssekamatte T, Nalugya A, Coker ES, Ssempebwa JC, Isunju JB (2022) Use of biomass fuels predicts indoor particulate matter and carbon monoxide concentrations; evidence from an informal urban settlement in Fort Portal city, Uganda. BMC Public Health 22(1):1–12. https://doi.org/10.1186/s12889-022-14015-w
50. Karanasiou A, Alastuey A, Amato F, Renzi M, Stafoggia M, Tobias A, Reche C, Forastiere F, Gumy S, Mudu P, Querol X (2021) Short-term health effects from outdoor exposure to biomass burning emissions: a review. Sci Total Environ 781:146739. https://doi.org/10.1016/j.scitotenv.2021.146739
51. Kaur-Sidhu M, Ravindra K, Mor S, John S (2020) Emission factors and global warming potential of various solid biomass fuel-cook stove combinations. Atmos Pollut Res 11(2):252–260. https://doi.org/10.1016/J.APR.2019.10.009
52. Kedar SA, Sonawale P, Valve P, Khujat P, Talekar P (2017) Thermal analysis of parabolic solar cooker with back reflection. Mater Today Proc 4(8):8035–8039. https://doi.org/10.1016/j.matpr.2017.07.141
53. Khadilkar P (2017) Formulation of a Framework for needs analysis and Stakeholders' behavioral Simulation for design for the Bop
54. Khodaei H, Guzzomi F, Patiño D, Rashidian B, Yeoh GH (2017) Air staging strategies in biomass combustion-gaseous and particulate emission reduction potentials. Fuel Process Technol 157:29–41. https://doi.org/10.1016/j.fuproc.2016.11.007
55. Kimambo CZM (2007) Development and performance testing of solar cookers. J Energy Southern Africa 18(3):41–51. https://doi.org/10.17159/2413-3051/2007/v18i3a3384
56. Koo CW, Gupta N, Baliff JP, Hudock K, Haas AR (2011) A tale of two sisters: biomass fuel exposure-related lung disease. Clin Radiol 66(2):190–193. https://doi.org/10.1016/j.crad.2010.08.006
57. Koraïem M, Assanis D (2021) Wood stove combustion modeling and simulation: technical review and recommendations. Int Commun Heat Mass Transfer 127(August):105423. https://doi.org/10.1016/j.icheatmasstransfer.2021.105423
58. Kristoferson LA, Bokalders V (1986) Wood and charcoal stoves. Renew Energy Technol 68–87. https://doi.org/10.1016/B978-0-08-034061-6.50013-5
59. Kshirsagar MP, Kalamkar VR (2014) A comprehensive review on biomass cookstoves and a systematic approach for modern cookstove design. Renew Sustain Energy Rev 30:580–603. https://doi.org/10.1016/j.rser.2013.10.039
60. Kumar S (2004) Natural convective heat transfer in trapezoidal enclosure of box-type solar cooker. Renew Energy 29(2):211–222. https://doi.org/10.1016/S0960-1481(03)00193-9
61. Lentswe K, Mawire A, Owusu P, Shobo A (2021) A review of parabolic solar cookers with thermal energy storage. Heliyon 7(10):e08226. https://doi.org/10.1016/j.heliyon.2021.e08226
62. Li PW, Chyang CS (2020) A comprehensive study on NO_x emission and fuel nitrogen conversion of solid biomass in bubbling fluidized beds under staged combustion. J Energy Inst 93(1):324–334. https://doi.org/10.1016/J.JOEI.2019.02.007
63. Lima FDM, Pérez-Martínez PJ, de Fatima AM, Kumar P, de Miranda RM (2020) Characterization of particles emitted by pizzerias burning wood and briquettes: a case study at Sao Paulo, Brazil. Environ Sci Pollut Res 27(29):35875–35888. https://doi.org/10.1007/s11356-019-07508-6
64. MacCarty N, Ogle D, Still D, Bond T, Roden C (2008) A laboratory comparison of the global warming impact of five major types of biomass cooking stoves. Energy Sustain Dev 12(2):56–65. https://doi.org/10.1016/S0973-0826(08)60429-9
65. Mannan M (1996) Women targeted and women negated. Dev Pract 6(2):113–120
66. Manjo Kumar, Sachin Kumar, Tyagi SK (2013) Design, development and technological advancement in the biomass cookstoves: a review. Renew Sustain Energy Rev 26:265–285. https://doi.org/10.1016/j.rser.2013.05.010

67. Mbodji N, Hajji A (2017) Modeling, testing, and parametric analysis of a parabolic solar cooking system with heat storage for indoor cooking. Energy Sustain Soc 7(1):32. https://doi.org/10.1186/s13705-017-0134-z
68. Miah MD, Al Rashid H, Shin MY (2009) Wood fuel use in the traditional cooking stoves in the rural floodplain areas of Bangladesh: a socio-environmental perspective. Biomass Bioenergy 33(1):70–78. https://doi.org/10.1016/j.biombioe.2008.04.015
69. Mishra A, Powar S, Dhar A (2019) Solar thermal powered bakery oven, pp 577–592
70. MNRE (2010) National Biomass Cookstoves Programme (NBCP). Special Project on Cookstove (SPC)
71. Muthusivagami RM, Velraj R, Sethumadhavan R (2010) Solar cookers with and without thermal storage—a review. Renew Sustain Energy Rev 14(2):691–701. https://doi.org/10.1016/j.rser.2008.08.018
72. Naeher LP, Brauer M, Lipsett M, Zelikoff JT, Simpson CD, Koenig JQ, Smith KR (2007) Woodsmoke health effects: a review. Inhal Toxicol 19(1):67–106. https://doi.org/10.1080/08958370600985875
73. Oischinger J, Steiner M, Meiller M, Hebauer M, Beer S, Daschner R, Hornung A, Kramb J (2020) Optimization of the fractional collection efficiencies for electrostatic precipitators used in biomass-fired boilers. Biomass Bioenergy 141:105703. https://doi.org/10.1016/J.BIOMBIOE.2020.105703
74. Omara AAM, Abuelnuor AAA, Mohammed HA, Habibi D, Younis O (2020) Improving solar cooker performance using phase change materials: a comprehensive review. Sol Energy 207:539–563. https://doi.org/10.1016/j.solener.2020.07.015
75. Ozgen S, Caserini S, Galante S, Giugliano M, Angelino E, Marongiu A, Hugony F, Migliavacca G, Morreale C (2014) Emission factors from small scale appliances burning wood and pellets. Atmos Environ 94:144–153. https://doi.org/10.1016/j.atmosenv.2014.05.032
76. Padma TV (2007) Biomass fuels blamed for premature deaths in rural settings. Nat Med 13(2):112. https://doi.org/10.1038/nm0207-112a
77. Pandey S, Goswami S, Saini P, Powar S, Dhar A (2021) Hybrid electrical-solar oven: a new perspective pp 237–255
78. Pei WC (1934) A preliminary report on the late-Palæolithic cave of Choukoutien 1. Bull Geol Soc China 13(1):327–358
79. Pham M, Pakrasi A (2019) Air pollution control technologies. In: Proceedings of the air and waste management association's annual conference and exhibition, AWMA, pp 377–428. https://doi.org/10.1016/B978-0-12-814934-8.00013-2
80. Pivello VR, Vieira I, Christianini AV, Ribeiro DB, da Silva ML, Berlinck CN, Melo FPL, Marengo JA, Tornquist CG, Tomas WM, Overbeck GE (2021) Understanding Brazil's catastrophic fires: causes, consequences and policy needed to prevent future tragedies. Perspect Ecol Conserv 19(3):233–255. https://doi.org/10.1016/j.pecon.2021.06.005
81. R Core Team K (2021) R: A language and environment for statistical computing
82. Regattieri A, Piana F, Bortolini M, Gamberi M, Ferrari E (2016) Innovative portable solar cooker using the packaging waste of humanitarian supplies. Renew Sustain Energy Rev 57:319–326. https://doi.org/10.1016/j.rser.2015.12.199
83. Sandoval J, Salas J, Martinez-Guerra ML, Gomez A, Martinez C, Portales A, Palomar A, Villegas M, Barrios R (1993) Pulmonary arterial hypertension and cor pulmonale associated with chronic domestic woodsmoke inhalation. Chest 103(1):12–20. https://doi.org/10.1378/chest.103.1.12
84. Saxena A, Varun PSP, Srivastav G (2011) A thermodynamic review on solar box type cookers. Renew Sustain Energy Rev 15(6):3301–3318. https://doi.org/10.1016/j.rser.2011.04.017
85. Schmidl C, Luisser M, Padouvas E, Lasselsberger L, Rzaca M, Ramirez-Santa Cruz C, Handler M, Peng G, Bauer H, Puxbaum H (2011) Particulate and gaseous emissions from manually and automatically fired small scale combustion systems. Atmos Environ 45(39):7443–7454. https://doi.org/10.1016/J.ATMOSENV.2011.05.006
86. SCI SCI (2023) Solar cookers international. https://www.solarcookers.org/about

87. Sedighi M, Salarian H (2017) A comprehensive review of technical aspects of biomass cookstoves. Renew Sustain Energy Rev 70:656–665
88. Sesan TA (2011) What's cooking? Participatory and market approaches to stove development in Nigeria and Kenya. Technol Soc 39(July):142–150
89. Shan M, Carter E, Baumgartner J, Deng M, Clark S, Schauer JJ, Ezzati M, Li J, Fu Y, Yang X (2017) A user-centered, iterative engineering approach for advanced biomass cookstove design and development. Environ Res Lett 12(9). https://doi.org/10.1088/1748-9326/aa804f
90. Shen G, Tao S, Wei S, Zhang Y, Wang R, Wang B, Li W, Shen H, Huang Y, Chen Y, Chen H, Yang Y, Wang W, Wei W, Wang X, Liu W, Wang X, Simonich SLM (2012) Reductions in emissions of carbonaceous particulate matter and polycyclic aromatic hydrocarbons from combustion of biomass pellets in comparison with raw fuel burning. Environ Sci Technol 46(11):6409–6416. https://doi.org/10.1021/es300369d
91. Smith KR (1989) Dialectics of improved stoves. Econ Polit Wkly 517–522
92. Smith KR, Mehta S, Maeusezahl-Feuz M (2004) Indoor air pollution from household use of solid fuels. In: Comparative quantification of health risks: global and regional burden of disease attributable to selected major risk factors, vol 2, pp 1435–1493
93. Smith KR, Samet JM, Romieu I, Bruce N (2000) Indoor air pollution in developing countries and acute lower respiratory infections in children. Thorax 55(6):518–532. https://doi.org/10.1136/thorax.55.6.518
94. Sulilatu WF (1984) Danger signals to human health
95. Sweileh WM (2018) Research trends on human trafficking: a bibliometric analysis using Scopus database. Global Health 14(1):106. https://doi.org/10.1186/s12992-018-0427-9
96. Taner S, Pekey B, Pekey H (2013) Fine particulate matter in the indoor air of barbeque restaurants: elemental compositions, sources and health risks. Sci Total Environ 454–455:79–87. https://doi.org/10.1016/j.scitotenv.2013.03.018
97. Troncoso K, Castillo A, Masera O, Merino L (2007) Social perceptions about a technological innovation for fuelwood cooking: case study in rural Mexico. Energy Policy 35(5):2799–2810
98. Ucella-Filho JGM, da Freire ASM, Carréra JC, Lucas FMF, Zucolotto SM, Dias Júnior AF, Mori FA (2022) Tannin-rich bark extract of plants as a source of antimicrobial bioactive compounds: a bibliometric analysis. South Afr J Botany 150:1038–1050. https://doi.org/10.1016/j.sajb.2022.09.018
99. Venkataraman C, Sagar AD, Habib G, Lam N, Smith KR (2010) The Indian national initiative for advanced biomass cookstoves: the benefits of clean combustion. Energy Sustain Dev 14(2):63–72. https://doi.org/10.1016/j.esd.2010.04.005
100. Vicente ED, Alves CA (2018) An overview of particulate emissions from residential biomass combustion. Atmos Res 199:159–185. https://doi.org/10.1016/J.ATMOSRES.2017.08.027
101. Vicente ED, Duarte MA, Calvo AI, Nunes TF, Tarelho L, Alves CA (2015) Emission of carbon monoxide, total hydrocarbons and particulate matter during wood combustion in a stove operating under distinct conditions. Fuel Process Technol 131:182–192. https://doi.org/10.1016/J.FUPROC.2014.11.021
102. Westhoff B, Germann D (1995) Stove images: a documentation of improved and traditional stoves in Africa, Asia and Latin America
103. Wu J, Cheng F, Zhang D (2021) Health effect of indoor $Pm_{2.5}$ and co emissions from coal and biomass fired domestic appliances in remote rural China. Int J Energy Clean Environ 22(5):33–49. https://doi.org/10.1615/INTERJENERCLEANENV.2021036424
104. Wu J, Zhao X, Xiao C, Xiong G, Ye X, Li L, Fang Y, Chen H, Yang W, Du X (2022) The role of lung macrophages in chronic obstructive pulmonary disease. Respir Med 205(August):107035. https://doi.org/10.1016/j.rmed.2022.107035
105. Xiangjum Y (1993) Chinese fuel saving stoves: a compendium
106. Xiliang Z, Smith KR (2005) Programmes promoting improved household stoves in China. Boiling Point 50:14–16
107. Yettou F, Azoui B, Malek A, Gama A, Panwar NL (2014) Solar cooker realizations in actual use: an overview. Renew Sustain Energy Rev 37:288–306. https://doi.org/10.1016/j.rser.2014.05.018

108. Zadravec T, Rajh B, Kokalj F, Samec N (2020) CFD modelling of air staged combustion in a wood pellet boiler using the coupled modelling approach. Therm Sci Eng Prog 20:100715. https://doi.org/10.1016/J.TSEP.2020.100715
109. Zhang R, Xu G, Li B, Wang Z, Gao J, Li J, Sun Y, Xu G (2023) Analysis of the pollution emission system of large-scale combustion of biomass briquette fuel in China. Process Saf Environ Prot 169:928–936. https://doi.org/10.1016/J.PSEP.2022.11.088

Wastes from Sustainable Forest Management as a Source of Biomass: The Case of Amazonia for Bioenergy Generation

Elvis Vieira dos Santos, Michael Douglas Roque Lima, Lina Bufalino, Paulo Ricardo Gherardi Hein, Paulo Fernando Trugilho, and Thiago de Paula Protásio

Abstract Knowledge about the availability and application of Amazonian wood wastes from sustainable forest management plans (SFMP) in energy generation is essential when considering a third of the world's population depends on wood as an energy source. In tropical countries such as Brazil, technological initiatives may be combined with scientific studies to add value to forest biomass and enhance its use as an energy input in the Legal Amazonia. The integral conversion of dense tropical forest vegetation to charcoal or supplying thermoelectric plants would not be admissible; however, forest wastes are sustainable and promising alternatives. This chapter aims to present: (i) the energy potential of wood wastes from SFMP in Amazonia and its importance in the sustainable expansion of energy systems and (ii) the potential of wood wastes to replace non-renewable sources, such as fossil fuels, to reduce the logging of tropical forests for energy generation. The biggest challenges

E. V. dos Santos · P. R. G. Hein · P. F. Trugilho
Federal University of Lavras (UFLA), Department of Forest Science,
Campus Universitário—Aquenta Sol, S/N, Lavras, Minas Gerais 37200-900, Brazil
e-mail: elvisvieiradossantos@gmail.com

P. R. G. Hein
e-mail: paulo.hein@ufla.br

P. F. Trugilho
e-mail: trugilho@ufla.br

M. D. R. Lima · L. Bufalino
Federal Rural University of Amazonia (UFRA), Institute of Agricultural Sciences,
Av. Tancredo Neves, 2501, Belém, Pará 66077-901, Brazil
e-mail: lima_florestal@outlook.com

L. Bufalino
e-mail: lina.bufalino@ufra.edu.br

T. de P. Protásio (✉)
Federal Rural University of Amazonia (UFRA), Av. Duane Silva, S/N, Parauapebas,
Pará 68515-000, Brazil
e-mail: thiago.protasio@ufra.edu.br

A. K. S. Pereira and A. F. Dias Júnior (eds.), *Impacts of Using Biomass as an Energy Source in Homes*, Green Energy and Technology,
https://doi.org/10.1007/978-3-031-38824-8_4

to producing sustainable energy from forest wastes in Amazonia are related to the low technological level of the kilns used to produce charcoal and the high variation in the wood characteristics concerning its dimensions and its physical, chemical, and energy properties. In that chapter, it was demonstrated that the diameter of the wastes varies from 0.123 to 0.760 m, and the basic density, for example, varies from 0.221 to 0.935 g/cm^3 between species. Studies were carried out to characterize these residual woods and solve these problems. This characterization promoted scientific development. The segregation of wastes according to their properties, mainly the basic density, increased the productivity of the kilns and the quality of the charcoal derived. Future research will address technological improvements in energy production through SFMP wastes to increase efficiency, quality, and sustainability.

Keywords Renewable energy · Wood quality · Residual biomass · Carajás pole · Bioreducer

1 Introduction

Using wood wastes from SFMP as fuel is a sustainable alternative for diversifying the national and even global energy matrix. In Amazonia, these lignocellulosic wastes are formed mainly by branches, buttresses, and stumps of felled trees, usually of large dimensions. Such wastes can represent up to two-thirds of the dry mass of the tree [29]. Therefore, they are highly available and economically viable. In addition, waste consumption is heading towards maximum efficiency in using natural resources.

Some studies have discussed interesting aspects of the environmental impact of extracting wastes from SFMP worldwide [1, 25] and found no negative effect on tree growth. On the contrary, they indicated positive effects of stump extraction on forest regeneration [25, 28]. Based on published data worldwide, the literature indicates that waste extraction increases nutrient export [1]. The export effect is complex because it is affected by the harvest cycle and specific characteristics of the logged forest ecosystem, such as the input of nutrients by weathering and output by leaching [32, 44]. Compared to fossil fuels, burning wood and charcoal emits low amounts of greenhouse gases [11]. The carbon emitted will be stored again during forest regeneration [9] promoted by techniques associated with SFMP [28].

Carbonization is an important conversion route for using by-products from forest management. In addition to its wide domestic use, charcoal is used in blast furnaces of steel mills as a biothermoreducer for pig iron, in the production of the called "green steel." The Carajás region, located in eastern Amazonia, has the largest pig iron production in northern Brazil [41]. Forest wastes partially supply the region's firewood demand for charcoal production [24].

The great challenge for the energy use of wastes from SFMP refers to the high variability of their characteristics. Lima et al. [19] found variation in the wood basic density from 0.525 to 0.895 g/cm^3. In addition, the wood presents high moisture, representing up to 45% of the wet mass without bark [24]. Basic density and moisture influence the productivity of the kilns [4]. This variability is increased by the lack of carbonization control [39] commonly carried out in brick kilns of low technologic level known as “hot tail.” The carbonization in these handmade kilns is controlled empirically, without temperature control or waste separation into quality classes.

Studies have been carried out to understand the variation in the quality of charcoal derived from waste [19, 23]. In general, the separation of wastes into groups of species according to the different wood characteristics significantly improves the productivity of the kilns, increases the yield, and reduces the variation in the charcoal quality compared to the conventional method [5]. The next steps involve conducting research that will contribute to more effective control of carbonization, increase its yield, improve the quality of charcoal derived from forest wastes in Amazonia, and mitigate gas emissions during carbonization. This chapter aims to describe the advances obtained so far related to the energy use of forest wastes arising from SFMP in Amazonia.

2 Background of Studies with Wastes from SFMP

The project entitled “Valuation of wood wastes from sustainable forest management for bioenergy in Legal Amazonia” (Public Selection Notice for Scientific and Technological Research—Edition 2018) was developed by the Federal Rural University of Amazonia (website: https://novo.ufra.edu.br/) and financed directly by the Banco da Amazônia. Several advances were achieved in understanding the characteristics of primary wastes from SFMP in Amazonia. It was possible to understand the variability of the main physical, chemical, and energy properties of the forest biomass used for bioenergy in Amazonia [19, 23]. Consequently, the data collected will allow decision-making in industrial units, such as steel mills and ceramics.

The various studies analyzed the wood characteristics of species from 22 genera. The waste characteristics that varied the most were basic density (0.525–0.895 g/cm^3), ash content (0.3–2.5%), and total extractives (1.7–17.9%). However, the elemental carbon content (49.18–52.16%), total lignin (30.2–38.1%), fixed carbon (16.5–22.0%), volatile matter (76.7–82. 8%), and higher heating value (19.4–20.4 MJ/kg) were characteristics that showed the smallest variations between the studied species [19]. Within the same species, wood properties have smaller ranges of variation. In the study of Lima et al. [19], *Manilkara elata* and *Dinizia excelsa* had a basic density of 0.900 and 0.890 g/cm^3, respectively. On the other hand, Lima et al. [24] reported values of 0.872 and 0.927 g/cm^3 of basic density for the same species. Between the two studies, there was a difference of about 3%. This variation can be considered low since this residual biomass has no genetic, age, or site control. These findings indicate the reliability and possibility of data extrapolation to represent the overall quality of waste from these species.

Based on the physical, chemical, and energy characterization of the wastes, it was demonstrated that the wood of the species has the quality to generate electric energy in boilers [19] and in charcoal production for domestic and industrial use. Lima et al. [19] verified a species effect on the technological properties of residual woods, making it necessary to qualify them for energy production. The use of wood wastes with up to 30% moisture (wet basis) has proven to be viable in modern cogeneration systems.

The species studied, such as *D. excelsa*, *M. elata*, *P. altissium*, and *G. glabra*, showed better energy properties than species and clones traditionally used in energy forests. *D. excelsa* had the highest mass of CO_2eq fixed in 1 m^3 of wood wastes (1687 kg), meaning that the use of 1 m^3 of wood wastes of this species would mitigate the emission of 1687 kg of CO_2eq. Finally, the wastes of this species showed the best properties for bioenergy.

The second step of the project proposed waste segregation for charcoal production. Principal component analysis (PCA), a multivariate statistical technique, enabled forming species groups with similar properties [19, 34].

Correlations of the physical and energetic properties of the wood with colorimetry parameters were found [20]. Near-infrared (NIR) spectroscopy was also evaluated with the same objective [21]. Both methods were effective in separating waste.

Lima et al. [20] reported that woods with greater red pigmentation (*M. elata* and *D. excelsa* had higher energy density. All groups formed by PCA can be recommended for bioenergy; however, the group formed by the species *M. elata* and *D. excelsa*, with a purplish-brown color, is the most promising.

In general, waste segregation increases carbonization efficiency, mainly due to the effect of wood basic density. High-density waste has less moisture, allowing to add a greater amount of wood mass in the kiln, increasing production and productivity per carbonization cycle. Figure 1 shows the productivity parameters of segregated and non-segregated woods in the different studies carried out during the execution of the project.

Lima et al. [22] reported a species effect on the physical, chemical, and energy properties of charcoal derived from SFMP wastes. The results showed that the content of extractives in the wood positively influenced the gravimetric yields of charcoal, the carbonization mass balance, the heating value, and the energy performance index of charcoal. Apparent relative density, gravimetric yields of charcoal, gravimetric yields of non-condensable gases, ash content, fixed carbon yield, energy density, energy yield, and retained carbon were the properties with a wide variation between species. The PCA separated charcoal samples from SFMP wastes of 48 species into five distinct groups. These groups can be used in carbonization in brick kilns to supply the Carajás Steel Pole and cogeneration systems in remote Amazonian communities. Finally, the *D. excelsa* species wastes showed the best charcoal properties and processes.

Table 1 presents the project outputs to date. Seven articles were published aiming to understand the waste properties and improve charcoal production through its segregation.

Currently, the project "Generation of bioenergy from wastes from sustainable forest management: decentralization of the energy matrix and socio-environmental impacts in Amazonia" (CNPq/MCTI/FNDCT No. 59/2022) is in effect. The project aims to clarify the combustibility processes of burning co-products from SFMP in

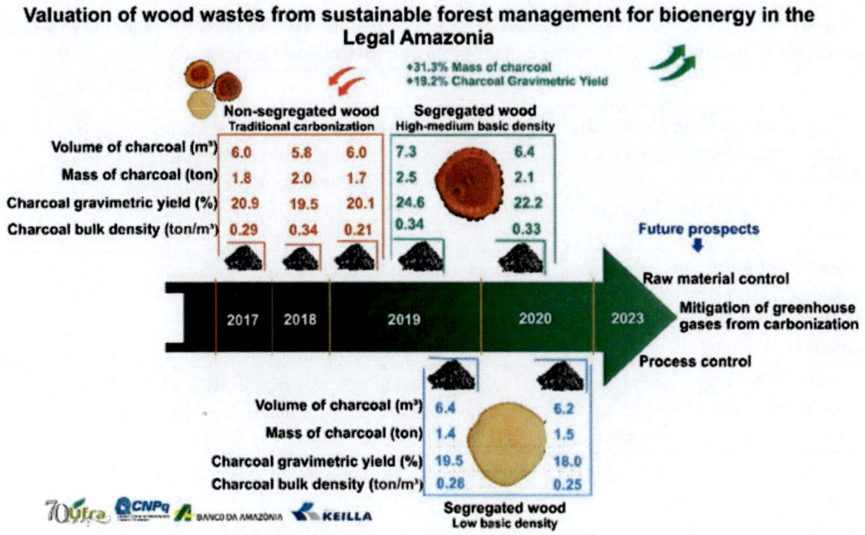

Fig. 1 Summary of results related to kiln productivity during project execution (*Source* The authors 2023)

Amazonia and mitigate the emission of greenhouse gases during the wood carbonization by implementing a furnace-kiln system, which allows the burning of gases formed during the carbonization.

3 Traditional Production of Charcoal from SFMP Wastes in Amazonia

The model of charcoal production in Brazilian Amazonia mainly includes rudimentary brick kilns of the hot tail type, which are semi-spherical kilns built by combining clay and sandy textured soil to avoid cracking. These relatively simple, handmade kilns present low conversion efficiency and are empirically controlled by experienced workers based on the color and amount of smoke expelled, in addition to touch. Consequently, these brick kilns have low productivity and gravimetric yields of charcoal between 15 and 25% on a wet basis [33, 34]. Typically, brick kilns do not include gas cleaning systems and may release unburned by-products into the environment [42]. A typical hot tail kiln is presented in Fig. 2.

The study conducted by Barros et al. [5] describes the carbonization model with wastes in Amazonia. The authors reported kilns with a base diameter of 3.20 m and a height of 2.5 m, with six openings at the base, six at the top, and one chimney, with a height of 70 cm for gas exhaustion. The kilns have a capacity of 10 m^3 of wood. The carbonization cycle normally ranges from 12 to 13 days, including filling, ignition, carbonization, cooling, and kiln discharge steps. The cooling stage refers

Table 1 Products of the project financed directly by the Banco da Amazônia "Valuation of wood wastes from sustainable forest management for bioenergy in Legal Amazonia"

Article	Goal	References
Logging wastes from sustainable forest management as alternative fuels for thermochemical conversion systems in Brazilian Amazon	Energy characterization of the wood wastes from twenty commercial Amazon species harvested in an SFMP and their energy equivalence to fossil fuels	[19]
Charcoal of logging wastes from sustainable forest management for industrial and domestic uses in the Brazilian Amazon	Characterization and production of charcoal from groups of wastes from SFMP in Amazonia according to their physical, chemical, and energy properties	[22]
Grouping of wood wastes from sustainable forest management aiming at bioenergy generation	Grouping of wood wastes from SFMP for firewood and charcoal production for steelmaking	[34]
Colorimetry as a criterion for segregation of logging wastes from sustainable forest management in the Brazilian Amazon for bioenergy	Use of wood color parameters to segregate SFMP wastes	[20]
Efficiency of near-infrared spectroscopy in classifying Amazonian wood wastes for bioenergy generation	Presentation of a method for classifying wood wastes from 12 Amazonian hardwoods based on near-infrared spectroscopy (NIR) and basic density	[24]
Classifying waste wood from Amazonian species by near-infrared spectroscopy (NIRS) to improve charcoal production	Classification of wood waste from 12 Amazonian species by near-infrared spectroscopy (NIRS) to improve charcoal production	[21]
Clarifying the carbonization temperature effects on the production and apparent density of charcoal derived from Amazonia wood wastes	Presentation of the effects of final carbonization temperature and different species of Amazonian wood on carbonization efficiency and apparent relative density of charcoal	[18]
Does the segregation of wood waste from Amazonia improve the quality of charcoal produced in brick kilns?	Comparison of the quality of charcoal derived from wood wastes previously segregated into four distinct and non-segregated groups of 23 Amazonian species carbonized in brick kilns	[5]

to the application of water mixed with clay on the kiln wall (Fig. 3a), with the aid of a tractor (Fig. 3b), intending to seal the kilns and reduce their internal temperatures to about 50 °C, which reduces the chances of charcoal reignition after opening the kiln.

Allied to the low technological apparatus used in the charcoal production model, the raw material used is very heterogeneous (Fig. 4). Typical wood wastes present low energy density, high moisture content, and variable diameters, hindering their use as energy sources, mainly for charcoal production with suitable quality for domestic and industrial applications. The empiricism of the activity is also related to raw material

Fig. 2 Rudimentary brick kiln commonly used in charcoal production with SFMP wastes in Brazilian Amazonia (*Source* The authors 2023)

Fig. 3 Cooling of the kilns through the application of water and clay (**a**) with the aid of a tractor, Massey Ferguson 275 model (**b**) (*Source* The authors 2023)

control, as it is carbonized without diametric standardization and moisture control. It is known that the higher the water content in wood, the lower its heating value, and the combustion will occur in an inadequate regime [12]. In addition, the charcoal yield reduces since the energy expenditure for the process's first stage, drying, is likely to be high. Canal et al. [7] demonstrated that the emissions of condensable gases (pyroligneous liquid) and non-condensable gases (CO_2, CO, CH_4, and H_2) increased, and the gravimetric yield of charcoal decreased with the increase in wood moisture during the carbonization processes.

Currently, there is no control over the moisture content of forest wastes for charcoal production in Legal Amazonia and, consequently, there is the excessive expenditure of energy, the appearance of cracks and internal fissures in the charcoal due to the sudden release of water in the form of steam and, consequent decrease

Fig. 4 SFMP wastes in the storage yard of a charcoal production unit in Brazilian Amazonia (*Source* The authors 2023)

in the mechanical properties of charcoal. Moreover, countless forest species are carbonized concomitantly; therefore, the wood's qualitative and quantitative aspects are not under control. Finally, it is necessary to improve carbonization by, for example, adjusting the process control ranges as a function of time and temperature, considering the thermal degradation of wood to maximize charcoal yield [8, 10, 30].

The effect of raw material heterogeneity and empirically controlled carbonization on charcoal production can be evidenced by data from a Brazilian Amazonia production unit (Table 2). SFMP wastes are carbonized in this production unit, especially branches. Table 2 shows the mass and volume balances obtained in twelve brick kilns in 2017, as well as the descriptive statistics associated with the analyzed variables.

The average values of the wet mass of wastes in the kiln (MW), estimated dry mass of wastes in the kiln (MSM), and estimated volume of wastes in the kiln (VW) obtained by the traditional carbonization model were 8.456 ± 0.371 t, 6.577 ± 0.295 t, and 17.16 ± 0.86 st, respectively. The evaluated kilns produce, on average, 1.762 ± 0.126 t of charcoal per cycle. The average volume of charcoal in the hot tail kilns was 6.01 ± 0.36 m^3.

The gravimetric and volumetric surveys performed were based on the operational conditions of the charcoal production unit and not necessarily on experimental conditions, which, in turn, are controlled. In this way, the factors inherent to the raw material, labor, and the control of carbonization can considerably affect the production and productivity of hot tail kilns. This effect can be verified by analyzing the mass of semi-carbonized pieces (MSP) produced per cycle, in which kiln 4 did not produce these by-products. In contrast, kiln 6 produced a high amount of semi-carbonized pieces (1.240 t) and, consequently, a low gravimetric yield of charcoal. Therefore, they cannot be considered wood or charcoal but a by-product of carbonization.

Table 2 Mass and volume balances per carbonization cycle of SFMP wastes in Brazilian Amazonia

Kiln	MW (t)	VW (st)	MSM (t)	Mch (t)	Vch (m^3)	MSP (t)
1	8.010	16.57	6.312	1.725	6.26	0.140
2	8.250	14.86	6.501	1.455	5.38	0.580
3	9.250	17.10	7.290	1.759	5.69	0.860
4	8.420	15.31	6.635	2.027	6.91	0.000
5	8.400	15.82	6.620	1.793	6.05	0.320
6	9.050	18.99	6.945	1.460	5.21	1.240
7	7.790	17.16	5.978	1.395	4.76	0.460
8	7.470	17.29	5.733	1.640	5.78	0.620
9	8.510	18.64	6.531	1.884	6.33	0.220
10	8.830	18.79	6.777	2.011	7.03	0.100
11	9.900	20.00	7.700	2.156	6.81	0.220
12	7.590	15.33	5.903	1.837	5.88	0.320
Average	8.456	17.16	6.577	1.762	6.01	423.3
CV (%)	8.46	9.70	8.64	13.75	11.68	83.98

MW wet mass of wastes in the kiln (in tons, t); *VW* estimated volume of wastes in the kiln (in stereo, st); *MSM* estimated dry mass of wastes in the kiln (t); *Mch* mass of charcoal (t); *Vch* volume of charcoal (in cubic meters, m^3); *MSP* mass of semi-carbonized pieces (t); *CV* coefficient of variation (%) (*Source* The authors 2017)

The following variations were reported for the gravimetric yield of charcoal—GYC (%, on a wet and dry mass basis) and gravimetric yield of semi-carbonized pieces—GYS (%, on a wet and dry mass basis): 16.13–24.20% (GYC, wet mass basis), 21.02–31.12% (GYC, dry mass basis), 0.00–13.70% (GYS, wet mass basis), and 0.00–17.85% (GYS, dry mass basis) (Table 3).

The production of semi-carbonized pieces is undesirable, as it decreases the kilns' production and productivity (Fig. 5a) and the gravimetric conversion coefficient (Fig. 5b). Kiln 6 presented GYC, wet mass basis, of 16.13%. Kilns 4 and 12, on the other hand, with lower production of semi-carbonized pieces, stood out in converting wood into charcoal. As already mentioned, the generation of semi-carbonized pieces is related to the raw material and the control of carbonization.

The kilns evaluated showed, on average, a GYC of 20.87 ± 1.35% (wet mass basis) and 26.84 ± 1.73% (dry mass basis). The GYS was 5.00 (based on wet mass) and 6.44% (based on dry mass), with high variation between kilns. These results indicate that the carbonization was performed differently in the kilns, probably due to the empirical character associated with charcoal production (lack of control based on specific temperature ranges and time).

Figure 6 shows the gravimetric yields of products and by-products generated in a charcoal production unit concerning the initial total wet mass (110.74 t). 67.51% of forest wastes are transformed into gases during carbonization, 19.09% result in

Table 3 Gravimetric yields of charcoal and semi-carbonized pieces per brick kiln in a charcoal production unit in Brazilian Amazonia

Kiln	GYC (%)		GYS (%)	
	Wet basis	dry basis	Wet basis	Dry basis
1	21.54	27.33	1.75	2.22
2	17.64	22.38	7.03	8.92
3	19.02	24.13	9.30	11.80
4	24.07	30.55	0.00	0.00
5	21.35	27.09	3.81	4.83
6	16.13	21.02	13.70	17.85
7	17.91	23.33	5.91	7.69
8	21.95	28.61	8.30	10.81
9	22.14	28.85	2.59	3.37
10	22.77	29.68	1.13	1.48
11	21.78	28.00	2.22	2.86
12	24.20	31.12	4.22	5.42
Average	20.87	26.84	5.00	6.44
CV (%)	12.46	12.42	80.10	80.53

GYC gravimetric yield of charcoal on a wet and dry basis (%); *GYS* gravimetric yield of semi-carbonized pieces on a wet and dry basis (%); *CV* coefficient of variation (%) (*Source* The authors 2017)

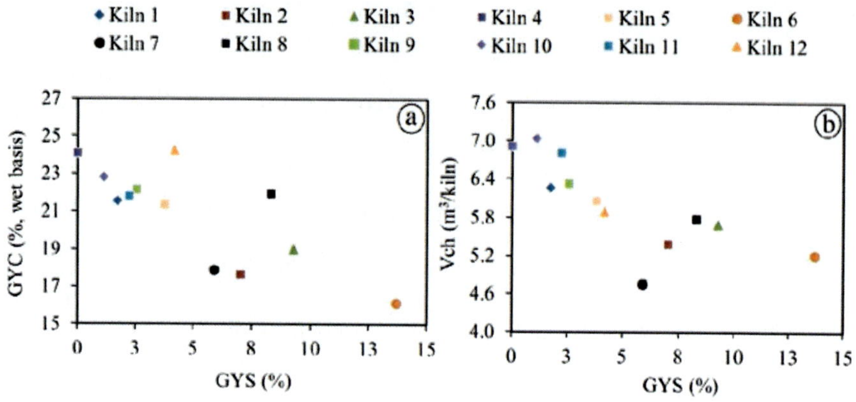

Fig. 5 Relationship between the gravimetric yield (GYC) (**a**) and volume of charcoal (**b**) with the gravimetric yield of semi-carbonized pieces (GYS) in the brick kilns of a charcoal production unit in Brazilian Amazonia (*Source* The authors 2017)

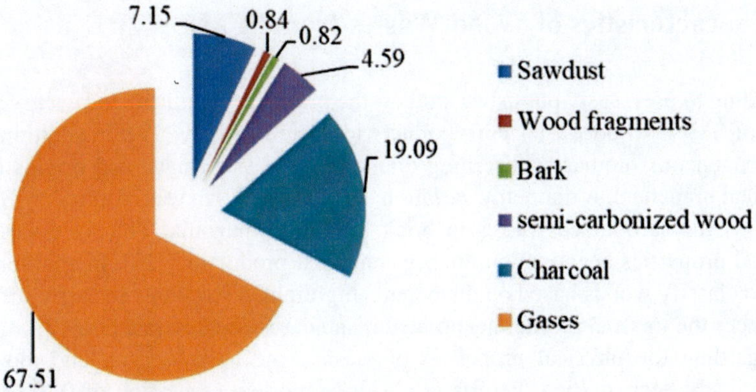

Fig. 6 Gravimetric yields of products and by-products generated in charcoal production, considering the initial total mass with 22.22% of moisture (*Source* The authors 2017)

charcoal, 4.59% in semi-carbonized pieces, and 8.81% are by-products generated in the sectioning wastes (sawdust, wood fragments, and bark).

The charcoal bulk density per kiln (CBD) evaluated in 2017 and the average value obtained for this initial prospecting can be seen in Fig. 7. On average, the CBD of waste was 0.293 t/m^3. The variation between the kilns was low; consequently, the minimum (0.270 t/m^3) and maximum (0.317 t/m^3) values were similar.

In summary, using different species with different physicochemical properties without prior separation for carbonization, combined with the diametric differences of the waste logs and empirical control of carbonization, help explain the previously reported results. Thus, alternatives are needed to maximize kiln productivity based on raw material and process control.

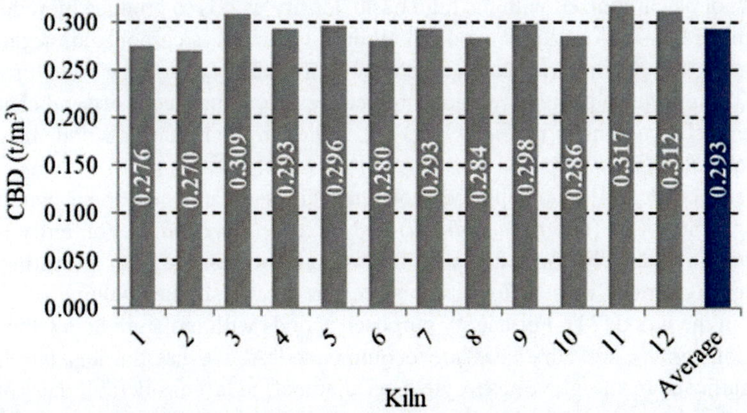

Fig. 7 Charcoal bulk density (CBD) per kiln evaluated in the production unit in Brazilian Amazonia (*Source* The authors 2017)

4 Characteristics of Wood Wastes from SFMP

According to previously published studies, the physical, chemical, and energy properties of SMFP wood wastes corroborate wide heterogeneity [19]. In addition, it is a heterogeneous biomass concerning diameter classes, formats, and origins (root, stem, and branches). A diametric variation of 0.123–0.760 m was reported by Barros et al. [5] for carbonized wastes in brick kilns in Amazonia. The wide variation in wood properties negatively influences charcoal production [23]. In addition, the need to classify woods based on diameter is highlighted since this property strongly influences the thermal profile, heat transfer, and carbonization rate [14, 17].

Regarding the physical properties of wastes, the basic density and moisture have already been studied. Pereira et al. [34] reported a variation of 0.221 (*Sterculia pruriens*) at 0.867 g/cm^3 (*Pseudopiptadenia psilostachya*), evaluating wood wastes of 18 tropical species from Brazilian Amazonia. Values ranging from 0.525 (*Couratari guianensis*) to 0.895 g/cm^3 (*M. elata*) were described by Lima et al. [19], evaluating wood wastes of 20 native species to the Amazonia. More recently, evaluating wood wastes with the near-infrared (NIR) spectroscopy technique, Lima et al. [24] reported basic density values between 0.354 (*Simara guianensis*) and 0.927 g/cm^3(*D. excelsa*).

A compilation of average waste basic density values from the previous studies and the basic density classes proposed by the International Association of Wood Anatomists [16] is shown in Table 4. Three classes were used, namely: low density ($Db \leq 0.500$ g/cm^3); medium density ($0.500 < Db \leq 0.720$ g/cm^3); and high density ($Db > 0.720$ g/cm^3). These are 55 tropical species, ranging from 0.221 to 0.935 g/cm^3.

The amplitudes reported for Amazonian wood wastes are within the variation range (0.140–1.210 g/cm^3) published by Fearnside [13], evaluating the wood properties of 268 tropical species from the same biome. The literature demonstrates that low-density woods (< 0.500 g/cm^3) are not suitable for energy purposes. Thus, they should not be carbonized with the high basic density woods to produce bioreducers.

In the reality of Amazonia, wood in all density ranges is carbonized, negatively influencing the gravimetric yield and charcoal productivity of brick kilns. Protásio et al. [37] reported a positive relationship between wood basic density and charcoal apparent relative density of *Eucalyptus* sp., indicating that denser woods result in dense charcoals.

Pereira et al. [34] also reported moisture values under operational conditions ranging from 24.99 (*Cordia goeldiana*) to 159.26% (*Sterculia* sp.) on a dry basis. Moisture is negatively correlated with the wood basic density [24], indicating that high-density woods present lower moisture due to the smaller volume of empty spaces in the wood [31]. For energy purposes, woods with moisture below the fiber saturation point (< 30%, dry basis) are recommended. Above that, the negative effects are significant in the gravimetric yield of charcoal, as a considerable part of the firewood is burned to release energy to meet the drying stage, as well as in the carbonization cycle, making it longer. In this sense, greater control of the moisture of wood wastes under operational conditions is necessary.

Table 4 Basic density of wood wastes of tropical species from Brazilian Amazonia

N	Species	Commercial name	BD (g/cm^3)	Class
1	*Sterculia pruriens*	Envira-quiabo	0.221 ± 0.016	Low
2	*Simaba guianensis*	Marupá-amarelo	0.354 ± 0.015	Low
3	*Simarouba amara*	Marupá	0.367 ± 0.015	Low
4	*Sterculia* sp.	Sucupira-babona	0.377 ± 0.052	Low
5	*Protium* sp.2	Breu-amesclim	0.392 ± 0.007	Low
6	*Tapirira guianensis*	Tapiririca	0.397 ± 0.017	Low
7	*Parkia gigantocarpa*	Fava-atanã	0.436 ± 0.033	Low
8	*Ocotea* sp.2	Louro-amarelo	0.452 ± 0.022	Low
9	*Cordia goeldiana*	Freijó	0.457 ± 0.023	Low
10	*Ocotea* sp.3	Louro-preto	0.470 ± 0.042	Low
11	*Anacardium giganteum*	Cajuaçu	0.486 ± 0.020	Low
12	*Parkia* sp.	Fava-branca	0.501 ± 0.010	Medium
13	*Couratari guianensis*	Tauarí-liso	0.525 ± 0.050	Medium
14	*Tetragastris altissima*	Amesclim	0.545 ± 0.042	Medium
15	*Couratari oblongifolia*	Tauarí-branco	0.545 ± 0.033	Medium
16	*Pouteria* sp.1	Guajará-bolacha	0.574 ± 0.028	Medium
17	*Ocotea* sp.1	Canela	0.586 ± 0.038	Medium
18	*Brosimum gaudichaudii*	Inharé	0.599 ± 0.006	Medium
19	*Peltogyne* sp.	Roxinho	0.641 ± 0.038	Medium
20	*Sclerolobium* sp.	Tachi	0.642 ± 0.042	Medium
21	*Pouteria* sp.5	Guajará-bolacha	0.667 ± 0.044	Medium
22	*Protium* sp.1	Breu-barrote	0.683 ± 0.057	Medium
23	*Pouteria oblanceolata*	Abiu	0.683 ± 0.038	Medium
24	*Vatairea sericea*	Angelim-amargoso	0.686 ± 0.051	Medium
25	*Lecythis* sp.	Sapucaia	0.690 ± 0.117	Medium
26	*Vantanea parviflora*	Uxirana	0.699 ± 0.070	Medium
27	*Pouteria* sp.4	Abiorana	0.701 ± 0.075	Medium
28	*Pouteria* sp.3	Abiorana	0.701 ± 0.061	Medium
29	*Caryocar* sp.2	Pequiarana	0.701 ± 0.029	Medium
30	*Caryocar glabrum*	Pequiarana	0.702 ± 0.074	Medium
31	*Caryocar villosum*	Pequiá	0.711 ± 0.048	Medium
32	*Protium altissimum*	Breu-barrote	0.721 ± 0.030	High
33	*Eschweilera pedicellata*	Matamatá	0.724 ± 0.045	High
34	*Eschweilera amazonica*	Jiboião	0.728 ± 0.050	High
35	*Pseudopiptadenia suaveolens*	Timborana	0.744 ± 0.086	High
36	*Eschweilera grandiflora*	Matamatá-preto	0.749 ± 0.057	High

(continued)

Table 4 (continued)

N	Species	Commercial name	BD (g/cm^3)	Class
37	*Goupia glabra*	Cupiúba	0.752 ± 0.035	High
38	*Pouteria* sp.2	Guajará-cinza	0.754 ± 0.011	High
39	*Lecythis lurida*	Jarana	0.755 ± 0.031	High
40	*Eschweilera* sp.1	Matamatá	0.779 ± 0.078	High
41	*Eschweilera coriacea*	Matamatá-branco	0.785 ± 0.012	High
42	*Eschweilera* sp.2	Matamatá	0.792 ± 0.029	High
43	*Parinari rodolphii*	Coco-pau	0.801 ± 0.046	High
44	*Caryocar* sp.1	Pequiá	0.802 ± 0.016	High
45	*Manilkara* sp.1	Maçaranduba	0.806 ± 0.079	High
46	*Hymenaea* sp.	Jatobá	0.811 ± 0.130	High
47	*Lecythis pisonis*	Sapucaia	0.812 ± 0.064	High
48	*Terminalia* sp.	Tanibuca	0.814 ± 0.021	High
49	*Enterolobium schomburgkii*	Orelha-de-macaco	0.836 ± 0.036	High
50	*Vantanea guianensis*	Uxirana	0.843 ± 0.114	High
51	*Licania canescens*	Casca-seca	0.858 ± 0.047	High
52	*Pseudopiptadenia psilostachya*	Timborana	0.867 ± 0.074	High
53	*Manilkara* sp.2	Maçaranduba	0.872 ± 0.010	High
54	*Manilkara elata*	Maçaranduba	0.903 ± 0.023	High
55	*Dinizia excelsa*	Angelim-vermelho	0.935 ± 0.044	High

N species number; *BD* wood basic density (g/cm^3). Commercial name in Brazil. Average ± standard deviation (*Source* The authors 2023)

Average values of the chemical properties of wood wastes, such as total lignin (LigT), total extractives (EXT), and elemental carbon (C), can be seen in Table 5. These are data on wood from branches of 20 logged tropical species by the reduced impact logging method in an SFMP certified by the Forest Stewardship Council (FSC) in Brazilian Amazonia, previously published by Lima et al. [19].

EXT (1.8–17.9%, dry mass basis), LigT (30.2–38.1, dry mass basis free of extractives), and C (49.2–52.4%, basis mass dry) demonstrated high variability in native tropical woods of Amazonia. Wood species with high levels of EXT, LigT, and C are promising for energy purposes, especially to supply the charcoal-producing complex in the Carajás region, located between Maranhão and Pará states, in Brazil. Lima et al. [23] demonstrated that the EXT had a positive relationship with the GYC, indicating that the species *D. excelsa, P. altissimum, M. elata,* and *G. glabra* showed the best carbonization mass balances.

The C value (42.82%) described by Haqiqi et al. [15] for *Eucalyptus pellita* and the variation range (47.23–48.80%) reported by Santos et al. [40] of four *Eucalyptus* hybrid clones (three *Eucalyptus urophylla* × *E. grandis* and one *Eucalyptus camaldulensis* × *E. grandis*) at seven years of age, were lower than the values found

Table 5 Total extractives, total lignin, and elemental carbon from SFMP wastes in Brazilian Amazonia

Species	Commercial name	EXT (%)*	LigT (%)**	C (%)*
Dinizia excelsa	Angelim-vermelho	17.9	37.6	51.7
Protium altissimum	Breu-barrote	12.6	31.0	50.9
Manilkara elata	Maçaranduba	11.9	30.2	51.0
Goupia glabra	Cupiúba	11.4	34.0	51.0
Pouteria sp. 2	Guajará-bolacha	9.0	33.3	50.3
Pouteria oblanceolata	Abiu	9.0	32.3	50.6
Lecythis lurida	Jarana	8.4	34.3	49.7
Pseudopiptadenia suaveolens	Timborana	8.0	32.9	52.0
Eschweilera grandiflora	Matamatá-preto	7.7	30.9	49.2
Caryocar glabrum	Pequiarana	7.7	32.6	50.7
Enterolobium schomburgkii	Orelha-de-macaco	6.0	33.0	51.8
Eschweilera pedicellata	Matamatá	6.0	32.4	52.4
Lecythis pisonis	Sapucaia	5.9	33.5	52.2
Couratari guianensis	Tauarí-liso	5.3	33.6	49.9
Caryocar villosum	Piquiá	4.9	34.5	51.0
Pouteria sp. 1	Abiorana	4.0	33.6	50.1
Couratari oblongifolia	Tauarí-branco	3.8	32.6	49.7
Licania canescens	Casca-Seca	3.6	36.6	49.7
Vantanea parviflora	Uxirana	2.4	33.3	50.0
Parinari rodolphii	Coco-pau	1.8	38.1	50.9

EXT total extractives (%); *LigT* total lignin (%); and *C* elemental carbon (%). *Based on dry wood mass. ** Based on dry wood mass free of extractives (*Source* Lima et al. [19])

for wood wastes, which ranged from 49.2 to 52.4%. This range of C indicates that the wastes are very promising since C is the main energetic element of biomass [2]. Woods of the *Eucalyptus genus* are the most used to compose energy forests, Brazil's main source of forest biomass. Thus, alternative renewable sources are important for diversifying the energy sector's raw materials.

Pereira et al. [35] discussed that woods with LigT above 28% are desired for charcoal production. All species shown in Table 5 showed values above the reference published by the authors, indicating that the species are suitable for this purpose. Wood species with high LigT values are more thermally stable and contribute positively to the gravimetric yield of charcoal of the production unit [27]. In addition, they have a high heating value, which indicates more energy is generated during combustion [43].

Although wastes from SFMP have suitable quality for charcoal production, the carbonization of wood with different physical and chemical characteristics negatively affects charcoal production unity, productivity in brick kilns, and the quality of the charcoal produced. Barros et al. [5] demonstrated that the carbonization of different woods together, which is the traditional model of carbonization in Amazonia, results in a reduction in the charcoal quality, negatively affecting friability, apparent relative density, ash content, volatile matter, fixed carbon, higher heating value, and energy density. The authors highlighted the need for better control of the raw material factor under operational conditions, especially with the wood segregation before carbonization, aiming to reduce the effect of heterogeneity on the production, productivity, and quality of the charcoal produced in traditional kilns.

5 Carbonization of Similar Wood Wastes

Several proposals were presented in the literature to reduce the heterogeneity of the residual raw material to produce charcoal based on the properties of the wood [19, 20, 34] and charcoal [23]. Carbonization should prioritize woods with similar properties to homogenize the process phases and the bioreducer quality. The wood basic density can be a criterion to separate wastes, as well as several combined properties, through multivariate statistical analyses of grouping.

Pereira et al. [34] proposed carbonization considering basic density classes. In this study, the authors separate wood wastes into classes of low (*Sterculia pruriens, Sterculia* sp., and *Cordia goeldiana*), medium (*Tetragastris altissima, Pouteria* sp., *Ocotea* sp., *Peltogyne* sp., *Sclerolobium* sp., *Protium* sp., *Lecythis* sp., and *Caryocar villosum*), and high (*Eschweilera amazonia, Lecythis lurida, Eschweilera* sp., *Manilkara* sp., *Hymenaea* sp., *Terminalea* sp., and *Pseudopiptadenia psilostachya*) density. They recommended wood of medium and high basic density for steel charcoal production, as it will result in a bioreducer with adequate apparent relative density.

Lima et al. [19] verified four groups of similar residual woods through the PCA technique, using physical (basic density, moisture, and maximum moisture content); chemical (proximate analysis: fixed carbon, volatile matter, and ash; and molecular analysis: total extractives, soluble, insoluble, and total lignin); and energy (higher heating value and energy density) properties; in addition to the specific consumption of firewood in charcoal production. This proposal promoted improvements of (+) 22, (−) 9.4, (+) 2.0, (−) 2.3, (+) 1.0, and (+) 23.6% in apparent relative density, ash content, fixed carbon, friability, higher heating value, and energy density of charcoal produced in brick kilns in Amazonia [5]. Table 6 presents the wood groups evaluated by the authors and the basic density variation ranges.

Lima et al. [23] used the same grouping analysis to separate species in charcoal production units in Amazonia based on physical, chemical, and energy properties. Furthermore, the authors considered the yields of the carbonization process under laboratory conditions. Colorimetric characteristics (lightness, green/red axis, blue/yellow axis, color saturation, and hue angle), physical (moisture and basic density), chemical (total extractives and total lignin), and energy density of these residual woods contributed to the formation of six species groups using the PCA technique

Table 6 Species groups segregated by PCA technique for carbonization in a charcoal production unit in Brazilian Amazonia

Group 1 (BD: 0.948–1.015 g/cm^3)	Group 2 (BD: 0.787–0.914 g/cm^3)	Group 3 (BD: 0.429–0.711 g/cm^3)	Group 4 (BD: 0.568–0.936 g/cm^3)	Group 5 (traditional) (BD: 0.352–1.015 g/cm^3)
D. excelsa	*P. rodolphii*	*A. giganteum*	*Pouteria* sp.1	*A. giganteum*
	L. canescens	*C. oblongifolia*	*V. sericea*	*C. glabrum*
		Ocotea sp.1	*P. altissimum*	*C. villosum*
		Ocotea sp.2	*G. glabra*	*C. oblongifolia*
		Pouteria sp.2	*L. lurida*	*D. excelsa*
			M. elata	*E. coriacea*
			E. coriacea	*G. glabra*
			C. glabrum	*L. lurida*
			C. villosum	*L. pisonis*
			L. pisonis	*L. canescens*
			P. suaveolens	*M. elata*
			V. guianensis	*Ocotea* sp.1
				Ocotea sp.2
				P. rodolphii
				P. gigantocarpa
				Pouteria sp.1
				Pouteria sp.2
				P. altissimum
				Protium sp.
				P. suaveolens
				S. amara
				V. guianensis
				V. sericea

BD basic density (g/cm^3) (*Source* [5])

[20]. In the study, the basic and energy densities were negatively correlated with lightness, blue/yellow axis, color saturation, and hue angle, demonstrating that darker woods, such as *D. excelsa* and *M. elata,* present greater energy potential.

Two methods were tested to segregate residual wood on an operational scale. The first was based on the basic density [34], and the second was based on several characteristics of the wood [19]. In this sense, positive effects are evident in the productivity of brick kilns with raw material control. Figure 8 shows the effects of separating wood into density classes (medium–high and medium–low) on charcoal production at an operational scale.

Medium–high basic density woods (0.739–0.993 g/cm^3) had lower average MC (41.1%, dry mass basis), while medium–low basic density woods (0.517–0.630 g/cm^3) had higher MC (67.5%, based on dry mass). The trend of water reduction in woods with high BD can be seen in Fig. 9a. Traditional carbonization brings

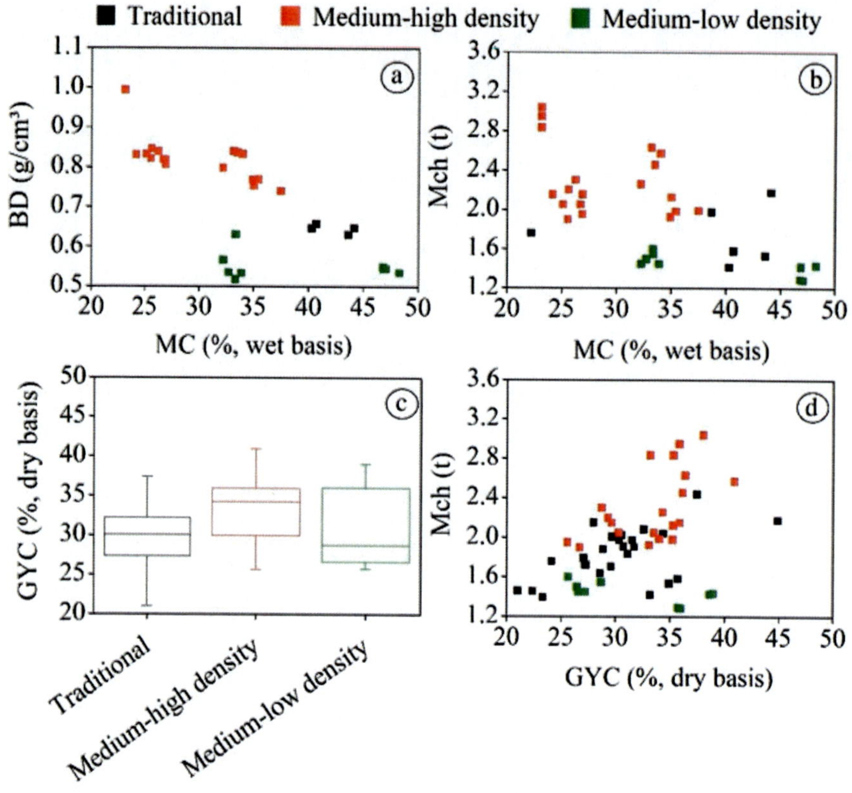

Fig. 8 Effects of waste segregation into density classes (medium–high and medium–low) on charcoal productivity at operational scale in Brazilian Amazonia. Relationship between basic density—BD (g/cm^3) and moisture content—MC (%) (**a**), mass of charcoal—Mch (t) and MC (%) (**b**), basic density—BD (g/cm^3) and gravimetric yield of charcoal—GYC (%) (**c**), and GYC (%) and Mch (t) (**d**) (*Source* The authors 2023)

together woods with a wide variation in BD, which is not interesting due to the different MCs of the woods, which makes it difficult to control the carbonization process, reducing Mch (Figs. 9b and d) and GYC (Fig. 9c). Traditional carbonization showed a GYC of 30.3% (based on dry mass), lower than the average values reported for the medium–high (33.4%, based on dry mass) and medium–low wood groups (31.6%, based on dry mass) density. It is important to note that traditional carbonization negatively affects bioreducer production, the unit's operational efficiency, and revenues. Segregation promotes greater production of charcoal with the same quantity of wood mass inserted in the kiln. From an operational and charcoal productivity point of view, BD as a criterion to separate residual wood is the most appropriate and simple method to be carried out by the employees.

Additional wood characteristics can be used as a separation criterion to improve the production and quality of the steel bioreducer. In this sense, the carbonization of the groups proposed by Lima et al. [19] was tested under operational conditions in brick kilns. Figure 9 presents weighted average data for moisture—MCw (Fig. 9a) and basic density—BDWw (Fig. 9b), wet mass—MW (Fig. 9c), and volume—VW (Fig. 9d) of wood per hot-tail kiln. Groups 2 (9.117 m^3) and 3 (9.500 m^3) filled the kilns with the highest average volumes of waste. Groups 1, 2, and 4 showed average MW, MCw, and BDWw above group 5 (conventional model).

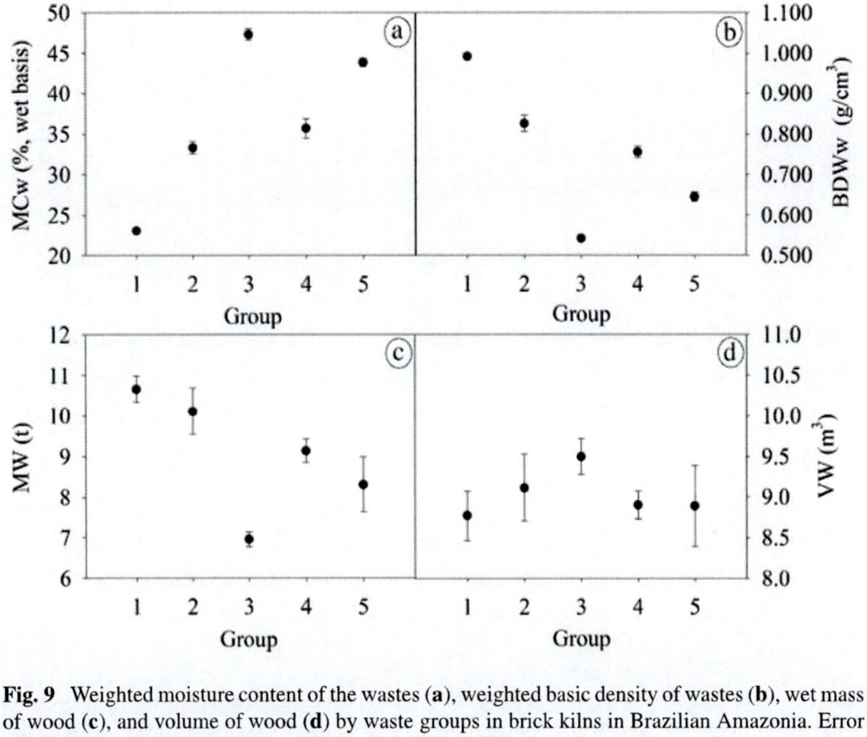

Fig. 9 Weighted moisture content of the wastes (**a**), weighted basic density of wastes (**b**), wet mass of wood (**c**), and volume of wood (**d**) by waste groups in brick kilns in Brazilian Amazonia. Error bars refer to standard deviations (*Source* The authors 2023)

Waste groups 1, 2, and 4, which combine medium and high-density species, had the highest MW. Consequently, they allowed better use of the kiln's internal space. In addition, such combined wood species had lower MCw and higher BDWw. The woods in group 3 presented a lower MW inside the kiln due to the lower BDWw. In summary, grouping similar tropical woods provided encouraging results related to the wood amount inserted in the kilns. It is known that the greater the dry wood mass in the kiln, the greater the mass of charcoal produced. The traditional model of carbonization promotes greater variation in the MW, and VW used in the process, which is not interesting, as it makes the monthly and annual planning of wood in the charcoal plant difficult.

Compared to the traditional model without wood separation, the charcoal productivity dataset from brick kilns demonstrates the positive effects of waste wood segregation in Amazonia (Fig. 10). In order, groups 1 (7.84 m^3), 2 (7.33 m^3), 4 (6.61 m^3), and 3 (6.43 m^3) showed better average values of charcoal volume (Vch), compared to the group without segregation, which produced 6.02 m^3. Regarding mass (Mch), gravimetric yield (GYC), and bulk density (CBD) of charcoal, the highest average values were reported for groups 1, 2, and 4.

The highest Mch and GYC results reported for group 1 (*D. excelsa*) are associated with the high content of extractives and lignin in the wood (see Table 5). The H/C

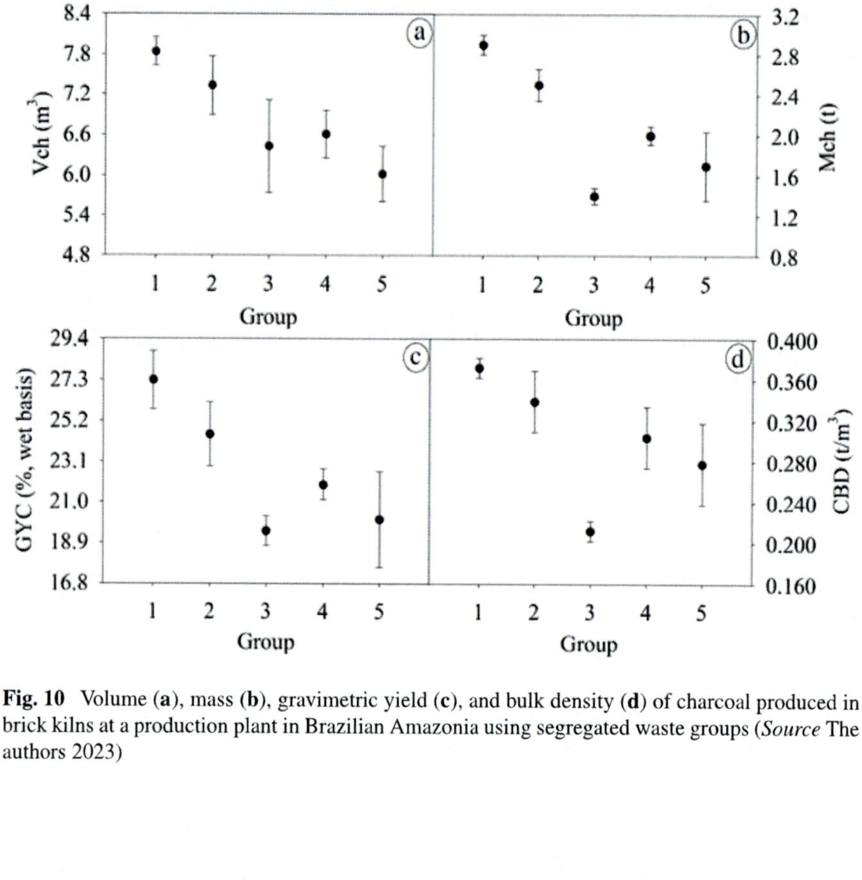

Fig. 10 Volume (**a**), mass (**b**), gravimetric yield (**c**), and bulk density (**d**) of charcoal produced in brick kilns at a production plant in Brazilian Amazonia using segregated waste groups (*Source* The authors 2023)

ratio and the high thermal stability of the extractives, mainly the phenolics, contribute to the increase in Mch and GYC [36]. In addition, this group will present reduced emissions of condensable and non-condensable gases into the atmosphere. Woods with higher densities explain the high CBD mean values of groups 1 (0.372 t/m^3), 2 (0.339 t/m^3), and 4 (0.304 t/m^3). In this sense, segregation will promote a greater mass of charcoal transported to the steel industries compared to group 5. On the other hand, the charcoal of group 3 is less dense and voluminous and can be considered as a basis for commercialization.

The production of charcoal must be better controlled, especially concerning raw materials and factors associated with the process. Clearly, the separation of wood promotes encouraging results related to the production and quality of charcoal, which can influence the operational and financial planning of the plant. Thus, the methodology to adequately control the quality of the consumed raw material must be adjusted. As charcoal is commercialized based on its mass, the basic density can be used as a qualitative index to discriminate the material by its quality. Furthermore, it is an easily determined property. The literature corroborates that dense woods generate charcoals with high apparent relative density [23]. On the other hand, groups with similar physical, chemical, and energy properties can clearly improve charcoal quality.

6 Future Prospects for Charcoal Production from Wood Wastes

The energy use of forest wastes contributes to reducing the environmental impact caused by improper disposal and the exacerbated use of "dirty" energy sources and, consequently, to the mitigation of greenhouse gas emissions. Forest biomass stands out among the various renewable materials used for energy purposes due to the numerous solid, liquid, and gaseous fuels obtained through thermochemical, biochemical, and mechanical routes and their conciliation. It is worth noting that Normative Instruction No. 5, of December 11, 2006, published by the Ministry of the Environment, mentions in Article 2, item XIV, that forest logging wastes (branches, buttresses, and remains of trunks, for example) can be used as secondary products of SFMP for wood and energy production [6]. However, there is still a lack of scientific research and technology transfer for the sustainable valorization of these resources for bioenergy generation on industrial and domestic scales.

Many advances have been obtained with the understanding of the quality of SFMP wastes destined for energy generation and steel charcoal production. However, there is a need to continue research to reach technological maturity for a real understanding of the energy potential of this biomass. Furthermore, converting this raw material into energy products is not sufficiently mastered. In this sense, thermochemical conversion routes must be sufficiently analyzed with integrated methodologies (energetic-environmental-economic). For example, the life cycle inventory will provide a technical–economic-environmental assessment of strategies for the sustainable use of residual biomass in SFMP associated with the carbon balance of conversion processes.

The studies carried out so far have promoted important advances in clarifying the quality parameters of forest wastes destined for energy production. Despite this, some aspects of its energy use need to be elucidated. As pointed out by Lima et al. [19], the thermal behavior and combustion kinetics of this fuel remains unclear. It is also necessary to know the energy performance of the wastes at different final carbonization temperatures. These studies will allow more efficient use of wood and subsidize new conversion routes, such as densification, torrefaction, gasification, and liquefaction. Biomass has been widely studied for producing electricity, bio-oil, and chemical compounds [26]. Although there is research on wood from Amazonian species [38] and other biomass types common in the region [3], there is no large-scale research on woody Amazonian wastes for this purpose.

From a strategic point of view, biomass energy conversion processes are economically viable techniques to generate energy for local communities and industrial production systems, even in remote locals, as in several regions of the Amazonia. In Brazil, direct combustion and carbonization are the two most widespread ways of using wood wastes from SFMP, the main legal tool for the rational use of timber and non-timber forest products in Brazilian Amazonia.

Previously reported studies were limited to a single charcoal production unit from waste wood. Future research needs to include more wooden species and Amazonian forest management regions in their studies. It is also necessary to know the amount of waste stored in the forests. Thus, efficient harvesting and storage strategies for these biomasses must be developed.

The sustainability of the entire production chain for the use of waste must become one of the main objects of future research, generating technological impacts for the full use of waste and improving sustainable technologies for bioenergy use of tropical forest resources; with new, creative, and viable solutions that contribute to technological, economic, and social developments, through research. Thus, the following questions must be answered: (i) How much carbon is emitted from the collection, transport, carbonization, and combustion of waste?; (ii) Does the carbon stored in the forest soil and the regeneration of trees after harvesting neutralize the carbon emitted by the production chain?; (iii) Are the strategies effective in minimizing environmental impacts?

The answers to these questions will support the development of new solutions in Amazonia for increasing the efficiency of waste use and minimizing polluting gas emissions during wood carbonization. The technologies used for planted forests burn polluting gases formed during carbonization [8] and can be improved, adapted, and applied to the carbonization of SFMP wood wastes.

7 Conclusions

SFMP wastes have suitable properties for charcoal production for domestic and industrial use. However, there are great challenges to producing sustainable energy with this Amazonian biomass, mainly due to the low technological level of the carbonization kilns and the high variation in the waste properties concerning their dimensions and physical, chemical, and energy properties.

For example, waste diameter and basic density show high variability among species, studies were carried out to characterize and classify these wood wastes into different groups to solve these problems. The results demonstrate an increase in the gravimetric yield of charcoal, the productivity of the kilns, and an improvement in the bioreducer quality.

Therefore, the importance of technological research is emphasized to enable the development of carbonization, direct heat generation, and other energy routes to adapt the industry in Brazil to the current requirements of the national and international markets. Future research should highlight technological improvements in energy production using SFMP wastes to increase efficiency, quality, and sustainability.

Acknowledgements The authors would like to thank the Coordination for the Improvement of Higher Education Personnel (CAPES—financing code 001), the National Council for Scientific and Technological Development (CNPq—process nº 306793/2019-9 and 406053/2022-7), the Amazonia Bank (financial support contract nº 2018/193), the Keilla Group, the Laboratory of Technology of Forest Products of the Federal Rural University of Amazonia (UFRA, Pará State, Brazil), the Multiuser Laboratory of Biomaterials and Biomass Energy of the Federal University of Lavras (UFLA, Minas Gerais State, Brazil), and the Embrapa Eastern Amazonia for the research support and funding.

We would like to specifically thank the following people who played an important role in the field stages of this project: Wilson Alan Santos do Rosário, Iara Alves Moreira, Alessandra Alves Pereira, Maria Kely Alves Gomes, Sindy Daniela Ferreira Sozinho, Raimunda Marcia Souza dos Santos, Irinaldo França Pimentel da Silva, Rita de Cássia Carvalho Silva, Patrícia Leonídia dos Santos, Evelym Poliana Santos Patrício, Romel da Costa Dias, Eidy Regina Oliveira da Silva, Eliana Martins de Sousa, Lanna Vitória Ribeiro Santos, Mirian Isis Demétrio Medeiros, Udson de Oliveira Barros Junior, Joseane Memória Ribeiro dos Santos, Carlos Alberto Oliveira Diniz, Amauri Roberto Müller, Joaquim Gomes da Silva Neto, Alexandre Rodrigues, Vanessa Spindola de Oliveira, Enoque Leda de Arruda, Edil Neves Gandra Junior, Marcos Victor da Conceição Paixão, Francisco de Assis Silva Matos, and Josué Evandro Ribeiro Ferreira.

References

1. Achat DL, Deleuze C, Landmann G, Pousse N, Ranger J, Augusto L (2015) Quantifying consequences of removing harvesting residues on forest soils and tree growth—a meta-analysis. For Ecol Manage 348:124–141. https://doi.org/10.1016/j.foreco.2015.03.042
2. Adamovics A, Platace R, Gulbe I, Ivanovs S (2018) The content of carbon and hydrogen in grass biomass and its influence on heating value. Eng Rural Dev 1(1):1277–1281. https://doi.org/10.22616/ERDev2018.17.N014

3. Araujo RO, Ribeiro FCP, Santos VO, Lima VMR, Santos JL, Vilaça JES, Chaar JS, Falcão NPS, Pohlit AM, de Souza LKC (2022) Renewable energy from biomass: an overview of the Amazon Region. Bioenergy Res 15(2):834–849. https://doi.org/10.1007/s12155-021-10308-x
4. Assis MR, Brancheriau L, Napoli A, Trugilho PF (2016) Factors affecting the mechanics of carbonized wood: literature review. Wood Sci Technol 50(3):519–536. https://doi.org/10.1007/s00226-016-0812-6
5. Barros DS, Lima MDR, Dias Junior AF, Bufalino L, Massuque J, Santos EV, Trugilho PF, Oliveira FA, Protásio TP (2023) Does the segregation of wood waste from Amazonia improve the quality of charcoal produced in brick kilns? Bioenergy Res. https://doi.org/10.1007/s12155-022-10551-w
6. Brazil (2006) Ministry of the Environment. Normative Instruction MMA No. 5 of December 11, 2006. Technical procedures for the preparation, presentation, execution, and technical evaluation of Sustainable Forest Management Plans—SFMP in primitive forests and their forms of succession in the Legal Amazon, and makes other arrangements. Brazil
7. Canal WD, Carvalho AMML, Carneiro ACO, Vital BR, Pereira BLC, Donato DB (2016) Effect of wood moisture on greenhouse gases emission from carbonization process. Sci For 44(112):831–840. https://doi.org/10.18671/scifor.v44n112.05
8. Cardoso MT, Damásio RAP, Carneiro ACO, Jacovine LAG, Vital BR, Barcelos DC (2010) Construction of a gas burning system resulting from carbonization to reduce pollutants emissions. Cerne 16:115–124
9. Crow SE, Reeves M, Turn S, Taniguchi S, Schubert OS, Koch N (2016) Carbon balance implications of land use change from pasture to managed eucalyptus forest in Hawaii. Carbon Manag 7(3–4):171–181. https://doi.org/10.1080/17583004.2016.1213140
10. Damásio RAP, Oliveira AC, Carneiro ACO, Barcelos DC, Pereira BLC, Magalhães MA, Silva MS (2015) Thermal profile and control of carbonization on circular kiln through the internal temperature. Ciência da Madeira 6(1):11–22. https://doi.org/10.12953/2177-6830/rcm.v61p11-22
11. Demirbas A (2005) Potential applications of renewable energy sources, biomass combustion problems in boiler power systems and combustion related environmental issues. Prog Energy Combust Sci 31(2):171–192. https://doi.org/10.1016/j.pecs.2005.02.002
12. Diniz VY (1981) Caldeiras a lenha. In: Gaseificação de madeira e carvão vegetal. pp 114–131
13. Fearnside PM (1997) Wood density for estimating forest biomass in Brazilian Amazonia. For Ecol Manage 90(1):59–87. https://doi.org/10.1016/S0378-1127(96)03840-6
14. Figueroa MJM, Moraes PD (2009) Wood behavior at high temperatures Manuel. Ambiente Construído 9(4):157–174
15. Haqiqi MT, Hudaya D, Septiana HA, Ramadhan R, Yuliansyah Y, Suwinarti W, Amirta R (2022) Short communication: analysis of the ultimate wood composition of a forest plantation species, *Eucalyptus pellita*, to estimate its bioelectricity potency. Biodiversitas 23(5):2389–2394. https://doi.org/10.13057/biodiv/d230516
16. International Association of Wood Anatomists - IAWA (1989) International association of wood anatomists: list of microscopc features for hardwood identification. IAWA J 10(3):219–232
17. Jesus MS, Carneiro ACO, Martinez CLM, Vital BR, Carneiro APS, Assis MR (2019) Thermal decomposition fundamentals in large-diameter wooden logs during slow pyrolysis. Wood Sci Technol 53:1353–1372. https://doi.org/10.1007/s00226-019-01133-9
18. Lima MDR, Massuque J, Bufalino L, Trugilho PF, Ramalho FMG, Protásio TP, Hein PRG (2022) Clarifying the carbonization temperature effects on the production and apparent density of charcoal derived from Amazonia wood wastes. J Anal Appl Pyrolysis 166:105636. https://doi.org/10.1016/j.jaap.2022.105636
19. Lima MDR, Patrício EPS, Barros Junior UO, Assis MR, Xavier CN, Bufalino L, Trugilho PF, Hein PRG, Protásio TP (2020) Logging wastes from sustainable forest management as alternative fuels for thermochemical conversion systems in Brazilian Amazon. Biomass Bioenergy 140:105660. https://doi.org/10.1016/j.biombioe.2020.105660
20. Lima MDR, Patrício EPS, Barros Junior UO, Silva RCC, Bufalino L, Numazawa S, Hein PRG, Protásio TP (2021) Colorimetry as a criterion for segregation of logging wastes from sustainable

forest management in the Brazilian Amazon for bioenergy. Renew Energy 163:792–806. https://doi.org/10.1016/j.renene.2020.08.078

21. Lima MDR, Ramalho FMG, Trugilho PF, Bufalino L, Dias Júnior AF, Protásio TP, Hein PRG (2022) Classifying waste wood from Amazonian species by near-infrared spectroscopy (NIRS) to improve charcoal production. Renew Energy 12(1):3224. https://doi.org/10.1016/j.renene.2022.05.048
22. Lima MDR, Simetti R, de Assis MR, Trugilho PF, de Carneiro ACO, Bufalino L, Hein PRG, de Protásio TP (2020) Charcoal of logging wastes from sustainable forest management for industrial and domestic uses in the Brazilian Amazonia. Biomass Bioenergy 142. https://doi.org/10.1016/j.biombioe.2020.105804
23. Lima MDR, Simetti R, Assis MRD, Trugilho PF, Carneiro ADCO, Bufalino L, Hein PRG, Protásio TDP (2020) Charcoal of logging wastes from sustainable forest management for industrial and domestic uses in the Brazilian Amazonia. Biomass Bioenergy 142:105804. https://doi.org/10.1016/j.biombioe.2020.105804
24. Lima MDR, Trugilho PF, Bufalino L, Dias Júnior AF, Ramalho FMG, Protásio TP, Hein PRG (2022) Efficiency of near-infrared spectroscopy in classifying Amazonian wood wastes for bioenergy generation. Biomass Bioenergy 166:106617. https://doi.org/10.1016/j.biombioe.2022.106617
25. Manolis EN, Zagas TD, Karetsos GK, Poravou CA (2019) Ecological restrictions in forest biomass extraction for a sustainable renewable energy production. Renew Sustain Energy Rev 110:290–297. https://doi.org/10.1016/j.rser.2019.04.078
26. Mednikov AS (2018) A review of technologies for multistage wood biomass gasification. Therm Eng 65(8):531–546. https://doi.org/10.1134/S0040601518080037
27. Moriana R, Vilaplana F, Ek M (2015) Forest residues as renewable resources for bio-based polymeric materials and bioenergy: chemical composition, structure and thermal properties. Cellulose 22(5):3409–3423. https://doi.org/10.1007/s10570-015-0738-4
28. Numazawa CTD, Krasovskiy A, Kraxner F, Pietsch SA (2020) Logging residues for charcoal production through forest management in the Brazilian Amazon: economic gains and forest regrowth effects. Environ Res Lett 15(11):114029. https://doi.org/10.1088/1748-9326/abb495
29. Numazawa CTD, Numazawa S, Pacca S, John VM (2017) Logging residues and CO_2 of Brazilian Amazon timber: two case studies of forest harvesting. Resour Conserv Recycl 122:280–285. https://doi.org/10.1016/j.resconrec.2017.02.016
30. Oliveira AC, Carneiro ACO, Pereira BLC, Vital BR, Carvalho AMML, Trugilho PF, Damásio RAP (2013) Optimization of charcoal production through control of carbonization temperatures. Revista Árvore 37(3):557–566. https://doi.org/10.1590/S0100-67622013000300019
31. Oliveira JTS, Hellmeister JC, Tomazello Filho M (2005) Variation of the moisture content and specific gravity in the wood of seven eucalypt species. Revista Árvore 29(1):115–127. https://doi.org/10.1590/S0100-67622005000100013
32. Paré D, Thiffault E (2016) Nutrient budgets in forests under increased biomass harvesting scenarios. Current Forestry Reports 2(1):81–91. https://doi.org/10.1007/s40725-016-0030-3
33. Peláez-Samaniego MR, Garcia-Perez M, Cortez LB, Rosillo-Calle F, Mesa J (2008) Improvements of Brazilian carbonization industry as part of the creation of a global biomass economy. Renew Sustain Energy Rev 12(4):1063–1086. https://doi.org/10.1016/j.rser.2006.10.018
34. Pereira AA, Lima MDR, Patrício EPS, Numazawa S, Goulart SL, Protásio TP (2020) Grouping of wood residues from sustainable forest management aiming at bioenergy generation. Sci For 48(127):1–14. https://doi.org/10.18671/scifor.v48n127.01
35. Pereira BLC, Carneiro ACO, Carvalho AMML, Colodette JL, Oliveira AC, Fontes MPF (2013) Influence of chemical composition of Eucalyptus wood on gravimetric yield and charcoal properties. Bioresources 8(3):4574–4592. https://doi.org/10.15376/biores.8.3.4574-4592
36. Pereira BLC, Oliveira AC, Carvalho AMML, Carneiro ACO, Santos LC, Vital BR (2012) Quality of wood and charcoal from Eucalyptus clones for ironmaster use. Int J For Res 2012:1–8. https://doi.org/10.1155/2012/523025

37. Protásio TP, Lima MDR, Scatolino MV, Silva AB, Figueiredo ICR, Hein PRG, Trugilho PF (2021) Charcoal productivity and quality parameters for reliable classification of Eucalyptus clones from Brazilian energy forests. Renew Energy 164:34–45. https://doi.org/10.1016/j.renene.2020.09.057
38. Rivera JEG, Merencio DO, Vistín ASR, Acosta RDL, Espinoza BDC, Abreu-Naranjo R (2022) Thermogravimetric characteristics and kinetic modeling of *Piptocoma discolor* pyrolysis and combustion processes to contribute to its use as a renewable energy source in the Ecuadorian Amazon region. Biomass Convers Biorefin. https://doi.org/10.1007/s13399-021-02178-2
39. Rodrigues T, Braghini Junior A (2019) Charcoal: a discussion on carbonization kilns. J Anal Appl Pyrolysis 143:104670. https://doi.org/10.1016/j.jaap.2019.104670
40. Santos RC, Carneiro ACO, Vital BR, Castro RVO, Vidaurre GB, Trugilho PF, Castro AFNM (2016) Effect of properties chemical and siringil/guaiacil relation wood clones of Eucalyptus in the production of charcoal. Ciência Florestal 26(2):657–669. https://doi.org/10.5902/1980509822765
41. SINDIFER. Sindicato da Indústria do Ferro no Estado de Minas Gerais (2021) Pig iron production in Minas Gerais and in Brazil. Statiscal yearbook reference: 2020. Minas Gerais
42. Surup GR, Trubetskaya A, Tangstad M (2020) Charcoal as an alternative reductant in ferroalloy production: a review. Processes 8(11):1432. https://doi.org/10.3390/pr8111432
43. Telmo C, Lousada J (2011) The explained variation by lignin and extractive contents on higher heating value of wood. Biomass Bioenergy 35(5):1663–1667. https://doi.org/10.1016/j.biombioe.2010.12.038
44. Vos MAE, den Ouden J, Hoosbeek M, Valtera M, de Vries W, Sterck F (2023) The sustainability of timber and biomass harvest in perspective of forest nutrient uptake and nutrient stocks. For Ecol Manage 530:120791. https://doi.org/10.1016/j.foreco.2023.120791

Renewable Energy Sources to Promote Food Sovereignty and Social Inclusion

Alfredo José dos Santos Junior, Paulo Renato Souza de Oliveira, João Marcelo Ribeiro Macedo, Allana Katiussya Silva Pereira, Daniel Saloni, Luis Filipe Cabral Cezario, José Otávio Brito, and Ananias Francisco Dias Júnior

Abstract Several examples in the literature and worldwide place biomass as an undervalued fuel, considering that it could have a more noble use than its use in combustion. However, is there anything nobler than eating? Forest biomass is used for cooking food in the wealthiest and most vulnerable social classes as a primary or secondary energy source. For underprivileged families, forest biomass is used in a raw form or with little processing in ineffective energy conversion technologies, which can bring health risks due to long-term exposure. For wealthy families, forest biomass is used with greater use of technology, for example, in densified fuels or efficient ovens and even, in some cases, in preparing "gourmet" foods. Therefore, as it is a

A. J. dos Santos Junior (✉) · P. R. S. de Oliveira · A. K. S. Pereira · J. O. Brito
"Luiz de Queiroz" College of Agriculture, University of São Paulo (ESALQ/USP), Department of Forests Sciences, Av. Pádua Dias, 11, Piracicaba, São Paulo 13418-900, Brazil
e-mail: alfredo.j.santos.junior@usp.br

P. R. S. de Oliveira
e-mail: renatosarievilo@usp.br

A. K. S. Pereira
e-mail: allana.florestal@gmail.com

J. O. Brito
e-mail: jobrito@usp.br

J. M. R. Macedo · L. F. C. Cezario · A. F. Dias Júnior
Federal University of Espírito Santo, Department of Forestry and Wood Sciences, Av. Governador Lindemberg, 316, Jerônimo Monteiro, Espírito Santo 29550-000, Brazil
e-mail: ribeirojoaomarcelomacedo@gmail.com

L. F. C. Cezario
e-mail: luisfilipecabral10@gmail.com

A. F. Dias Júnior
e-mail: ananias.dias@ufes.br

D. Saloni
North Carolina State University (NCSU), College of Natural Resources, Department of Forest Biomaterials, Raleigh, NC 27695, USA
e-mail: desaloni@ncsu.edu

A. K. S. Pereira and A. F. Dias Júnior (eds.), *Impacts of Using Biomass as an Energy Source in Homes*, Green Energy and Technology,
https://doi.org/10.1007/978-3-031-38824-8_5

fuel for the daily use of many populations, forest biomass becomes directly related to food security issues. In addition to participating in broader contexts, forest biomass can promote economic and social development in different locations, enabling food production with added value by families, for example, in rural settlements. Thus, this chapter presents an approach to how other family groups from different social classes use biomass for cooking and maintaining their survival or enjoying exclusive pleasures provided by biomass energy.

Keywords Food security · Sustainable food production · Cooking technologies · Food policy · Clean cooking

1 Introduction

How can we promote using forest biomass in food preparation without harming the environment and people? The answer to this question goes through many dimensions. These issues permeate discussions about food security and sovereignty and broader discussions, such as: economic, social, political, and technological. Thus, it becomes necessary to understand these aspects to use forest biomass for cooking efficiently.

Food and Nutritional Security (FNS) is the realization of everyone's right to regular and permanent access to quality food in sufficient quantity without compromising access to other essential needs, based on health-promoting food policies that respect diversity culturally and that are socially, economically, and environmentally sustainable [74]. It is observed that the FNS concept is a mixture of issues that bring up the challenges of having access to quality food, consuming it in healthy ways, and producing it in sustainable practices.

The conceptualization of FNS is interpreted in different ways around the world. Wealthy countries and large agricultural producers claim to be fighting for their food and nutritional security when they impose barriers to imports and, in this way, increase food prices artificially [72, 152]. On the other hand, in poor or developing countries governed by leaders with a populist position, the concept of Food and Nutrition Security is used to tabulate costs and to establish a villainous image of agricultural producers [107, 110]. With these statements, one arrives at the thought that defining food security is difficult, considering factors ranging from culture and regionality to industrialization and politics. This definition becomes even more complicated when there is a diverse social and cultural panorama, where several peoples and cultures coexist in the same space and, according to the FNS concept, have the same right [13]. In Italy, efforts are to ensure food security and sustainable food production by shifting from conventional to organic farming and encouraging diet changes to reduce meat consumption [92]. In Uganda and Tanzania, efforts are being made to ensure food security through research, development, and agricultural extension for using high-efficiency nitrogen fertilizers in maize crops so that less favored families can access adequate food [42].

To deal with an issue such as Food and Nutritional Security in a country the size of Brazil, considering factors such as the abundance of natural, territorial, and cultural resources, one must first deal with a concept contrary to FNS, the idea of hunger. To define the concept of hunger, dividing a single word into three meanings is necessary. In English, we have three words that can lead to a better understanding: *hunger*, *starvation*, and *famine*. *Hunger* is a desire, a momentary and biological need, and the rich can feel this hunger as much as the poor. *Starvation* is the increase in hunger; it ceases to be individual and becomes collective; we can give the example of students who only eat at school; it is prolonged hunger until a certain point. Finally, *famine* is understood as hunger imposed as a condition on a social group; it is the complete lack of access to food and the means of producing it [84].

In Brazil, there is a belief that the issue of hunger in the country begins in the formation process of Brazilian society, starting in the colonial period. However, on the contrary, it is observed that the first studies on eating habits only began to appear in the nineteenth century (1880–1890), and the first measures to combat hunger were instituted during the Vargas Dictatorship (1937–1945), that is, the political emergency of tackling the scourge of hunger in Brazil was perceived very late, leaving the problem to settle down almost wholly [139]. Nowadays, Federal Government programs aim to establish food security in Brazil in a way adapted to people's social, cultural, and economic reality. The Brazilian federal government, during the term of President Luiz Inácio Lula da Silva, to combat hunger by focusing on its structural causes to guarantee Food and Nutritional Security for Brazilians on three fronts: a set of public policies, the participatory construction of a National Food and Nutrition Security Policy; and a great effort against hunger, involving the three spheres of Government (Federal, State, and Municipal) and all Ministries.

The program's initiatives range from financial aid to the poorest families (with the Bolsa Família card) to the creation of cisterns in the northeastern hinterland, passing through the construction of popular restaurants, instruction on eating habits, distribution of vitamins and food supplements, microcredit loan for poorer families, among others [134]. It should also be mentioned measures that are not directly linked to the issue of Food Security but which have the consequence of providing the poorest regions and those affected by hunger with access and the possibility of changing this reality, such as the transposition of the São Francisco River and construction of roads in the north of the country. Measures like these facilitate food production in areas where there was no production before and contribute to family farming, which many see as the solution to hunger and food insecurity in Brazil.

When discussing the territorial area and cultivable areas, Brazil is undoubtedly interpreted as synonymous with abundance. After all, 851,0345.540 km^2 of extension is a privileged number concerning other countries considered "rich," such as France (55,1695 km^2) and Germany (35,7588 km^2), bringing this to the food production scene. Brazil has 55 million hectares of food grown [62]. However, at the same time as the country deals with super crops and the international recognition of the strength of its power to produce agricultural commodities and food, a significant portion of its people deal with the ghost of hunger, despite Brazil not currently being considered a "poor" country social injustices and the dynamics of the construction process country's economy have contributed to the social problems being experienced by the contemporary Brazilian population [95].

Issues such as land distribution, slavery, European immigration, the unplanned growth of urban areas, lack of access to knowledge, and lack of incentive for those who plant to survive, are problems that link the FNS concept with the concept of food sovereignty [33, 139]. The term sovereignty has a widespread meaning: Superior authority that any other power cannot restrict. Therefore, food sovereignty is connected mainly to the concept of FNS. It is correct to say that the portrait of a country where the two issues are not treated with a singularity is the loss of focus on a primary objective and, consequently, hunger. Although in Brazil, there are model programs that act almost entirely against the issue of Food Insecurity, it is still a challenge to deliver Food Sovereignty to all. Most critics believe it creates a place of passivity and complacency between government and beneficiary since none frees the people from hunger. Whoever kills the need of the people is the government and not the people themselves [7, 34].

Observing history, changes in government, political and cultural bias, between growth and economic setbacks, Food and Nutritional Security in Brazil, and Food Sovereignty is and has been interpreted in different ways over time. However, in all kinds of situational diversity, it has had There is and still is a clear objective, the realization of the right of all to regular and permanent access to quality food, in sufficient quantity, without compromising access to other essential needs, based on health-promoting food policies that respect cultural diversity and that are socially, economically, and environmentally sustainable.

2 Forest Biomass in the World

It is undeniable that biomass is one of the most important fuels for cooking. Yet, according to the [149], about 2.4 billion people still lack access to clean cooking technologies. Forest biomass for household use generally comes in the following forms: firewood, charcoal, briquettes, and pellets. Their use cuts across all social classes, from the poorest to the richest [5, 133].

Firewood is the raw form of wood and requires less technology. Therefore, it is used primarily in South America, sub-Saharan Africa, and South Asia for cooking over open fires and low-technology stoves [16, 79, 86]. Another way to use firewood is to prepare food in bakeries and pizzerias [77] or in restaurants to prepare grilled and smoked food [101, 122].

Charcoal can be produced from forest biomass through the pyrolysis process and has about twice the calorific value of raw wood [115]. As firewood, charcoal is used for grilling and smoking food, especially in countries in South America and sub-Saharan Africa [88, 97]. Families in developing countries also use charcoal for heating and lighting [109].

Forest biomass is the raw material for producing briquettes and pellets through a densification process. The result is a fuel with more homogeneous particle sizes, high heating value, and low greenhouse gas and particulate emissions [26, 50, 124]. Briquettes and pellets undergo a more technological manufacturing process. They require more efficient stoves for combustion, which is why they are more commonly used in developed countries in North America, Europe, and China [45, 90, 105]. Generally, pellets and briquettes are usually used to heat and light homes because they have a high heating value, more controlled combustion, and low gas emissions [85, 144].

We can understand that people widely use forest biomass to meet their cooking, heating, and lighting needs. Firewood is most used for cooking and is common among both the poorest and wealthiest segments of the population. Charcoal is also used primarily for cooking at home. Still, it is less common compared to firewood because it must go through a pyrolysis process. Briquettes and pellets are the least used forest biomass fuels because they must undergo a densification process and require more technically sophisticated stoves, primarily in European and North American countries.

3 What Does Forest Biomass Mean for Households?

Forest biomass is one of the renewable energy sources. With wind, water, solar, geothermal, and ocean energy, it is essential to transition away from the fossil-fuel-based global economic model. Today, the production of heat, electricity, fuels, chemicals, and other petroleum-based materials contributes to climate change by releasing greenhouse gases (GHGs) and deficits in carbon sequestration. In this context, we face increasing pressure on natural ecosystems, changing land-use patterns, intensive agriculture, biodiversity loss, and energy crisis [112]. In addition, extreme weather events are likely to become more frequent [63], and human and non-human populations face heat waves, storms, floods, and droughts. This results in losses in agricultural productivity, generating food shortages and favoring population displacement, promoting numerous conflicts [127]. In general, the use of forest biomass among renewable sources can mitigate environmental impacts because it does not alter the carbon cycle in the long term and helps optimize CO_2 levels in the atmosphere [14]. It is also a feedstock for local and decentralized businesses, helping to reduce dependence on fossil fuels [142]. In addition, research on vegetable biomass is diverse and has recently been summarized. Among the possible uses are conversion into biofuels or bioproducts and electricity generation [119].

Moreover, what is biomass? It is a biological and renewable material derived from living or recently dead organisms, including plants and animals. There are two typical applications of the term. The first refers to the ecological sense of biomass, which usually indicates the carbon stock of ecosystems from living or dead matter below or above ground [47]. The second relates to energy practices, including traditional heat production for cooking, domestic heating, modern conversion to biofuels, and combined heat and power generation [65]. Biomass of plant origin has a broad scope, as it comprises materials formed through photosynthesis or generated in its use [19], including wood and firewood, energy crops, charcoal, agroforestry, rural, industrial, domestic, and commercial waste (Fig. 1).

In this context, we can classify it as a heterogeneous raw material. Moreover, forest biomass is distributed in different spaces and times, so its volume and characteristics are influenced by local climatic and geographic factors [142]. Therefore, production,

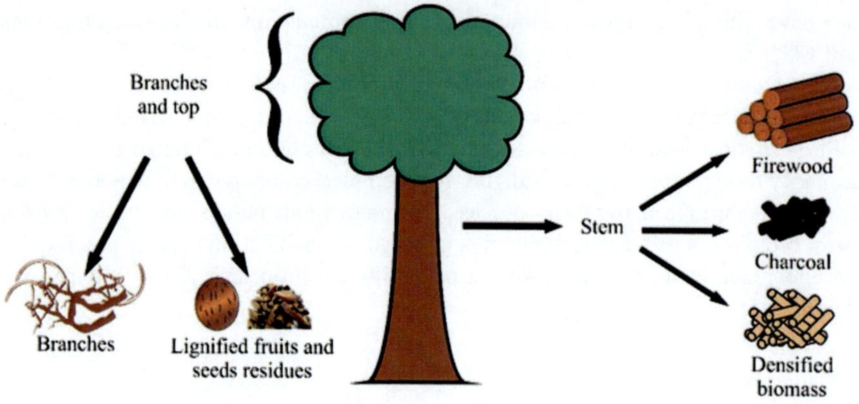

Fig. 1 Possibilities for using forest biomass (*Source* The authors 2023)

stock, and transport data are essential in building a forest biomass production chain [54]. Thus, regional solutions for biomass use for energy are thought in countries from the south [18, 59] to the global north [48, 58, 99].

Humanity has always had its socioeconomic progress linked to using biomass for energy. However, the separation between traditional and modern use, or bioenergy, has opened space to depreciate biomass for such a noble purpose as cooking. Cooking is a necessary process for humankind. Proper cooking techniques make food more digestible and increase food's bioavailability and flavor [69]. It is further estimated that raw or unprocessed biomass and charcoal in cooking food occur in 39.5% of households worldwide [147, 148]. However, this high number is affiliated with reality in emerging and developing countries [116]. In these nations, large populations use energy sources that contribute to household air pollution by emitting carbon monoxide (CO) and particulate materials [98]. The term clean cooking refers to the use of efficient fuels or equipment inside homes, seeking a lower emission of pollutants. According to the WHO Global Air Quality Guidelines [148], emissions should meet the annual average air quality guideline level (AQG, 5 $\mu g\ m^{-3}$) or the Interim Target-1 level (IT1, 35 $\mu g\ m^{-3}$) for $PM_{2.5}$, and either the 24-h average air quality guideline level (AQG, 4 mg m^{-3}) or the Interim Target-1 level (IT-1, 7 mg m^{-3}) for CO [148]. As WHO [146] described, solar-based, electric cooking systems that use biogas, natural gas, liquefied petroleum gas (LPG), or alcohol fuels, including ethanol, are considered clean. Solid fuels can also fall into this category, provided they meet adequate levels confirmed in laboratory tests following an international laboratory testing protocol.

Expanding access to clean cooking is a strong ally for advances in at least 5 of the United Nations (UN) Sustainable Development Goals (SDGs), these being (3) Good health and well-being, (5) Gender equality, (7) Affordable and clean energy, (13) Climate action, (15) Life on land [118]. When forcing on SDG 7, which aims at universal provision of modern, reliable, and affordable energy services by 2030, including services of electricity and clean cooking facilities, we notice progress over time. However, the historical series given by the World Health Organization [150] indicates a slow pace of growth, not compatible with the targets set. Hence, there is a double question, one of technical nature and the other of social interest: are there efficient ways to use traditional biomass for cooking food for these populations? At the same time, are there ways to mitigate the effects of domestic air pollution? Finally, having overcome these material questions, what is the role of biomass in the food security of these populations?

4 Forest Biomass in Food Preparation by Different Socioeconomic Classes

In addition to carbon–neutral products and energy, the appropriate use of forest biomass is necessary to develop robust production models that transform linear production chains into circular ones. Thus, we could advance the transition to a circular economy [76] and strengthen the United Nations' 17 SDGs, which aim to motivate and facilitate the world's economies to address serious social and environmental problems. In practice, products necessary for modern life could be from renewable sources. Therefore, it is common that biomass applications such as sophisticated biofuels and other bioproducts are always required and encouraged. However, the traditional use of biomass in food preparation is given low prominence in socioeconomic and technical studies. It is even treated as a less noble material that needs to be avoided because it offers health risks (Fig. 2).

There are examples of studies that prove the association of the use of biomass for cooking with symptoms of depression and anxiety in the elderly [37], loss in cognitive

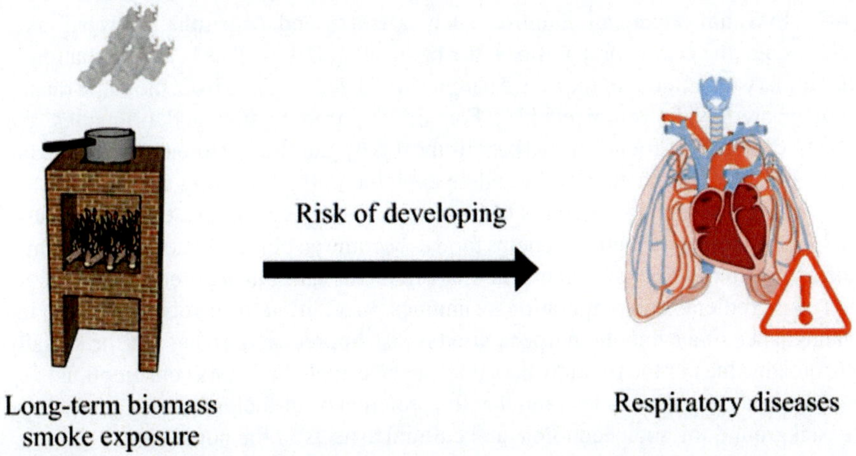

Fig. 2 Exposure to biomass smoke may increase the risk of developing respiratory diseases (*Source* The authors 2023)

skills in children up to 4 years old [83], with respiratory health risks in pregnant women [138] and even consequences with pre-menopausal depression [10]. Such cited researches are naturally observational, so they compared different populations and investigated a sole factor: the employment of biomass in households. There is robustness in this type of verification, and we can obtain data without purposely subjecting people to a risk factor for their health. In this sense, it is necessary to remember that populations that use biomass are often poor and dependent on its frequent use for food.

By 2020, sub-Saharan Africa and South Asia countries culminated in the list of nations with the lowest share of the population with access to clean cooking fuels and technologies [147]. People in these countries are also more likely to be deprived of adequate nutrition, sanitation, housing, or access to clean water and electricity [136]. Thus, we must consider that poverty brings with it personal material disadvantages, systemic and environmental constraints, daily experiences of stress, cognitive demands, and situations of scarcity that together act to worsen people's health [6], with the use of biomass for cooking being one more conditioning factor among them. There is a technical bias that transforms traditional biomass into a more efficient biofuel. At the same time, it is possible to modify the conditions of a household and its equipment to receive this energy source for cooking food without serious emissions of pollutants. Thus, we must understand that forest biomass is a potential source and instrument of an energy chain that should be broad and diverse. With it, we can start an integrated approach to overcome the systemic deprivations conditioned to the lower social classes, in which policies for improving conversion systems into biofuels and domestic equipment with low pollutant emissions are included.

Another relevant context is utilizing forest biomass as an alternative and not exclusive fuel in food preparation. In these cases, the use of charcoal is ordinary in food preparation due to the characteristics it gives to food in terms of flavor and texture [141]. In South American countries such as Brazil and Argentina, charcoal takes center stage in specific preparations: the barbecue [28, 43]. This is a combination of the culinary techniques of the native peoples with beef obtained from the cattle raising implemented by the colonizers [52]. For similar purposes, firewood, briquettes, and pellets can still be found in markets in the USA, European Union, Australia, and Japan [68, 128]. Such cited fuels will be explored in the following session.

Thus, going beyond the limits of food security for homes, forest biomass is also fuel for the culinary tradition. It helps food to become an object of adequate consumption by attributing physicochemical characteristics and flavor through the interaction of ingredients and preparation techniques. In addition to corroborating with the maintenance of a symbolic purpose, food is also appreciated because of the socially attributed values or the relationships that can be established in its consumption [52]. Finally, it is crucial to understand that the preferences of fuel for cooking food have as background the socioeconomic and cultural aspects of the populations.

5 Conversion of Forest Biomass into Clean Energy for Cooking

5.1 Solid Biofuels to Replace Traditional Firewood

Densification is the first example of technology to generate modern and efficient solid biofuels. It is a biomass compaction process that allows the production of a new product with higher apparent density than the raw material that originated it. Its application in residual biomass is common, bringing as an advantage the energy use associated with the reduction of deforestation for the removal of firewood and the use of renewable sources for fuels [135]. Densification occurs under different operating conditions of temperature and pressure, in natural binders' presence exclusively or not [17]. The type of biomass used influences the production parameters employed. In general, the compaction pressure creates plastic and elastic deformities that reduce the voids between particles, favoring the mechanical strength of the biofuel [106] and the energy increment [111]. Temperature acts by modifying lignin and extractives to the plastic condition, which is necessary for the connection and adhesion of biomass particles [24, 71]. Under these conditions, the binding agents inherent in biomass (e.g., lignin, starch, and proteins) corroborate with interparticle adhesion. However, adding binders may be necessary [123]. Densification is integrated with preparation steps, which may include drying, biomass reduction, and grading of the particles obtained.

In general, we can obtain two different products in the form of briquettes or pellets. Both have higher volumetric energy concentration, lower humidity, and homogeneity than the source feedstock [51]. The dimensions of pellets vary between 6 and 16 mm. At the same time, briquettes have diameters starting at 50 mm, varying according to market demand or standards [38]. Both products are direct substitutes for wood in furnaces, heaters, and boilers. According to FAO [96], pellets are the forest biomass fuels with the greater economic importance globally, moving approximately 4.4 billion dollars in imports and exports. The same authors estimated that the production and consumption of pellets worldwide increased by 49% and 41%, respectively, between 2016 and 2020.

The use of these biofuels in domestic furnaces has been the subject of research. In Nigeria, briquettes produced with corn cob and bark of palm oil trunks showed suitable thermal properties for combustion and GHG emissions aligned with environmental standards [78]. Studies in Uganda have emphasized that household use of briquettes made from rice husks, coffee, and binders translates into energy savings [93]. In the Bolivian context, briquettes with waste cardboard and sawdust showed better performance through environmental Life Cycle Assessment (LCA) within a 100–130 km radius of the production plant, being superior to all fossil fuels used in Andean areas for heating and cooking food [44]. Finally, we point out that replacing raw biomass with briquettes can benefit health. This was indicated by the research of [108], elucidating the similarities and differences in $PM_{2.5}$ (particulate matter with diameter ≤ 2.5 μm) emissions and cytotoxicity between these two fuels. The emission characteristics of briquettes decrease inflammatory response and cell membrane damage, which contributes to reducing adverse pulmonary effects. In this context, studies that perform on-site analysis of densified fuels associated with traditional and efficient kilns are still required.

Furthermore, thermochemical processes can convert biomass into another biofuel suitable for a domestic environment without pollution if they meet the above recommendations. The pyrolysis or carbonization of woody materials occurs in high-temperature environments with low presence or absence of oxygen, and its main product is charcoal. This material is rich in carbon and has higher specific energy, resulting from removing water and other volatile compounds [41]. However, we can tense a debate about using environmentally and socially responsible sources of charcoal production. In sub-Saharan African countries such as Ghana, charcoal activity results in vegetation loss and ecological impacts while meeting household energy needs, supporting the livelihoods of different social groups, and transforming local economies [2]. In this sense, there are social risks to extinguishing charcoal production. However, it is recommended to incorporate interventions such as switching from the more efficient production models that earth mound for charcoal production, given that producers are flexible in adapting to any new technology indicated [8]. For [12] decision-makers should look at production methods, creating policies to make it economically viable and environmentally friendly, especially to reduce GHG emissions. A similar narrative is supported by [70], who indicated that this conversion technology is a valuable tool in the energy diversification process in Mozambique. Even though these same authors raise questions about applying laws disfavoring already weakened social groups that depend on coal for subsistence activities, finally, local and context-specific planning and evaluation should be considered for projects that aim at rural development so that the actors and stakeholders can contribute to the process and the production cycle [57].

In Brazil, in a different context from the production for homes and subsistence purposes, authors have indicated that the substitution of traditional kilns by the furnace-kiln system (Fig. 3) has shown excellent performance in reducing the emissions of pollutants, following the targets established by international agreements [120].

As for the source of woody material, this can also be diversified and thus mitigate pressures on native biomes. Studies in the Brazilian Amazon, as seen in Chap. 4 of this book, indicate that using residues from forest harvesting activities can supply domestic systems with quality charcoal [88]. Furthermore, due to the great diversity of residues, it is also possible to control the carbonization process in brick kilns to obtain a high-quality product [87]. Pruning residues are other potential sources of charcoal in nearby or urban areas [102].

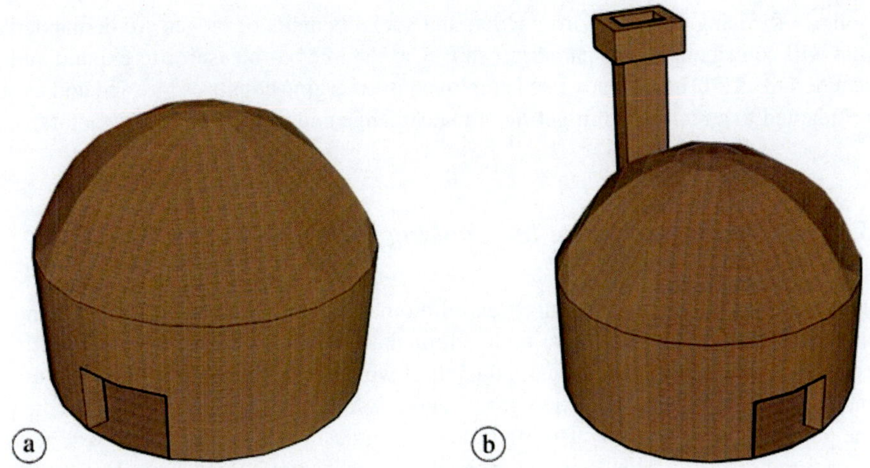

Fig. 3 Types of kilns for charcoal production, where: **a** traditional kiln; **b** furnace-kiln system) (*Source* [120])

5.2 *Gaseous Biofuels Replacing Cooking Gas*

Households with access to clean cooking facilities have increased in recent years in favor of liquefied petroleum gas (LPG). According to the International Energy Agency [64], this fuel is essential to significantly reduce indoor air pollution, bringing health benefits to several households. Crucial policy instruments for this reality involve informing about the LPG market and clarifying the consequences of using traditional biomass for cooking [66]. Alternatively, biogas fulfills the function of clean fuel, with a smaller carbon footprint and not dependent on fossil sources. It is a product of the biochemical conversion of biomass in an anaerobic environment. The elements that makeup biogas are primarily composed of methane (CH_4) in a proportion ranging from 50 to 75%, along with carbon dioxide (CO_2), which comprises 25–50% of the composition [55]. Small amounts of other gases, such as oxygen (O_2), hydrogen sulfide (H_2S), nitrogen (N_2), and water vapor, are also present. In addition, biogas can be converted into biomethane using techniques to purify CH_4 and recover CO_2 simultaneously [89], which is employed in industry for various purposes.

Biogas from household digesters could improve the standard of living of families. In rural areas, this biogas contributes to reducing fuel costs and time to collect firewood, besides mitigating $PM_{2.5}$ emissions in the domestic environment [113]. We emphasize that its use applies to homes with sufficient outdoor space to store the equipment. In this context of rural homes, LPG usually has higher costs due to supply logistics, transportation, and storage requirements [60]. In this sense, biogas can contribute to the energy transition of cooking fuels in areas with land availability and near the feedstock. It is necessary to (i) overcome technologies poorly adapted to local conditions, (ii) overcome barriers to financing facilities [125], (iii) study available biomass; and (iv) manage waste professionally and in the long term [91]. Sound policies focusing on the environmental and social benefits of biogas are demanded that will contribute to this paradigm shift. Case studies of measures to expand and democratize LPG has already been employed in emerging countries [15, 75] and can be adapted to a reality of mitigating climate change and improving social factors.

5.3 *Efficient Equipment for Cooking Food*

Traditional stoves using raw/unprocessed biomass have low energy efficiency and high pollutant emissions due to their rudimentary structure [36]. Some geometric configurations of the stove are essential, all of which will affect the efficiency of the biomass burning, the combustion gases, and consequently, the energy that reaches the pot with the food [104]. During biomass combustion in traditional stoves, heat is released diffusely. However, the optimal is to concentrate this energy in a single direction, allowing food cooking with less fuel [94]. Thus, a combustion chamber is necessary, in adequate proportions, and made of material that limits energy loss to the environment. The size of combustion chambers will also affect energy efficiency [36]. So larger heights favor the reaction time of the combustion gases with oxygen, releasing more heat into the system. However, there are limits so that the height does not affect the flame temperature, which is responsible for heating the pots used to cook the food. In the study by [61], the ideal measurements for the combustion chamber of 16 cm in height and 13 cm in diameter were obtained through modeling.

Another critical component of an efficient stove is the chimney. Its existence mitigates the emissions of particles and pollutants in the indoor environment, favoring people's health [82]. Moreover, it dictates the stove's power supply since it tends to change the gas temperature and the combustion efficiency [114]. It is worth clarifying that excess air is mandatory in complete combustion [35, 143]. Therefore, building systems that favor airflow is appropriate, as this burns the biomass and gains in convective heat transfer [4]. However, care must be taken with high airflow. The combustion gases are expected to be removed from the system with improving energy efficiency, instead, it only favors biomass consumption [67].

The biomass source used will also modify the efficiency of the stove, as well as its layout, size, moisture content, and heating value. [94] Therefore, in practice, only some models can replace traditional stoves. It is valid to ensure suitable configurations for different stoves and families according to their needs [80]. So, the uptake of efficient stoves by poorer populations occurs through programs that consider people's preferences. Experiments show that stove dissemination rates are low when these factors are ignored and programs fail [40]. Finally, it should be remembered that efficient equipment is no long-term substitute for universal access to affordable, reliable, and modern energy services.

6 Promoting the Sustainable Use of Forest Biomass

In addition to security and food sovereignty issues, promoting forest biomass with low environmental and human health impacts also involves economic, social, political, and technological discussions.

From an economic perspective, using forest biomass is a cost-effective way to obtain energy for cooking food because it is a fuel that can be easily found anywhere in the world and can often be obtained from nature without much effort [16]. The use of raw biomass in the form of firewood is usually associated with low-income families who do not have access to another energy source for cooking and rely on obtaining wood from the nearest forest for survival [100, 101]. The data on the use of forest biomass for cooking can be alarming when we put them in context, such as in two large emerging economies, Brazil and India, where about 25% and 50% of households, respectively, rely on fuelwood [53, 75]. Indiscriminate use of forest biomass for cooking can cause problems such as high gas emissions, suppression of native vegetation, and loss of biodiversity. However, public policies, technological development, and financial incentives for properly using forest biomass can restrain this practice.

From a social perspective, using biomass may be based on regional or cultural reasons, making the transition from raw biomass to cleaner energy sources challenging. Most families that use forest biomass inefficiently to cook food live either in remote regions or on the outskirts of cities and are usually composed of poor people [56, 96, 131]. The geographic location of this population is often a determining factor in the choice of energy source used for cooking, so the closest option is firewood [1, 101]. Another social aspect of using forest biomass for cooking is the culture of a particular population. Often the preparation way of food is passed down from generation to generation, so the use of this energy source continues. In the case of Brazil, grilling is a food embedded in the country's culture, and the most common energy source used is charcoal, which directly affects the outcome of the meal [39, 103]. Thus, the social aspect is essential in promoting the sustainable use of forest biomass.

From a political perspective, the use of forest biomass should be encouraged by public actions in the form of laws, either reducing the social insecurity of people who need to use this source of energy or creating protection mechanisms against the exploitation of native forests and for the implementation of forests for energy purposes. The uncontrolled extraction of wood can lead to the destruction of ecosystems, soil deterioration, and change of ownership of watercourses, among others [1, 137]. Many countries have laws that combat illegal logging. However, more than enforcement is needed to prevent environmental impacts [11, 121, 140]. In China, there are public policies that aim at the transition from the use of traditional biomass to the use of modern biomass in rural homes [29, 151]. In Ethiopia, public policy development focuses on educating rural households about using biomass and more efficient cooking equipment [73]. In Brazil, the National Biofuels Policy presents as one of its foundations the importance of adding value to biomass as a form of environmental preservation and promotion of development and economic and social inclusion [22]. In developing countries such as Brazil, Nepal, and Ethiopia, the use of forest biomass is linked to the collection distance, in which low-income families prefer to use the closer material [49, 83, 126]. In developed countries such as Portugal, Italy, and Germany, the use of forest biomass is regulated by more specific norms, mainly regarding the technical characteristics of the fuel and the cooking technology [129]. Another point to be addressed politically is creating social assistance mechanisms that allow the poorest population to use forest biomass for cooking more efficiently.

From a technological perspective, the use of forest biomass can be improved on two points: (I) technologies for burning with greater thermal efficiency and (II) more efficient forest-based fuels. More efficient cooking appliances provide lower greenhouse gas (GHG) emissions and greater energy efficiency so that human exposure to harmful gases is reduced and the fuel needed is lower [20, 130]. The emission of GHG, nitrous oxides (NO_X), polycyclic aromatic hydrocarbons (PAHs), and particulate matter from the inefficient burning of biomass can generate serious problems for public health and the environment [21, 145]. Another critical point that can be approached from a technological perspective to encourage the use of forest biomass for cooking is the development of more efficient fuels through physical–mechanical or thermochemical routes. Concrete evidence suggests biomass densification techniques can increase its physical properties and energy efficiency [9, 27, 36]. Briquettes and pellets are the most common forms of densified biomass. The main characteristics that make them more efficient than firewood are higher energy density, a lower percentage of moisture, a better combustion rate, and lower emission of particulate matter [25, 81].

In addition to densification processes, there are also thermochemical routes for improving forest biomass for cooking, such as torrefaction and pyrolysis. Torrefaction is a thermochemical process that aims to improve the characteristics of biomass, taking place in an environment lacking or with little oxygen at temperatures ranging between 200 and 300 °C [30, 96]. Pyrolysis is a thermochemical process that aims at the complete transformation of forest biomass into new products: charcoal, pyroligneous liquid, and non-condensable gases, occurring at temperatures above 350 °C, with heating rates and residence time in the reactors varying depending on the type of pyrolysis [31, 32]. As it is a process that depends on technology, the pyrolysis product commonly used for cooking is charcoal, as it does not need any treatment after production [3, 15, 23]. Charcoal is a fuel that may have its origin in forest biomass. It has some characteristics superior to firewood, such as higher calorific value, less volatile materials, and higher energy density [39, 115]. In domestic environments, charcoal is used in stoves or barbecues with low technology employed, usually without a ventilation system (Fig. 4a) [117].

On the other hand, steakhouses, pizzerias, or restaurants use charcoal to prepare "gourmet" dishes with high added value in ovens with an exhaust system and liquid collection (Fig. 4b). Therefore, the use of charcoal as a source of energy for cooking permeates all social classes, from less favored families in rural areas to families with greater purchasing power in urban centers. In this way, developing technologies for

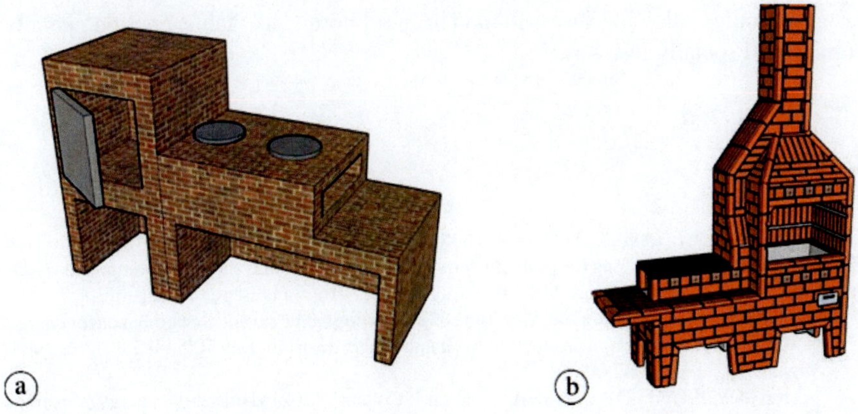

Fig. 4 Illustration of ovens for cooking food in which: **a** rural oven with low technology, without air circulation system and liquid collection; **b** oven in a commercial environment, with an exhaust system and collection channels (*Source* The authors 2023)

stoves and forest biomass fuels can guarantee access to more efficient renewable energy sources for cooking food.

Therefore, understanding the economic, social, political, and technological perspectives can sustainably promote the use of forest biomass to ensure its ecologically correct, socially just, and economically viable use. In this way, it contributes to developing food sovereignty in countries. Furthermore, it provides the population's food security by promoting the efficient use of forest biomass in the household energy matrix.

7 Conclusions

In a world with frequent discussions about food distribution, public policies, and sustainability, the role of people, governments, and non-governmental entities becomes more evident when solving these problems directly bound to human survival. This chapter addressed concerns about how food security and food sovereignty are correlated and what actions all actors involved can take to produce, consume, and distribute food sufficiently and fairly without negatively affecting future generations. Another point addressed was how families from different social classes relate to forest biomass in food preparation. There is the perception that wealthier families are more likely to achieve cleaner and more efficient cooking technologies. However, in comparison, poorer families use natural forest fuel in more rustic equipment, generating public health problems. Thus, to mitigate this food inequality, public policies are needed to expand the access of poorer people to more efficient cooking technologies and the development of equipment and fuel that are cheaper and simpler to use. Only then will a food transition be possible for less favored peoples, allowing access to food in an economically viable, environmentally correct, and socially just way.

References

1. Abdu N, Tinch E, Levitt C, Volker PW, Hatton MacDonald D (2022) Illegal firewood collection in Tasmania: approaching the problem with the Institutional Analysis and Development (IAD) framework. Land use policy 118. https://doi.org/10.1016/j.landusepol.2022.106130
2. Ablo AD, Osei L, Jarawura FX, Yaro JA (2022) Sacrificing the savannah woodlands for energy and livelihoods? Charcoal production in Ghana. Energy Sustain Dev 70:549–559. https://doi.org/10.1016/j.esd.2022.09.001
3. Abreu MW, Ferreira DV, Pereira AO, Cabral J, Cohen C (2021) Household energy consumption behaviors in developing countries: a structural decomposition analysis for Brazil. Energy Sustain Dev 62:1–15. https://doi.org/10.1016/j.esd.2021.03.001
4. Agenbroad J, DeFoort M, Kirkpatrick A, Kreutzer C (2011) A simplified model for understanding natural convection driven biomass cooking stoves—Part 2: With cook piece operation and the dimensionless form. Energy Sustain Dev 15(2):169–175. https://doi.org/10.1016/j.esd.2011.04.002
5. Antar M, Lyu D, Nazari M, Shah A, Zhou X, Smith DL (2021) Biomass for a sustainable bioeconomy: An overview of world biomass production and utilization. Renew Sustain Energy Rev 139
6. Appelhans BM (2023) The cognitive burden of poverty: a mechanism of socioeconomic health disparities. Am J Prev Med 64(2):293–297. https://doi.org/10.1016/j.amepre.2022.08.012
7. Aranha AV (2010) Fome Zero - Uma História Brasileira
8. Asare F, Owusu FW, Gazo R (2022) Sustainable charcoal production drive in rural communities in Ghana, West Africa. Energy Sustain Dev 68:364–372. https://doi.org/10.1016/j.esd.2022.04.013
9. Bajwa DS, Peterson T, Sharma N, Shojaeiarani J, Bajwa SG (2018) A review of densified solid biomass for energy production. Renew Sustain Energy Rev 96:296–305
10. Banerjee M, Siddique S, Dutta A, Mukherjee B, Ranjan Ray M (2012) Cooking with biomass increases the risk of depression in pre-menopausal women in India. Soc Sci Med 75(3):565–572. https://doi.org/10.1016/j.socscimed.2012.03.021
11. Bär R, Reinhard J, Ehrensperger A, Kiteme B, Mkunda T, Wymann von Dach S (2021) The future of charcoal, firewood, and biogas in Kitui County and Kilimanjaro Region: scenario development for policy support. Energy Policy 150. https://doi.org/10.1016/j.enpol.2020.112067
12. Bekele B, Kemal AW (2022) Determents of sustainable charcoal production in AWI zone; the case of Fagita Lekoma district Ethiopia. Heliyon 8(12):e11963. https://doi.org/10.1016/j.heliyon.2022.e11963
13. Belik W (2003) Perspectivas para segurança alimentar e nutricional no Brasil Prospects for food and nutricional safety in Brazil. Saúde e Sociedade 12(1):12–20
14. Benebere P, Appiah DO, Aabeyir R, Abass K, Guodaar L (2023) Climate change mitigation potential of agroforestry farms in West African Savanna. Environ Challenges 10:100678. https://doi.org/10.1016/j.envc.2023.100678
15. Bezerra P, Cruz T, Mazzone A, Lucena AFP, De Cian E, Schaeffer R (2022) The multidimensionality of energy poverty in Brazil: a historical analysis. Energy Policy 171:113268. https://doi.org/10.1016/j.enpol.2022.113268
16. Bharadwaj B, Pullar D, To LS, Leary J (2021) Why firewood? Exploring the co-benefits, socio-ecological interactions and indigenous knowledge surrounding cooking practice in rural Nepal. Energy Res Soc Sci 75. https://doi.org/10.1016/j.erss.2021.101932

17. Bhattacharya SC, Sett S, Shrestha RM (1989) State of the art for biomass densification. Energy Sour 11(3):161–182. https://doi.org/10.1080/00908318908908952
18. Bowd R, Quinn NW, Kotze DC, Guilfoyle MJ (2018) A systems approach to risk and resilience analysis in the woody-biomass sector: a case study of the failure of the South African wood pellet industry. Biomass Bioenerg 108:126–137. https://doi.org/10.1016/j.biombioe.2017.10.032
19. Brand MA (2010) Energia de biomassa florestal. Editora Interciência, Rio de Janeiro
20. Brandelet B, Rose C, Landreau J, Druette L, Rogaume Y (2021) Toward a cleaner domestic wood heating by the optimization of firewood stoves? J Clean Prod 325. https://doi.org/10.1016/j.jclepro.2021.129338
21. Brandelet B, Rose C, Rogaume C, Rogaume Y (2018) Impact of ignition technique on total emissions of a firewood stove. Biomass Bioenergy 108:15–24. https://doi.org/10.1016/j.biombioe.2017.10.047
22. Brasil (2017) Política Nacional de Biocombustíveis (RenovaBio). Presidência da República. Disponível em: https://www.planalto.gov.br/ccivil_03/_ato2015-2018/2017/lei/l13576.htm. Acesso em: 07 mar 2023
23. Broto VC, Arthur FSRM, Guibrunet L (2020) Energy profiles among urban elite households in Mozambique: Explaining the persistence of charcoal in urban areas. Energy Res Soc Sci 65. https://doi.org/10.1016/j.erss.2020.101478
24. Cao Z, Zhang S, Huang X, Liu H, Sun M, Lyu J (2020) Correlations between the compressive strength of the hydrochar pellets and the chemical components: evolution and densification mechanism. J Anal Appl Pyrol 152:104956. https://doi.org/10.1016/j.jaap.2020.104956
25. Carter E, Shan M, Zhong Y, Ding W, Zhang Y, Baumgartner J, Yang X (2018) Development of renewable, densified biomass for household energy in China. Energy Sustain Dev 46:42–52. https://doi.org/10.1016/j.esd.2018.06.004
26. Cen K, Chen F, Chen D, Gan Z, Zhuang X, Zhang H (2022) Life cycle assessment of torrefied cornstalk pellets combustion heating system. Fuel 320. https://doi.org/10.1016/j.fuel.2022.123968
27. Cencin A, Zanetti M, Urso T, Crivellaro A (2021) Effects of an innovative densification process on mechanical and physical properties of beech and Norway spruce veneers. J Wood Sci 67(1). https://doi.org/10.1186/s10086-021-01948-w
28. Champredonde M (2008) The source and market development of a premium product—beef from the Argentine Pampas. Meat Sci 79(3):534–540. https://doi.org/10.1016/j.meatsci.2007.10.021
29. Chen Q, Yang H, Liu T, Zhang L (2016) Household biomass energy choice and its policy implications on improving rural livelihoods in Sichuan, China. Energy Policy 93:291–302. https://doi.org/10.1016/j.enpol.2016.03.016
30. Chen WH, Peng J, Bi XT (2015) A state-of-the-art review of biomass torrefaction, densification and applications. Renew Sustain Energy Rev 44:847–866
31. Chen X, Che Q, Li S, Liu Z, Yang H, Chen Y, Wang X, Shao J, Chen H (2019) Recent developments in lignocellulosic biomass catalytic fast pyrolysis: strategies for the optimization of bio-oil quality and yield. Fuel Process Technol 196
32. Choi YS, Lee KH, Zhang J, Brown RC, Shanks BH (2015) Manipulation of chemical species in bio-oil using in situ catalytic fast pyrolysis in both a bench-scale fluidized bed pyrolyzer and micropyrolyzer. Biomass Bioenergy 81:256–264. https://doi.org/10.1016/j.biombioe.2015.07.017
33. Chonchol J (2005) A soberania alimentar. Estudos Avançados 19(55):33–48. https://doi.org/10.1590/S0103-40142005000300003
34. de Freitas Coca EL, Santos LLM, Souza RDP, Salvaterra JR (2020) A soberania alimentar na Geografia Agrária brasileira. Terra Livre 1(54):586–615
35. Deng M, Li P, Shan M, Yang X (2020) Optimizing supply airflow and its distribution between primary and secondary air in a forced-draft biomass pellet stove. Environ Res 184:109301. https://doi.org/10.1016/j.envres.2020.109301

36. Deng M, Zhang P, Yang H, Ma R (2023) Directions to improve the thermal efficiency of household biomass cookstoves: A review. Energy Build 278:112625. https://doi.org/10.1016/j.enbuild.2022.112625
37. Deng Y, Gao Q, Yang T, Wu B, Liu Y, Liu R (2021) Indoor solid fuel use and incident arthritis among middle-aged and older adults in rural China: a nationwide population-based cohort study. Sci Total Environ 772:145395. https://doi.org/10.1016/j.scitotenv.2021.145395
38. Dias, JMCS, Souza DT, Braga M, Onoyama MM, Miranda, CHBM, Rocha JD (2012) Produção de briquetes e péletes a partir de resíduos agrícolas, agroindustriais e florestais. Embrapa Agroenergia, Brasília
39. Dias Júnior AF, Andrade CR, Milan M, Brito JO, de Andrade AM, de Souza ND (2020) Quality function deployment (QFD) reveals appropriate quality of charcoal used in barbecues. Sci Agric 77(6). https://doi.org/10.1590/1678-992x-2019-0021
40. Ekouevi K, Tuntivate V (2012) Household energy access for cooking and heating: lessons learned and the way forward. A World Bank study. World Bank, Washington, DC. Available in: http://hdl.handle.net/10986/9372. Accessed 08 Mar 2023
41. El Bassam N (2021) Energy resources, global contribution, and applications. In: Distributed renewable energies for off-grid communities. Elsevier, pp 165–211
42. Falconnier GN, Leroux L, Beillouin D, Corbeels M, Hijmans RJ, Bonilla-Cedrez C, van Wijk M, Descheemaeker K, Zingore S, Affholder F, Lopez-Ridaura S, Malézieux E, Makowski D, Rurinda J, van Ittersum MK, Vanlauwe B, Giller KE, Lammoglia SK, Waha K (2023) Increased mineral fertilizer use on maize can improve both household food security and regional food production in East Africa. Agric Syst 205. https://doi.org/10.1016/j.agsy.2022.103588
43. Fernandes AM, Teixeira OS, Revillion JP, Souza ÂRL (2022) Beef as a socio-cultural identity: Rural and urban consumers' attitudes from Rio Grande do sul, Brazil, facing cultured beef. J Rural Stud 95:438–448. https://doi.org/10.1016/j.jrurstud.2022.09.035
44. Ferronato N, Baltrocchi APD, Romagnoli F, Calle Mendoza IJ, Gorritty Portillo MA, Torretta V (2023) Environmental life cycle assessment of biomass and cardboard waste-based briquettes production and consumption in Andean areas. Energy Sustain Dev 72:139–150. https://doi.org/10.1016/j.esd.2022.12.005
45. Ferronato N, Calle Mendoza IJ, Ruiz Mayta JG, Gorritty Portillo MA, Conti F, Torretta V (2022) Biomass and cardboard waste-based briquettes for heating and cooking: thermal efficiency and emissions analysis. J Clean Prod 375. https://doi.org/10.1016/j.jclepro.2022.134111
46. Food and Agriculture Organization of The United Nations—FAO (2022) FAO yearbook of forest products. FAO, Rome. https://doi.org/10.4060/cc3475m
47. Food and Agriculture Organization of The United Nations—FAO. Global Forest Resources Assessment 2020. FAO, Rome. www.fao.org/forestry. Accessed 5 Jan 2023
48. Furubayashi T, Nakata T (2021) Analysis of woody biomass utilization for heat, electricity, and CHP in a regional city of Japan. J Clean Prod 290:125665. https://doi.org/10.1016/j.jclepro.2020.125665
49. Gallagher M, Beard M, Clifford MJ, Watson MC (2016) An evaluation of a biomass stove safety protocol used for testing household cookstoves, in low and middle-income countries. Energy Sustain Dev 33:14–25. https://doi.org/10.1016/j.esd.2016.03.008
50. Gan Q, Xu J, Peng S, Yan F, Wang R, Cai G (2021) Effects of heating temperature on pore structure evolution of briquette coals. Fuel 296. https://doi.org/10.1016/j.fuel.2021.120651
51. Gendek A, Aniszewska M, Malaťák J, Velebil J (2018) Evaluation of selected physical and mechanical properties of briquettes produced from cones of three coniferous tree species. Biomass Bioenerg 117:173–179. https://doi.org/10.1016/j.biombioe.2018.07.025
52. Gimenes MH (2009) O uso turístico das comidas tradicionais: algumas reflexões a partir do Barreado, prato típico do litoral paranaense (Brasil). Turismo e Sociedade 2(1):8–24
53. Gioda A, Tonietto GB, de Leon AP (2019) Exposure to the use of firewood for cooking in Brazil and its relation with the health problems of the population. Ciencia e Saude Coletiva 24(8):3079–3088. https://doi.org/10.1590/1413-81232018248.23492017

54. Gold S, Seuring S (2011) Supply chain and logistics issues of bio-energy production. J Clean Prod 19(1):32–42. https://doi.org/10.1016/j.jclepro.2010.08.009
55. Gomez CC (2013) Biogas as an energy option: an overview. In: The Biogas Handbook. Elsevier, pp 1–16
56. Gould CF, Schlesinger SB, Molina E, Bejarano ML, Valarezo A, Jack DW (2020) Household fuel mixes in peri-urban and rural Ecuador: Explaining the context of LPG, patterns of continued firewood use, and the challenges of induction cooking. Energy Policy 136. https://doi.org/10.1016/j.enpol.2019.111053
57. Hazelton JA, Windhorst K, Amezaga JM (2013) Forest based biomass for energy in Uganda: stakeholder dynamics in feedstock production. Biomass Bioenerg 59:100–115. https://doi.org/10.1016/j.biombioe.2013.04.014
58. He L, English BC, Menard RJ, Lambert DM (2016) Regional woody biomass supply and economic impacts from harvesting in the southern U.S. Energy Economics 60:151–161. https://doi.org/10.1016/j.eneco.2016.09.007
59. Hoffmann BS, Szklo A, Schaeffer R (2012) An evaluation of the techno-economic potential of co-firing coal with woody biomass in thermal power plants in the south of Brazil. Biomass Bioenerg 45:295–302. https://doi.org/10.1016/j.biombioe.2012.06.016
60. Hollands AF, Daly H (2023) Modelling the integrated achievement of clean cooking access and climate mitigation goals: an energy systems optimization approach. Renew Sustain Energy Rev 173:113054. https://doi.org/10.1016/j.rser.2022.113054
61. Honkalaskar VH, Sohoni M, Bhandarkar UV (2014) Thermo-chemical modelling of a village cookstove for design improvement. Combust Theor Model 18(3):414–453. https://doi.org/10.1080/13647830.2014.921730
62. Instituto Brasileiro de Geografia e Estatística—IBGE (2021) Áreas Territoriais. https://www.ibge.gov.br/geociencias/organizacao-do-territorio/estrutura-territorial/15761-areas-dos-municipios.html?=&t=acesso-ao-produto. Accessed 18 Feb 2023
63. Intergovernmental Panel on Climate Change—IPCC (2022) Summary for policymakers. In: Pörtner HO, Roberts DC, Tignor M, Poloczanska ES, Mintenbeck K, Alegría A, Craig M, Langsdorf S, Löschke S, Möller V, Okem A, Rama B (eds) Climate change 2022: impacts, adaptation, and vulnerability. Cambridge University Press, Cambridge and New York. https://doi.org/10.1017/9781009325844.001
64. International Energy Agency—IEA (2017) Energy access outlook 2017: from poverty to prosperity. World Energy Outlook Special Report
65. International Renewable Energy Agency—IRENA and International Energy Agency—IEA (2015) Biomass for heat and power. www.irena.org/Publications. Accessed 13 Jan 2023
66. Ishengoma EK, Igangula NH (2021) Determinants of household choice of cooking energy-mix in a peri-urban setting in Tanzania. Energy Sustain Dev 65:25–35. https://doi.org/10.1016/j.esd.2021.09.004
67. Jain T, Sheth PN (2019) Design of energy utilization test for a biomass cook stove: formulation of an optimum air flow recipe. Energy 166:1097–1105. https://doi.org/10.1016/j.energy.2018.10.180
68. Jelonek Z, Drobniak A, Mastalerz M, Jelonek I (2021) Emissions during grilling with wood pellets and chips. Atmos Environ X 12:100140. https://doi.org/10.1016/j.aeaoa.2021.100140
69. Joardder MUH, Islam MdK (2023) Thermophysical aspects of cooking processes. In: High-temperature processing of food products. Elsevier, pp 47–57
70. Jones D, Ryan CM, Fisher J (2016) Charcoal as a diversification strategy: the flexible role of charcoal production in the livelihoods of smallholders in central Mozambique. Energy Sustain Dev 32:14–21. https://doi.org/10.1016/j.esd.2016.02.009
71. Kaliyan N, Vance Morey R (2009) Factors affecting strength and durability of densified biomass products. Biomass Bioenerg 33(3):337–359. https://doi.org/10.1016/j.biombioe.2008.08.005
72. Kazungu I, Kumburu NP (2023) Agripreneurship as a panacea for food security in Tanzania: a systematic review. Heliyon 9

73. Kedir MF, Bekele T, Feleke S (2019) Problems of Mirt, and potentials of improved Gonzie and traditional open cook stoves in biomass consumption and end use emission in rural wooden houses of Southern Ethiopia. Sci Afr 3:64. https://doi.org/10.1016/j.sciaf.2019.e0
74. Kepple AW, Segall-Corrêa AM (2011) Conceituando e medindo segurança alimentar e nutricional Conceptualizing and measuring food and nutrition security. Cien Saude Colet 16(1):187–199. https://doi.org/10.1590/S1413-81232011000100022
75. Khanwilkar S, Gould CF, DeFries R, Habib B, Urpelainen J (2021) Firewood, forests, and fringe populations: Exploring the inequitable socioeconomic dimensions of liquefied petroleum gas (LPG) adoption in India. Energy Res Soc Sci 75:102012. https://doi.org/10.1016/j.erss.2021.102012
76. Kirchherr J, Reike D, Hekkert M (2017) Conceptualizing the circular economy: an analysis of 114 definitions. Resour Conserv Recycl 127:221–232. https://doi.org/10.1016/j.resconrec.2017.09.005
77. Kosemani BS, Ilori AT, Atere AO (2021) Modification and optimization of a baking oven for small scale bread production. Agric Sci 12(06):630–644. https://doi.org/10.4236/as.2021.126041
78. Kpalo SY, Zainuddin MF, Manaf LA, Roslan AM (2021) Evaluation of hybrid briquettes from corncob and oil palm trunk bark in a domestic cooking application for rural communities in Nigeria. J Clean Prod 284:124745. https://doi.org/10.1016/j.jclepro.2020.124745
79. Krecl P, Targino AC, Lara C, Oukawa GY, Soares J, Mollinedo EM (2023) Detecting local and regional air pollution from biomass burning at a suburban site. Atmos Environ 297. https://doi.org/10.1016/j.atmosenv.2023.119591
80. Kshirsagar MP, Kalamkar VR (2016) User-centric approach for the design and sizing of natural convection biomass cookstoves for lower emissions. Energy 115:1202–1215. https://doi.org/10.1016/j.energy.2016.09.048
81. Kuai B, Wang Z, Gao J, Tong J, Zhan T, Zhang Y, Lu J, Cai L (2022) Development of densified wood with high strength and excellent dimensional stability by impregnating delignified poplar by sodium silicate. Constr Build Mater 344. https://doi.org/10.1016/j.conbuildmat.2022.128282
82. Kumar N, Phillip E, Cooper H, Davis M, Langevin J, Clifford M, Stanistreet D (2021) Do improved biomass cookstove interventions improve indoor air quality and blood pressure? A systematic review and meta-analysis. Environ Pollut 290:117997. https://doi.org/10.1016/j.envpol.2021.117997
83. Kvestad I, Chandyo RK, Schwinger C, Ranjitkar S, Hysing M, Ulak M, Shrestha M, Shrestha L, Strand TA (2022) Biomass fuel use for cooking in Nepalese families and child cognitive abilities, results from a community-based study. Environ Res 212:113265. https://doi.org/10.1016/j.envres.2022.113265
84. Leme AS (2021) Josué de Castro and the metamorphoses of hunger in Brazil, 1932–1946. História, Ciências, Saúde - Manguinhos 28(4):1115–1135. https://doi.org/10.1590/S0104-59702021000400010
85. Li G, Hu R, Hao Y, Yang T, Li L, Luo Z, Xie L, Zhao N, Liu C, Sun C, Shen G (2023) CO_2 and air pollutant emissions from bio-coal briquettes. Environ Technol Innov 29. https://doi.org/10.1016/j.eti.2022.102975
86. Li X, Guo Y, Xiao J, Liu T, Zeng W, Hu J, He G, Rong Z, Zhu Z, Wu F, Ma W (2022) The effect of polluting cooking fuels on depression among older adults in six low- and middle-income countries. Sci Total Environ 838. https://doi.org/10.1016/j.scitotenv.2022.155690
87. Lima MDR, Massuque J, Bufalino L, Trugilho PF, Ramalho FMG, de Paula Protásio T, Hein PRG (2022) Clarifying the carbonization temperature effects on the production and apparent density of charcoal derived from Amazonia wood wastes. J Anal Appl Pyrol 166:105636. https://doi.org/10.1016/j.jaap.2022.105636
88. Lima MDR, Simetti R, de Assis MR, Trugilho PF, Carneiro ACO, Bufalino L, Hein PRG, de Paula Protásio T (2020) Charcoal of logging wastes from sustainable forest management for industrial and domestic uses in the Brazilian Amazonia. Biomass Bioenergy 142. https://doi.org/10.1016/j.biombioe.2020.105804

89. Lins LP, Martinez DG, Furtado AC, Padilha JC (2022) Biomethane generation and CO_2 recovery through biogas production using brewers' spent Grains. Biocatalysis Agric Biotechnol 102579. https://doi.org/10.1016/j.bcab.2022.102579
90. Liu Y, Li Z, Floess E, Zhang Y, Lam N, Mawusi SK, Shrestha P, Li X, Xue C, Liu G (2023) Field assessment of straw pellet combustion in improved heating stoves in rural Northeast China. J Environ Sci (China) 127:295–307. https://doi.org/10.1016/j.jes.2022.05.046
91. Lohani SP, Dhungana B, Horn H, Khatiwada D (2021) Small-scale biogas technology and clean cooking fuel: assessing the potential and links with SDGs in low-income countries—a case study of Nepal. Sustainable Energy Technol Assess 46:101301. https://doi.org/10.1016/j.seta.2021.101301
92. Lombradi GV, Parrini S, Atzori R, Stefani G, Romano D, Gastaldi M, Liu G (2021) Sustainable agriculture, food security and diet diversity. The case study of Tuscany, Italy. Ecol Model 458:109702. https://doi.org/10.1016/j.ecolmodel.2021.109702
93. Lubwama M, Yiga VA (2018) Characteristics of briquettes developed from rice and coffee husks for domestic cooking applications in Uganda. Renew Energy 118:43–55. https://doi.org/10.1016/j.renene.2017.11.003
94. Machado GO, Christoforo AL, Araujo VA, Lahr FAR (2016) Química da Madeira no Contexto Energético. EESC/USP, São Carlos
95. Maciel MDA, Troian A, Oliveira SV (2022) Brasil do agro, país da fome: pensando estratégias para o desenvolvimento sustentável. Espac Abierto 31(3):23–41
96. Mainimo EN, Okello DM, Mambo W, Mugonola B (2022) Drivers of household demand for cooking energy: a case of Central Uganda. Heliyon 8(3). https://doi.org/10.1016/j.heliyon.2022.e09118
97. Makonese T, Ifegbesan AP, Rampedi IT (2018) Household cooking fuel use patterns and determinants across southern Africa: evidence from the demographic and health survey data. Energy Environ 29(1):29–48. https://doi.org/10.1177/0958305X17739475
98. Mannucci P, Franchini M (2017) Health effects of ambient air pollution in developing countries. IJERPH 14(9):1048. https://doi.org/10.3390/ijerph14091048
99. Marques A, Rasinmäki J, Soares R, Amorim P (2018) Planning woody biomass supply in hot systems under variable chips energy content. Biomass Bioenerg 108:265–277. https://doi.org/10.1016/j.biombioe.2017.11.016
100. Martínez GJ (2015) Cultural patterns of firewood use as a tool for conservation: a study of multiple perceptions in a semiarid region of Cordoba, Central Argentina. J Arid Environ 121:84–99. https://doi.org/10.1016/j.jaridenv.2015.05.004
101. Mazzone A, Cruz T, Bezerra P (2021) Firewood in the forest: social practices, culture, and energy transitions in a remote village of the Brazilian Amazon. Energy Res Soc Sci 74. https://doi.org/10.1016/j.erss.2021.101980
102. Meira AM, Nolasco AM, Klingenberg D, de Souza EC, Dias Júnior AF (2021) Insights into the reuse of urban forestry wood waste for charcoal production. Clean Techn Environ Policy 23(10):2777–2787. https://doi.org/10.1007/s10098-021-02181-1
103. Mencarelli A, Cavalli R, Greco R (2022) Variability on the energy properties of charcoal and charcoal briquettes for barbecue. Heliyon 8(8). https://doi.org/10.1016/j.heliyon.2022.e10052
104. Menghini D, Marra FS, Allouis C, Beretta F (2008) Effect of excess air on the optimization of heating appliances for biomass combustion. Exp Thermal Fluid Sci 32(7):1371–1380. https://doi.org/10.1016/j.expthermflusci.2007.11.018
105. Monteiro H, Soares N (2022) Integrated life cycle assessment of a southern European house addressing different design, construction solutions, operational patterns, and heating systems. Energy Rep 8:526–532. https://doi.org/10.1016/j.egyr.2022.02.101
106. Muazu RI, Stegemann JA (2015) Effects of operating variables on durability of fuel briquettes from rice husks and corn cobs. Fuel Process Technol 133:137–145. https://doi.org/10.1016/j.fuproc.2015.01.022
107. Neufeld LM, Tolentino L (2012) Nutritional surveillance: developing countries. In: Encyclopedia of Human Nutrition. Elsevier Inc., pp 289–302

108. Niu X, Liu X, Zhang B, Zhang Q, Xu H, Zhang H, Sun J, Ho K-F, Chuang H-C, Shen Z, Cao J (2023) Health benefits from substituting raw biomass fuels for charcoal and briquette fuels: in vitro toxicity analysis. Sci Total Environ 866:161332. https://doi.org/10.1016/j.scitotenv.2022.161332
109. Okore MKL, Koske J, Letema S (2022) Household-based determinants of cooking and heating fuel mixes in informal settlements of Kisumu City, Kenya. Energy Sustain Dev 71:64–72. https://doi.org/10.1016/j.esd.2022.09.002
110. Olesen RS, Hall CM, Rasmussen LV (2022) Forests support people's food and nutrition security through multiple pathways in low- and middle-income countries. One Earth 5:1342–1353
111. Oliveira PRS, Trugilho PF, Oliveira TJP (2022) Briquettes of acai seeds: characterization of the biomass and influence of the parameters of production temperature and pressure in the physical-mechanical and energy quality. Environ Sci Pollut Res 29(6). https://doi.org/10.1007/s11356-021-15847-6
112. Pfau S, Hagens J, Dankbaar B, Smits A (2014) Visions of sustainability in bioeconomy research. Sustainability 6(3):1222–1249. https://doi.org/10.3390/su6031222
113. Pizarro-Loaiza CA, Antón A, Torrellas M, Torres-Lozada P, Palatsi J, Bonmatí A (2021) Environmental, social and health benefits of alternative renewable energy sources. Case study for household biogas digesters in rural areas. J Cleaner Prod 297:126722. https://doi.org/10.1016/j.jclepro.2021.126722
114. Prapas J, Baumgardner ME, Marchese AJ, Willson B, DeFoort M (2014) Influence of chimneys on combustion characteristics of buoyantly driven biomass stoves. Energy Sustain Dev 23:286–293. https://doi.org/10.1016/j.esd.2014.08.007
115. Protásio TP, Roque Lima MD, Scatolino MV, Silva AB, Rodrigues de Figueiredo IC, Gherardi Hein PR, Trugilho PF (2021) Charcoal productivity and quality parameters for reliable classification of Eucalyptus clones from Brazilian energy forests. Renew Energy 164:34–45. https://doi.org/10.1016/j.renene.2020.09.057
116. REN21 (2020) Renewables 2020 Global Status Report. REN21 Secretariat
117. Rose J, Bensch G, Munyehirwe A, Peters J (2022) The forgotten coal: charcoal demand in sub-Saharan Africa. World Dev Perspect 25. https://doi.org/10.1016/j.wdp.2022.100401
118. Rosenthal J, Quinn A, Grieshop AP, Pillarisetti A, Glass RI (2018) Clean cooking and the SDGs: integrated analytical approaches to guide energy interventions for health and environment goals. Energy Sustain Dev 42:152–159. https://doi.org/10.1016/j.esd.2017.11.003
119. Saleem M (2022) Possibility of utilizing agriculture biomass as a renewable and sustainable future energy source. Heliyon 8(2):e08905. https://doi.org/10.1016/j.heliyon.2022.e08905
120. Schettini BLS, Jacovine LAG, Torres CMME, Carneiro ACO, Villanova PH, Rocha SJSS, Rufino MPMX, Silva LB, Castro RVO (2022) Furnace-kiln system: how does the use of new technologies in charcoal production affect the carbon balance? Ind Crops Prod 187:115330. https://doi.org/10.1016/j.indcrop.2022.115330
121. Schueftan A, Sommerhoff J, González AD (2016) Firewood demand and energy policy in south-central Chile. Energy Sustain Dev 33:26–35. https://doi.org/10.1016/j.esd.2016.04.004
122. Seko T, Ishihara K, Suzuki T, Takagi S, Taga K, Iida Y, Shigematsu Y, Itabashi Y, Nakamichi Y, Fujiwara Y, Inada A, Yamashita Y (2022) Effects of moisture content of firewood used in the manufacture of Japanese traditional smoked-dried bonito, katsuobushi, on polycyclic aromatic hydrocarbon (PAH) generation. J Food Compos Anal 111. https://doi.org/10.1016/j.jfca.2022.104630
123. Setter C, Costa KLS, Oliveira TJP, Mendes RF (2020) The effects of kraft lignin on the physicomechanical quality of briquettes produced with sugarcane bagasse and on the characteristics of the bio-oil obtained via slow pyrolysis. Fuel Process Technol 210:106561. https://doi.org/10.1016/j.fuproc.2020.106561
124. Shrestha P, Zhang W, Mawusi SK, Li J, Xu J, Li C, Xue C, Liu G (2021) In-use emissions and usage trend of pellet heating stoves in rural Yangxin, Shandong Province. Environ Pollut 280. https://doi.org/10.1016/j.envpol.2021.116955

125. Silaen M, Taylor R, Bößner S, Anger-Kraavi A, Chewpreecha U, Badinotti A, Takama T (2020) Lessons from Bali for small-scale biogas development in Indonesia. Environ Innov Soc Trans 35:445–459. https://doi.org/10.1016/j.eist.2019.09.003
126. Simões GMF, Leder SM (2022) Energy poverty: the paradox between low income and increasing household energy consumption in Brazil. Energy Build 268. https://doi.org/10.1016/j.enbuild.2022.112234
127. Smith Kirk R, Woodward A Campbell-Lendrum D, Chadee D, Honda Y, Iu Q, Olwoch Jane M, Revich B, Sauerborn R (2014) Human health: impacts, adaptation, and co-benefits. In: FIeld Cb, Barros V, Dokken Dj, Mach Kj, Mastrandrea Md, Bilir Te, Chatterjee M, Ebi Kl, Estrada Yo, Genova Rc, Kissel Es, Maccracken S, White L. (org) Climate change 2014: impacts, adaptation, and vulnerability. Intergovernmental Panel on Climate Change, Cambridge. Cambridge University Press, Cambridge and New York
128. Sotannde OA, Oluyege AO, Abah GB (2010) Physical and combustion properties of charcoal briquettes from neem wood residues. Int Agrophys 24(2):189–194
129. Song C, He J, Zhang H (2021) Comprehensive zoning of biomass energy heating in EU countries reference for China from European experience. Chin J Popul Resour Environ 19(4):321–329. https://doi.org/10.1016/j.cjpre.2022.01.005
130. Srichat A, Vengsungnle P, Bootwong A (2017) A study to increasing the thermal efficiency of the improvements salt boiling stove which the fuel was firewood in Ban Dung area, Udon Thani province by CFD. In: Energy Procedia. Elsevier Ltd., pp 446–451
131. Stabridis O, van Gameren E (2018) Exposure to firewood: consequences for health and labor force participation in Mexico. World Dev 107:382–395. https://doi.org/10.1016/j.worlddev.2018.03.009
132. Thengane SK, Kung KS, Gomez-Barea A, Ghoniem AF (2022) Advances in biomass torrefaction: parameters, models, reactors, applications, deployment, and market. Prog Energy Combust Sci 93
133. Toka A, Iakovou E, Vlachos D, Tsolakis N, Grigoriadou AL (2014) Managing the diffusion of biomass in the residential energy sector: an illustrative real-world case study. Appl Energy 129:56–69. https://doi.org/10.1016/j.apenergy.2014.04.078
134. Tomazini CG, da Silva Leite CK (2016) Programa Fome Zero e o paradigma da segurança alimentar: Ascensão e queda de uma coalizão? Revista de Sociologia e Politica 24(58):13–30. https://doi.org/10.1590/1678-987316245801
135. Ullah S, Noor RS, Sanaullah GT (2021) Analysis of biofuel (briquette) production from forest biomass: a socioeconomic incentive towards deforestation. Biomass Conv Bioref. https://doi.org/10.1007/s13399-021-01311-5
136. United Nations Development Programme—UNDP and Oxford Poverty and Human Development Initiative—OPHI (2022). 2022 Global multidimensional poverty index (MPI): unpacking deprivation bundles to reduce multidimensional poverty. New York
137. Vasile M, Iordăchescu G (2022) Forest crisis narratives: illegal logging, datafication and the conservation frontier in the Romanian Carpathian Mountains. Polit Geogr 96. https://doi.org/10.1016/j.polgeo.2022.102600
138. Van Vliet EDS, Kinney PL, Owusu-Agyei S, Schluger NW, Ae-Ngibise KA, Whyatt RM, Jack DW, Agyei O, Chillrud SN, Boamah EA, Mujtaba M, Asante KP (2019) Current respiratory symptoms and risk factors in pregnant women cooking with biomass fuels in rural Ghana. Environ Int 124:533–540. https://doi.org/10.1016/j.envint.2019.01.046
139. Vasconcelos FAG (2005) Combate à fome no Brasil: uma análise histórica de Vargas a Lula Fighting hunger in Brazil: a historical analysis from Presidents Vargas to Lula. Revista de Nutrição 18(4):439–457. https://doi.org/10.1590/S1415-52732005000400001
140. Vásquez Lavin F, Barrientos M, Castillo Á, Herrera I, Ponce Oliva RD (2020) Firewood certification programs: key attributes and policy implications. Energy Policy 137. https://doi.org/10.1016/j.enpol.2019.111160
141. Vicente ED, Vicente A, Evtyugina M, Carvalho R, Tarelho LAC, Oduber FI, Alves C (2018) Particulate and gaseous emissions from charcoal combustion in barbecue grills. Fuel Process Technol 176:296–306. https://doi.org/10.1016/j.fuproc.2018.03.004

142. Vijay V, Kapoor R, Singh P, Hiloidhari M, Ghosh P (2022) Sustainable utilization of biomass resources for decentralized energy generation and climate change mitigation: a regional case study in India. Environ Res 212:113257. https://doi.org/10.1016/j.envres.2022.113257
143. Wander PR, Bianchi FM, Caetano NR, Klunk MA, Indrusiak MLS (2020) Cofiring low-rank coal and biomass in a bubbling fluidized bed with varying excess air ratio and fluidization velocity. Energy 203:117882. https://doi.org/10.1016/j.energy.2020.117882
144. Wang D, Liu L, Yuan Y, Yang H, Zhou Y, Duan R (2020) Design and key heating power parameters of a newly-developed household biomass briquette heating boiler. Renew Energy 147:1371–1379. https://doi.org/10.1016/j.renene.2019.09.081
145. Welfle DA, Chingaira S, Kassenov A (2020) Decarbonising Kenya's domestic and industry Sectors through bioenergy: an assessment of biomass resource potential and GHG performances. Biomass Bioenergy 142. https://doi.org/10.1016/j.biombioe.2020.105757
146. World Health Organization—WHO (2014) WHO guidelines for indoor air quality: household fuel combustion. World Health Organization, Geneva. https://apps.who.int/iris/handle/10665/141496. Accessed 9 Jan 2023
147. World Health Organization—WHO (2021a) Cooking fuels and technologies (database). https://www.who.int/publications/m/item/database-primary-reliance-on-fuels-and-technologies-for-cooking. Accessed 6 Feb 2023
148. World Health Organization—WHO (2021b) WHO global air quality guidelines. Particulate matter (PM2.5 and PM10), ozone, nitrogen dioxide, sulfur dioxide and carbon monoxide. World Health Organization, Geneva
149. World Health Organization (2022) WHO publishes new global data on the use of clean and polluting fuels for cooking by fuel type. https://www.who.int/news/item/20-01-2022-who-publishes-new-global-data-on-the-use-of-clean-and-polluting-fuels-for-cooking-by-fuel-type. Accessed 15 Feb 2023
150. World Health Organization—WHO (2023) Household air pollution data. https://www.who.int/data/gho/data/themes/air-pollution/household-air-pollution. Accessed 6 Feb 2023
151. Zhang Y, Zhi G, Jin W, Xu P, Li Z, Kong Y, Zhang H, Shen Y, Hu J (2023) Identifying the fundamental drives behind the 10-year evolution of northern China's rural household energy and emission: implications for 2030 and beyond. Sci Total Environ 865. https://doi.org/10.1016/j.scitotenv.2022.161053
152. Zhao J, Cao Y, Yu L, Liu X, Yang R, Gong P (2022) Future global conflict risk hotspots between biodiversity conservation and food security: 10 countries and 7 biodiversity hotspots. Glob Ecol Conserv 34. https://doi.org/10.1016/j.gecco.2022.e02036

Springer Series in Reliability Engineering

Series Editor

Hoang Pham, Industrial and Systems Engineering, Rutgers, The State University of New Jersey, Piscataway, NJ, USA

Today's modern systems have become increasingly complex to design and build, while the demand for reliability and cost effective development continues. Reliability is one of the most important attributes in all these systems, including aerospace applications, real-time control, medical applications, defense systems, human decision-making, and home-security products. Growing international competition has increased the need for all designers, managers, practitioners, scientists and engineers to ensure a level of reliability of their product before release at the lowest cost. The interest in reliability has been growing in recent years and this trend will continue during the next decade and beyond.

The Springer Series in Reliability Engineering publishes books, monographs and edited volumes in important areas of current theoretical research development in reliability and in areas that attempt to bridge the gap between theory and application in areas of interest to practitioners in industry, laboratories, business, and government.

Now with 100 volumes!

Indexed in Scopus and EI Compendex

Interested authors should contact the series editor, Hoang Pham, Department of Industrial and Systems Engineering, Rutgers University, Piscataway, NJ 08854, USA. Email: **hopham@soe.rutgers.edu, or Anthony Doyle, Executive Editor, Springer, London. Email: anthony.doyle@springer.com**.

509

7

OK

M

System Dynamics for Performance Management & Governance

Volume 7

Robinson Salazar Rua

Improving Consistency in Performance Measurement System Design

The Case of the Colombian Public Schools

CED4
System Dynamics Group

Springer

Robinson Salazar Rua
The Center for Community Health Integration
Case Western Reserve University
Cleveland, OH, USA

ISSN 2367-0940 ISSN 2367-0959 (electronic)
System Dynamics for Performance Management & Governance
ISBN 978-3-031-24698-2 ISBN 978-3-031-24699-9 (eBook)
https://doi.org/10.1007/978-3-031-24699-9

This Springer imprint is published by the registered company Springer Nature Switzerland AG
The registered company address is: Gewerbestrasse 11, 6330 Cham, Switzerland

Preface

The purpose of this book is to analyze behavioral distortions in public schools and outline outcome-oriented performance measurement systems that may prevent and mitigate them. An instrumental view of dynamic performance management (DPM) has been used to support that endeavor by (1) identifying how performance drivers impact on outcome and output end results, (2) determining how end results affect strategic resources, and (3) understanding how strategic resources and benchmarks define the dynamics of performance drivers. This approach is also used to foster a shift from an output to an outcome-based view in performance management, intending to pursue sustainable results in the long term.

More specifically, this book provides an answer to the following questions:

- Can performance measurement systems trigger behavioral distortions?
- What are the characteristics of such performance measurement systems that generate dysfunctional behaviors?
- In which fields can behavioral distortions be found from the use of performance measurement systems? Why?
- What is the contribution of dynamic performance management to deal with behavioral distortions?
- How can behavioral distortions be framed in the Colombian public schools?
- Why is boundary spanning relevant to enhance school performance?

The case of the Colombian public schools has been analyzed to frame behavioral distortions associated with performance measurement systems. Test-based accountability systems are often used to assess school performance. An inconsistent design of these measurement systems may provoke the emergence of dysfunctional behaviors, which may jeopardize school outcomes such as educational quality. The case study presented in this research illustrates how an excessive focus on improving test scores caused narrowing of the school curricula. In particular, a DPM-based model has been built to show how an outcome-oriented view may support policymakers and system designers in dealing with behavioral distortions from the use of school assessment systems.

Reading this book is beneficial to (1) graduate students in public management, (2) researchers in performance management, system dynamics, and education, and (3) decision-makers in public schools.

This book is not designed to make the reader an experienced system dynamics modeler in the field of education. Instead, it aims to develop the reader's capability to frame dysfunctional behaviors from inconsistent school performance measurement systems and propose solutions by means of an outcome-oriented view. This book also includes a comprehensive literature review at the end of each chapter, intending to further strengthen the readers' knowledge on the discussions.

The book consists of five chapters. The first chapter introduces the reader to the topic of performance measurement systems and behavioral distortions. Basic concepts are clarified to set the foundations for posterior analyses. The second chapter describes the use of performance measurement systems to measure school outcomes such as educational quality and civic-mindedness. Likewise, this chapter makes a distinction between assessment and evaluation and provides the reader with a historical background on the administration of standardized tests. Finally, it explains the public school system in Colombia and common criticisms associated with the school measures. The third chapter explores the theoretical background of outcome-oriented performance measurement systems. In particular, the design of control systems and the theory of street-level bureaucracy are discussed. The fourth chapter illustrates how dynamic performance management (DPM) can be used to outline outcome-oriented performance measurement systems in the school context. The case of the Colombian public schools is used for such purpose. The fifth chapter broadens the boundaries of the system with the purpose of enhancing school performance. To this end, the importance of boundary spanning initiatives in education is recognized, and the adoption of an "outside-in" view of policymaking—through DPM—is proposed. Lastly, contributions to the body of knowledge and research limitations are detailed.

I trust that this work will make a contribution to the domain of system dynamics applied to performance management, with an eye on the behavioral effects triggered by output-oriented school performance measurement systems.

I would like to thank Professor Carmine Bianchi from the University of Palermo (Italy) as his ideas have strongly influenced the process leading to this book. I feel a profound respect and admiration for him because he has been my main mentor in academia during the last few years, and he has also taught me innumerable lessons that have enriched my personal and professional life.

I would also like to extend my gratitude to my current colleagues at Case Western Reserve University (USA). I really appreciate our close interactions and the assistance I have received since I am working here. In particular, I would like to thank Peter Hovmand for the trust he placed in me, and for his encouragement during the writing process of the last chapter. This book would not have been possible without his patience and support. I would also like to express my sincere thanks to Kurt Stange, James Bindas, and Robin Gotler. I will never forget the help they offered me before and after my arrival in the USA.

I am also grateful to my new and old friends because they have emotionally supported me by sharing quality time. Even though some of them live far away, they are always in my thoughts. I know that we will meet again and we will share wonderful experiences in the future. Each of them represents the family that I chose for myself.

Last but not least, I would like to thank God, my parents, my grandmas, and my aunt for being always with me. I love you.

Cleveland, OH, USA Robinson Salazar Rua

Contents

About the Author

Robinson Salazar Rua is a researcher and policy analyst with several years of experience in the fields of system dynamics and performance management. He is skilled in analyzing data, designing, and implementing innovative strategies through modeling and simulation techniques. He possesses a strong ability to apply scientific frameworks to support strategic decision-making in public and private organizations. He has worked in a broad number of projects ranging from public education to healthcare. He is also experienced in participatory systems modeling to promote conflict resolution and foster strategic dialogue.

Robinson received his bachelor's degree from the National University of Colombia, his master's degree in system dynamics from the University of Bergen (Norway), his MBA from Radboud University (the Netherlands), and his PhD in model-based public planning, policy design, and management from the University of Palermo (Italy). He currently works as a research associate in the Center for Community Health Integration at Case Western Reserve University (USA).

Email: rssalazarr@gmail.com, rss188@case.edu

Chapter 1
Affecting Organizational Behavior Through Performance Measurement Systems

1.1 Exploring the Concept of Performance Measurement

Performance measurement is central for business success (Platts and Sobotka 2010). This is the reason why different frameworks, methodologies, and tools for performance measurement have been developed (Kaplan and Norton 1996; Bourne et al. 2002; De Waal 2004). Using measurement systems aimed at measuring and tracking performance gained support in the private sector in 1980. Several scholars were pioneers in this movement by introducing the use of scientific methods and techniques into the business sector to support goal attainment (Kaplan and Norton 1993; Deming 1986). Such methods and techniques allowed system designers to identify quantifiable indicators related to organizational tasks, intending to make mission and vision statements more explicit. In addition, these indicators enabled decision-makers to determine performance over time and compare current values with benchmarks. In the 1980s, this was an important innovation for assessing organizational performance, as the use of inspections was mostly bounded to a financial perspective. In particular, the quantitative approach related to the measurements contrasted with the subjective and intuitive nature of the traditional inspections (McGinnes and Elandy 2012).

Radnor (2005, p.10) defines performance measurement as "quantifying, either quantitatively or qualitatively the output of an activity or process." Similarly, other scholars conceive the performance measurement system as the formal routines and procedures used by managers to keep or change organizational patterns. Throughout the years, this practice—aimed at increasing efficiency and effectiveness—has gained growing acceptance among policymakers, consultants, and academics (Simons 1995; Brigham and Fitzgerald 2001).

De Lancer Julnes (2006) and Hatry (2006) described the major components of performance measurement systems as follows.

R. Salazar Rua, *Improving Consistency in Performance Measurement System Design*, System Dynamics for Performance Management & Governance 7,
https://doi.org/10.1007/978-3-031-24699-9_1

- Inputs: These are the resources required to produce organizational outputs. In public education, classroom materials, teachers' experience, and expenditures per student are examples of this category.
- Outputs: They make up the final products or services that should lead to the achievement of the desired outcomes. In the education sector, scores in standardized tests are a popular output that is used to assess schools' performance and design public policies. It is widely assumed that high test scores should lead to an improvement in outcomes such as educational quality.
- Outcomes: These are the long-term consequences of the outputs in the whole system. They are often difficult to quantify due to their broad and complex nature. Examples of school outcomes are student placement, family involvement in the learning process, educational quality, and civic-mindedness.

Ridgway (1956) developed a classification system based on single, multiple, and composite criteria to analyze the impact of using different measures on organizational performance. The criteria are explained below.

Single criteria imply that only one indicator is used to gauge performance. For instance, in the 1950s the Soviet industries measured organizational success as a function of monthly production targets (Berliner 1956). Other performance measures were not significantly taken into account.

Multiple criteria denote the use of simultaneous variables to evaluate organizational progress. This type of criteria is based on the premise that organizational performance goes beyond the achievement of a single output. It aims at increasing the investment of efforts on a range of tasks that will allow the organization to grow balanced.

Composite criteria are used to calculate merged measures by assigning weights to each performance variable. For example, in Colombia, the Synthetic Education Quality Index (ISCE) is a merged measure used to assess public schools' performance. Such performance is calculated as an average result from four dimensions (i.e., progress, performance, efficiency, and school environment), which have different weights (Zambrano 2015).

In addition, when measuring organizational performance, the concept of "trade-offs" (Skinner 1969) gains relevance. It implies that the choice to improve performance in a certain organizational area may cause a performance reduction in other areas, as multiple conflicting goals may exist. Several scholars assert that trade-offs are inherent to organizational contexts (Santos et al. 2002). Therefore, decision-makers must comprehend how to select among diverse scenarios that might have positive or negative impacts on performance. Thus, a proper choice may lead to a consistent investment of resources, which may help organizations improve the delivery of outcomes.

In the same line, several frameworks aimed at supporting the assessment of organizational performance have been identified. Some of them include the balanced scorecard (Kaplan and Norton 1992), the performance pyramid (Lynch and Cross 1991), and the performance prism (Neely et al. 2001). Three decades ago, these frameworks represented an important innovation in the performance management field because performance was mainly associated with financial measures before their introduction. The inclusion of non-financial variables—such as those related to customers, internal process, and innovation—represented a different view to understand the rationales behind a high-quality service delivery. However, societies all over the world have rapidly evolved, and in parallel, the complexity and magnitude of their problems have grown up. This has made that those priceless frameworks, which initially were an insightful answer to understand organizational performance, have lost their inner glow. One of the most important causes related to this phenomenon is their static nature (Todd 2000), which contrasts with the dynamism of the current world. In these circumstances, novel research in performance measurement systems has arisen to support the decision-making process in modern organizations (Santos et al. 2002; Youngblood 2003; Li and Maani 2011; Castellano et al. 2004; Bianchi 2016).

Performance indicators are prevalently used to assist decision-making in both public and private organizations. However, several scholars have claimed that such indicators are linear and ignore the effect of trade-offs in the system (Bianchi 2016; Santos et al. 2002). Others argue that performance measures induce individuals to be driven by a short-term view (Youngblood 2003), which lacks a meaningful purpose for the entire organization. In the same line, further research suggests that indicators may (1) neglect the contribution of interactive control systems, (2) fail to support the process view, and (3) make people misconceive the role of benchmarking (Castellano et al. 2004). As a result, unexpected over-reactions might be triggered and over-interventions might be implemented in organizational environments (Li and Maani 2011). In this regard, Smith (1990) denotes how the extensive use of performance indicators to hold public institutions accountable may push managers to work through a narrow perspective. Under this position, measures are often linked to political goals according to individual—rather than group—interests. The author also emphasizes that performance indicators may lead to a fragmental view on managing public services, which may hinder the perception of a whole picture for decision-makers. This may reduce the chances of properly dealing with complex social problems.

In the next section, the previous ideas will be expanded by discussing in detail criticisms associated with the use of performance measures in public organizations.

1.2 Limitations and Side Effects of Performance Measures

Public organizations are expected to work in the delivery of community outcomes (e.g., quality of life, local attractiveness, educational quality, trust in government, and civic-mindedness). The results of such organizations are made accountable by performance indicators. However, criticisms associated with the use of measurements have arisen as described below.

In the first instance, the relation between the use of performance measures and successful performance in an organization is not entirely clear. Some studies have found positive effects of using performance measurement systems in organizations (Lingle and Schiemann 1996; Gubman 1998; Evans 2004). However, other studies have found partial or contrary effects (Neely et al. 2004; Ittner et al. 2003; Austin and Gittel 2002; Bourne and Franco 2003; Smith 1993). For instance, Neely et al. (2004) carried out a comparative analysis in two divisions of an energy distribution company. In one division, performance was assessed through the use of formal performance measurement systems. In the other division, flexible rules for performance evaluation were used. The authors of this study did not find real advantages in the level of sales and profits from using a particular approach to gauge performance. Similarly, Bourne and Franco (2003) found that managers perceived activities related to performance measurement as negative to organizational performance because of its highly bureaucratic nature. Some managers even thought their employees should use time at work to develop new projects rather than spending it on data reporting. In education, the use of performance measures based on school inspections is controversial since positive and negative effects on the achievement of learning outcomes have been denoted in the literature (Klerks 2013). On the one hand, several findings support the fact that using school inspections as an evaluation mechanism may positively influence the improvement of educational quality (Hanushek and Raymond 2005; Luginbuhl et al. 2009). On the other hand, some findings in the field do not relate inspections to an enhancement of school performance (Cullingford and Daniels 1999; Shaw et al. 2003; Rosenthal 2004).

In the second instance, performance indicators often neglect contextual factors that impact on organizational performance. Therefore, performance comparisons are often perceived as unfair because of external influences out of the managerial control. For example, Van Thiel and Leeuw (2002) found that Dutch police officers were erroneously judged to have low performance levels compared to past statistics. However, the scholars discovered that crime patterns in the Netherlands had changed in the last few decades, and such information had not been taken into account in the current measurement systems. In healthcare, Sipkoff (2007) found that using performance results to contrast hospitals from the British National Health Service (NHS) was not a wise decision, as demographical differences and typology of patients were not appropriately considered by the indicators. Similarly, in education, Brown (2005) stated that ranking public schools through performance statistics is arbitrary since these measures favor institutions in wealthy areas. In the same line, Bosker and Scheerens (2000) asserted that school league tables overlook students' socio-economic conditions. The previous two investigations denote that variations among school performance may reflect not only the use of distinct teaching strategies but also the existence of contextual factors that go beyond teachers and principals' responsibilities.

In the third instance, difficulties in selecting and agreeing on the most appropriate measures to be implemented in each organizational unit are recurrent, as conflicting goals among agencies represent the norm rather than the exception. Consequently, it is complex to determine whether or not the organizational efforts are being addressed toward the achievement of community outcomes (Adab et al. 2002; Hodgson et al. 2007). This point of view is supported by research in the field that has detected a misalignment of performance measures used in different institutional hierarchies across public agencies (Micheli and Neely 2010). As a result, such misalignment may lead to ambiguous goals which make it difficult to have a clear picture of how individual actions are linked to organizational performance (Buchanan 1974; Lan and Rainey 1992; Perry and Porter 1982; Rainey 1993). Likewise, implementing reward mechanisms might contravene the improvement of the service delivery, as weak performance measures might be used (Han Chun and Rainey 2006).

In the fourth instance, outcomes are difficult to quantify. Therefore, they are often operationalized as a set of short-term outputs (e.g., solved crimes per year, the number of fines per month, and students' scores in the standardized tests). Problems arise as politicians—in the attempt to improve results during their mandates—put excessive pressure over street-level bureaucrats to maximize outputs. It fosters the development of strategies that prioritize outputs over outcomes, which may be translated into dysfunctional behaviors in the long term such as gaming and goal displacement (Bohte and Meier 2000).

Lastly, dysfunctional effects—caused by performance measurement systems—have been detected in multiple domains (Goddard and Mannion 2004; Pidd 2005; Bird et al. 2005). For instance, in mobility, the Shadow Strategic Rail Authority (SSRA) informed that British train operators missed stations to attain performance measures based on punctuality (SSRA 2000). In safety, Her Majesty's Inspectorate of Constabulary (HMIC) stated that in England the number of crimes is even four times higher than that reported by the police (HMIC 2000). This report (1) provided evidence of crimes that were not registered in the database until they were solved and (2) brought to light intentional misclassifications of crimes in order to meet quotas. In healthcare, a study revealed that waiting lists are reduced in specific months by increasing the initial time to have an appointment with doctors (McCartney and Brown 1999). Similarly, Werner and Asch (2005) found that some doctors prescribed cancer screening procedures to reach targets, even if such procedures were not advisable. In education, experiments in schools illustrated how feedbacks may paradoxically cause detrimental effects on the achievement of long-term outcomes. Such effects have principally been observed during the execution of cognitive tasks (Visscher and Coe 2003; Kluger and DeNisi 1996).

Performance measurement systems are based on the premise that holding individuals and organizations accountable leads to an improvement in the quality of services. However, this assumption is arguable because people tend to change their behavior to meet the policy intended goals. Such a phenomenon is called—throughout this book—as *behavioral distortion*. For instance, in the mid-1990s, Argyris (1952) conducted a study on behavioral distortions and its impact on the achievement of organizational outcomes. In this research, the effects of using budgets to measure organizational performance were analyzed through a case study. Argyris reported that employees in a factory prioritized "easy orders" to meet quotas at the end of a financial period. This strategy triggered delays in the delivery of previous orders, as they were not finished in the same sequence in which they were received. This implied a violation of the procedures and rules in the organization (Jaworski and Young 1992).

The above example—of dysfunctional behaviors associated with performance measurement systems—denotes the importance of (1) identifying common side effects caused by the adopted indicators in organizations (Smith 1993) and (2) designing consistent performance measures to prevent and mitigate such detrimental effects (Bianchi 2016; Bianchi and Williams 2015). This may help policymakers understand the reasons that induce people to manipulate the system to maximize their pay-off, and consequently to formulate strategies that align individual and organizational interests.

More specifically, Smith (1993) identified side effects of using performance measures in organizations. These effects are described below.

Myopia refers to the focus on short-term activities at the expense of long-term goals. *Ossification* means a reluctance to experiment with new approaches. *Tunnel vision* is the preference for "easy to measure" targets. *Sub-optimization* implies a fragmental approach that leads to an unbalanced growth of different organizational areas. *Measure fixation* refers to a shift in the focus of attention from the intended outcome to the measured output. *Misrepresentation* occurs when results are made up to produce an impression of outstanding performance.

In a similar vein to Smith, other scholars have noted further behavioral distortions associated with the use of performance indicators. These are listed below.

Illegal acts and falsification refer to an intentional modification in the data to satisfy particular standards and requirements (Mars 1982; Vaughn 1983; Simon and Eitzen 1986). *Biasing and focusing* imply that data reporting is aimed at emphasizing favorable circumstances for the manager and the organization (Birnberg et al. 1983). *Smoothing* is a deliberate alteration of the data flow that does not affect the achievement of outputs in a given period (Ronen and Sadan 1981). *Filtering* happens when information is withheld to avoid negative reactions from superiors (Read 1962; O'Reilly and Roberts 1974; Birnberg et al. 1983).

The above behavioral distortions have been categorized by Jaworski and Young (1992) as "strategic information manipulations" because the performance measurement system is affected directly. This author has also defined a further category of dysfunctional practices called "gaming performance indicators." This category implies that the processes are perversely affected but the rules are not broken. It leads to maximize personal benefits at the expense of organizational goals. "Gaming of a performance measure is said to exist when the subordinate knowingly selects his activities so as to achieve a more favorable measure on the surrogate used by the superior for evaluation at the expense of selecting an alternative course of action that would result in a more desirable level of performance as far as the superior's true goal is concerned" (Birnberg et al. 1983, p.123, as cited in Soobaroyen 2005).

As gaming the system is an important concept in this book, the next section will be devoted to deepening the discussions on it and highlighting its relationship with output-oriented rewards.

1.3 Gaming Performance Measures

The concept of gaming has been widely debated in the scientific literature. For instance, Smith (1995) describes gaming as an inclination to take advantage of the system of rules. This behavior results from using the wrong variables to assess performance. Similarly, Courty and Marschke (2007, p.905) label gaming as "undesirable responses that the designers of performance measures did not foresee." This definition emerged as the result of empirical analyses carried out in federal job training programs. In this research, the scholars found that policy designers developed program-compliance strategies aimed at improving performance outputs significantly (Courty and Marschke 2004). In addition, Kelman and Friedman (2009, p.924) define gaming as a "behavior that consumes real resources but produces no genuine performance improvement even on a measured dimension." This definition implies the creation of a false image of organizational progress.

Gaming has also been conceived as a set of unintended managerial responses to accountability systems designed to support policy implementation (Terman and Yang 2016). Such unintended responses are described in the literature as follows.

The threshold effect suggests that performance measurement systems encourage people to meet organizational goals but discourage them to exceed such goals (Bird et al. 2005). *The ratchet effect* results from setting future targets through an increase of previous targets (Hood 2006). Under this effect, people repeatedly report low performance levels to prevent managers from setting higher goals (Goddard and Mannion 2004). This effect also makes people lack organizational commitment in the long term, as they gradually look for reducing their effort investments. *Output distortion* is denoted as "hitting the target and missing the point" (Hood 2006, p.516). This type of gaming implies that organizational outputs are achieved at the expense of community outcomes, which compromises the real purpose of the measures. For instance, Oppel and Shear (2014) found that the United States Department of Veterans Affairs kept some veterans on a secret waiting list for months—before registering them into the patient tracking system—to meet the 14 days target.

In a similar vein, Bohte and Meier (2000) described the concept of gaming as an attempt to manipulate performance measures. These system manipulations are detailed below.

Cutting corners refers to a managerial approach that prioritizes quantity over quality to deliver outcomes. In this category, decision-makers place their efforts in the production of outputs regardless of the impact of their actions on the whole system (Levine et al. 1990). For instance, sometimes public schools allow students—who have not acquired an adequate knowledge level—to graduate, with the purpose of increasing the graduation rates. These dysfunctional behaviors often emerge in contexts where the achievement of outcomes is crucial but resource availability is constrained. As a result, organizations spread their efforts across a wide range of outputs rather than in a few of them, intending to be perceived as productive (Bohte and Meier 2000). *Biasing sample* occurs when individuals and agencies only select and report cases with a high probability of producing positive results. This phenomenon is also known as "cream-skimming," and it implies a perverse alteration of organizational efforts (Wilson 1989). For instance, it has been found that some teachers focus their efforts on increasing the number of white people who pass university entrance exams while they reduce attention on minority pass rates, as the latter group usually scores lower in the tests (Bohte and Meier 2000). *Lying* is a high-risk method of organizational gaming, as it implies a loss of trust between managerial and political divisions (Bohte and Meier 2000). This dysfunctional strategy is often used to hide disadvantageous information and portray results that favor particular interests (Downs 1967). In this regard, Meier (1985) claimed that the manager of "the Environmental Protection Agency's Superfund toxic waste cleanup program" intentionally lied about her reports to show an impartial image of the inspections that were carried out in toxic waste sites during the 1980s. Such reports were based on personal rather than objective criteria.

According to the scientific literature, the most common causes of gaming behaviors in public and private organizations are the design of poor performance measures (Greener 2005), along with the use of output-oriented rewards (Heinrich 2002; Courty and Marschke 2004). These two causes may induce people to turn away tasks that require high investment of resources and devote more time and energy to those "easy to do" activities. For instance, Figlio and Rouse (2006) found that schoolteachers focused their efforts on the lowest performing students in the standardized tests, as their scores were easier to increase than the highest performing students. This strategy allowed them to receive performance bonuses.

Heinrich and Marschke (2010) declared that distinct reward mechanisms—aimed at aligning individual actions with organizational goals—are often administered during the design of performance measurement systems. Social and financial benefits are linked to the attainment of outputs for the purpose of improving organizational performance. In particular, Rothstein (2008) carried out a detailed review on the use of compensation models in different organizational and geographical contexts. The scholar analyzed the use of performance indicators and rewards in Soviet manufacturing enterprises, several transportation companies in Chile, the Britain's National Health Service (NHS), and test-based accountability systems.

In parallel, psychological research has also been conducted on the use of incentives and its impact on motivation (Skinner 1938; Deci et al. 1999). In particular, the scientific literature distinguishes two types of motivation. Intrinsic motivation, which implies the enjoyment of an activity for its own sake, is increased by using internal rewards (e.g., public recognition). Alternatively, people may have an extrinsic motivation that is amplified by external rewards such as financial benefits (Platts and Sobotka 2010). For instance, in healthcare, Davies et al. (2000) claimed that practitioners, from the British National Health Service (NHS), were primarily driven by internal rewards to improve the quality of the services. In education, several scholars argue that the use of incentives may boost performance in public schools (Hanushek et al. 1994; Kemmerer and Windham 1997; Odden and Kelley 1997). Further research suggests that schoolteachers are mainly motivated by internal rewards and students' satisfaction (Kelley 1999; Heneman 1999). However, it does not imply a null preference for financial benefits. In fact, it has also been found that payment schemes related to performance may positively influence schoolteachers' behaviors (Ladd 1999). The above studies show how the use of reward mechanisms has been a widespread practice in multiple policy domains during the last few years. Thus, rewards can be used to address individual behaviors toward the achievement of organizational and community outcomes.

On the other hand, side effects from the use of rewards in organizations have also been evidenced. Kohn (1999) suggests the existence of a perverse effect of using rewards to assess performance, as individuals may get used to them. As a result, people may develop tactics to improve the "numbers." It provokes a loss of meaning in the adopted indicators. Similarly, Pollitt (2005) found that public sector employees learnt how to increase their scores in the measures for which they were made accountable. This scholar concluded that the power of indicators falls over time, and therefore they should be replaced periodically. Furthermore, Kohn (1999) also demonstrated how pressure—generated by incentive programs—may induce salespeople to game the system by prioritizing the sale of products for which they receive greater financial benefits. Likewise, this researcher evidenced that the chances of unethical behaviors tend to increase due to external incentives. Kohn's research outlined the dangers associated with performance measurement systems that strongly rely on rewards to address individual behaviors. These reward mechanisms may make people prefer "easy to reach" outputs at the expense of long-term outcomes.

In a similar vein, Platts and Sobotka (2010) argued that dysfunctional behaviors from the use of performance measurement systems were tightly related to reward systems and staff motivation. According to research in the field, performance measurement systems that rely on external rewards may reduce intrinsic motivation in the long term (Deci 1971, 1972, 1976; Weiner 1980; Deci et al. 1981, 2001; Levine and Broderick 1983; Jordan 1986; Shalley et al. 1987; Kreps 1997). Therefore, the design of performance measures tied to external rewards may not be enough to achieve targets (Kunz and Pfaff 2002).

In addition, gaming the system may emerge as a consequence of linking performance to both targets and budgets. It can eventually produce a counterproductive effect that may affect organizational performance (Jensen 2003). Jensen argues that performance measurement systems—based on rewards and sanctions—make people inclined to lie and cheat, as punitive actions are forecasted if targets are not achieved. He also asserts that once people were forced to cheat due to the system of rules, they will rarely change their mindsets and they will even spread this dysfunctional behavior in further organizational dimensions. Other scholars argue that gaming behaviors may be conceived as a rational, rather than irrational, response to restrictive performance measurement systems that influence human relations. Exerting pressure—through control mechanisms based on rewards—may lead to an increase in organizational stress and tension.

Moreover, if rewards are not distributed fairly, people will adjust their actions to favor non-cooperative settings (Ouchi 1979). It may increase the likelihood of dysfunctional behaviors from the adoption of performance measures that are erroneously aligned to reward systems. For instance, Bianchi and Williams (2015) analyzed the behavioral distortions associated with the CompStat program of the New York Police Department. They highlighted the unintended effects on reported crime caused by performance measures and reward systems focused on the achievement of short-term goals. In particular, they demonstrated how CompStat and similar programs from other geographical contexts could be the reason for *the performance paradox* (Leeuw 2003; Meyer and O'Shaughnessy 1993). This phenomenon suggests that the adopted indicators lose the capability to distinguish between outstanding and poor performance because people adapt their actions to reach the targets without producing real improvements in the overall system (Van Dooren et al. 2010).

On the other hand, the emergence of behavioral distortions could imply an unrecognized problem. For instance, gaming behaviors may arise as a response to irrational goals set by politicians (Terman and Yang 2016). In these circumstances, efforts to achieve short-term outputs are symbolic gestures to please a political agenda rather than to enhance the delivery of outcomes. However, this topic is beyond the scope of this research.

In the next section, behavioral distortions will be framed in the school context, in an effort to provide the reader with enough knowledge to understand the Colombian case study that will be discussed in a posterior stage.

1.4 Behavioral Distortions in Public Schools and Recommendations to Deal with Them

The emergence of behavioral distortions in public schools—due to an inconsistent design of performance measures—is the main concern of this book. Hence, this section has been devoted to discuss previous research about dysfunctional behaviors in education.

Over the last few decades, performance measurement systems have been used around the world to hold schools and teachers accountable for student achievement. In particular, the use of performance statistics has become an extended practice in public education (Goddard and Mannion 2004). These statistics are reported regularly to the audience by means of league tables, intending to rank school performance. This implies that public schools are classified according to their results in standardized tests (McGinnes and Elandy 2012; Goldstein and Spiegelhalter 1996). In particular, school decision-makers and politicians often design and implement educational policies based on such tests and tables (Hamilton et al. 2002).

School performance indicators are often assumed to be self-evident. It suggests that decisions made from their use should not require an exploratory view (Strand 1997). In addition, in the school context, measures recurrently embody a "carrot and stick" approach by which institutions are rewarded or sanctioned based on their results (Jacobs et al. 2006). This approach is grounded on the idea that schoolteachers will improve service delivery if their efforts are linked to outputs.

Public schools contribute to the development of society by a high-quality education service. Schools are expected to promote cultural unity, enhance intellectual and economic life, and turn out "well-educated" citizens who support social and political initiatives (Tyack and Cuban 1995; Smith 1998). Therefore, public schools are held accountable for a range of outputs such as graduation rates, test scores, and pass rates on the college entrance exams. However, the existence of multiple goals and performance targets raises the probability of an inversion between means and ends. This phenomenon is reinforced by limited financial conditions and high-pressure levels from external entities such as the government (Bohte and Meier 2000). Likewise, perception of unfairness regarding how the system measures the organizational performance and how compensation schemes distribute social and financial benefits may rationally induce people to engage in behaviors that maximize their pay-off to the detriment of the service delivery (McGinnes and Elandy 2012). Particularly, motivation to game the system may be an expected consequence of designing inconsistent measures (Bianchi and Williams 2015). On this subject, Berman (2002) found that people learnt to use performance indicators as a means to get political and financial advantages.

It has been observed that dysfunctional behaviors—such as gaming and goal displacement—emerge because of a disproportionate focus on improving scores in standardized tests. Bohte and Meier (2000) asserted that low-income public schools in the USA were prompt to game the system, with the purpose of producing comparable results to those schools with better financial conditions. Moreover, these scholars claimed that incentive structures—aligned with performance measures—may lead people to cheat the system as an attempt to maximize their benefits. In the same line, Lynn (1996) affirmed that public agencies may intentionally execute actions that will enhance output measures to the detriment of desirable community outcomes. It occurs due to performance and reward criteria that make people lose their sense of purpose (Blau and Meyer 1971; Downs 1967). As a result, reward mechanisms divert the attention away from achieving organizational goals and foster an individualistic culture. For instance, Garcia (1995) detected an "alarming relationship" between correct answers and erasures on students' reading tests in an American public school. This behavior was an unintended consequence of a school performance program that offered financial benefits to schoolteachers and principals based on score improvements in standardized tests. Therefore, the nature of such reward system influenced the school responses.

In a similar vein, in the 1990s, an improvement in student performance was observed after implementing reward mechanisms (Elmore et al. 1996). However, similar effects occurred in other schools, while a suspicious shift in school strategies was implemented. This "strategic adaptation" resulted from the desire of increasing bonus payments (Kelley 1999). Analogously, further side effects in public schools (e.g., increase in stress and pressure levels, cheating in standardized tests, gaming the system, and narrowing the curricula) have also been framed by the literature (Clotfelter and Ladd 1996; Kelley and Protsik 1997; Koretz 1996).

The organizational culture may also influence the emergence of gaming behaviors as observed by Vasquez Heilig and Darling-Hammond (2008). The scholars carried out interviews to principals and teachers from Texas public schools, and they found that dysfunctional strategies (e.g., holding back and suspending students) were performed to boost scores in standardized tests. Using such strategies was usually justified through a culture based on assessments, evaluations, and performance indicators. In addition, the scholars found that some public schools refused the enrollment of low-performing students as they could affect the accountabilities negatively.

In order to mitigate the previous detrimental behaviors to the organizational performance, several recommendations should be considered.

Performance indicators should reflect both tangible and non-tangible resources that affect policy implementation in an institution. Likewise, stakeholders' interests should be made explicit during the design of indicators by using a participatory approach. Moreover, pressure levels associated with the accomplishment of tasks and the attainment of specific outputs should be controlled strategically (Van Thiel and Leeuw 2002).

Other research has also emphasized the need to improve the integration between performance measures and compensation schemes (Courty and Marschke 2004). Similarly, several scholars (Heinrich 2007; Frey and Benz 2005) have recommended aligning performance programs with symbolic recognitions—rather than monetary benefits—as it may foster the development of an intrinsic motivation without side effects. Moreover, experts in the performance management field have suggested using "enabling controls" for designing performance measurement systems, as they may increase transparency and flexibility by means of a multi-player approach (Cuganesan et al. 2014; Adler and Borys 1996; Ahrens and Chapman 2004). According to these experts, "enabling controls" will make people feel less constrained by measures and more prompt to improve performance. For instance, in 2008, a study found that performance measurement systems that involved the participation of multiple stakeholders—during the stages of design and implementation—mitigated the emergence of behavioral distortions (Wouters and Wilderom 2008).

In the same vein, the judgment of performance—only based on indicators—may not reflect the real contributions of people in the whole system, as inherent characteristics to social issues (e.g., unpredictability) may not be considered correctly (Bosker and Scheerens 2000). Therefore, making comparisons among players in a single system appears to be an obsolescent approach. Perhaps, using dynamic indicators—from a combined institutional and inter-institutional view—could be an insightful answer to improve service delivery and collaboration among stakeholders (Bianchi 2016).

This book proposes the use of the dynamic performance management (DPM) as an approach to support the design of consistent performance measurement systems aimed at preventing and counteracting the emergence of behavioral distortions. In particular, the case study of the Colombian public schools has been used to illustrate the benefits of implementing an outcome-oriented view in performance management to deal with dysfunctional behaviors. In order to set the foundations for analyzing such a case study, school performance measurement systems need to be discussed in depth. The second chapter of this book has been written with that purpose in mind.

References

Adab P, Rouse AM, Mohammed MA, Marshall T (2002) Performance league tables: the NHS deserves better. BMJ 324(7329):95–98

Adler PS, Borys B (1996) Two types of bureaucracy: enabling and coercive. Adm Sci Q 41(1): 61–89

Ahrens T, Chapman CS (2004) Accounting for flexibility and efficiency: a field study of management control systems in a restaurant chain. Contemp Account Res 21(2):271–301

Argyris C (1952) The impact of budgets on people. School of Business and Public Administration, Cornell University, Ithaca

Austin R, Gittel JH (2002) Anomalies of measurement: when it works but should not. Business performance measurement: theory and practice. Cambridge University Press, Cambridge, p xiii, 366

Berliner JS (1956) A problem in Soviet business management. Adm Sci Q 1:86–101

Berman E (2002) How useful is performance measurement? Public Perform Manag Rev 25(4): 348–351

Bianchi C (2016) Dynamic performance management. Springer International Publishing, Zurich

Bianchi C, Williams DW (2015) Applying System Dynamics Modeling to foster a cause and effect perspective in dealing with behavioral distortions associated with a city's performance measurement programs. Public Perform Manag:395–425

Bird SM, Cox D, Farewell VT, Goldstein H, Holt T, Smith PC (2005) Performance indicators: good, bad, and ugly. J R Stat Soc Ser A Stat Soc 168(1):1–27

Birnberg JG, Turopolec L, Young SM (1983) The organizational context of accounting. Acc Organ Soc 8(2):111–129

Blau PM, Meyer MW (1971) Bureaucracy in modern society, 2nd edn. Random House, New York

Bohte J, Meier K (2000) Goal displacement: assessing the motivation for organizational cheating. Public Adm Rev 60(2):173–182

Bosker RJ, Scheerens J (2000) Publishing school performance data. Eur Educ 32(3):12–30

Bourne M, Neely A, Platts K, Mills J (2002) The success and failure of performance measurement initiatives. Perceptions of participating managers. Int J Oper Prod Manag 22(11)

Bourne M, Franco M (2003) Corporate performance management. SAS Institute Inc., Marlow, Centre for Business Performance, Cranfield University School of Management

Brigham BH, Fitzgerald L (2001) Controlling managers and organisations: the case of performance measurement in a regulated water company. Centre for Management under Regulation, Warwick Business School, Warwick

Brown A (2005) Implementing performance management in England's primary schools. Int J Product Perform Manag 54(5/6):468–481

Buchanan B (1974) Government managers, business executives, and organizational commitment. Public Adm Rev 35:339–347

Castellano JF, Young S, Roehm HA (2004) The seven fatal flaws of Performance Measurement. CPA J 74(6):32

Clotfelter CT, Ladd H (1996) Recognising and rewarding success in public schools. In: Ladd HF (ed) Holding schools accountable: performance based reform in education. Brookings, Washington DC, pp 23–64

Courty P, Marschke G (2004) An empirical investigation of gaming responses to explicit performance incentives. J Labor Econ 22(1):23–56

Courty P, Marschke G (2007) Making government accountable: lessons from a federal job training program. Public Adm Rev 67(5):904–916
Cuganesan S, Guthrie J, Vranic V (2014) The riskiness of public sector performance measurement: a review and research agenda. Financ Account Manag 30(3):279–302
Cullingford C, Daniels S (1999) Effects of OFSTED inspections on school performance. In: Cullingford C (ed) An inspector calls. Kogan Page, London, pp 59–69
Davies H, Mannion R, Goddard M, Smith PC (2000) How health care providers use comparative outcomes data to monitor and improve care: a cross-national study. Paper presented at the Academy for Health Services Research and Health Policy Meeting
De Lancer JP (2006) Performance measurement: an effective tool for government accountability? The debate goes on. Evaluation 12(2):219–235
Deci EL (1971) Effects of externally mediated rewards on intrinsic motivation. J Pers Soc Psychol 18:105–115
Deci EL (1972) The effects of contingent and noncontingent rewards and controls on intrinsic motivation. Organ Behav Human Perform 18:217–229
Deci EL (1976) The hidden costs of rewards. Organ Dyn 4(3):61–72
Deci EL, Betlly G, Kahle J, Abrams L, Porac J (1981) When trying to win: competition and intrinsic motivation. Personal Soc Psychol Bull 7:79–83
Deci EL, Koestner R, Ryan RM (1999) A meta-analytic review of experiments examining the effects of extrinsic rewards on intrinsic motivation. Psychol Bull 125:627–668
Deci EL, Koestner R, Ryan RM (2001) Extrinsic rewards and intrinsic motivation in education: reconsidered once again. Rev Educ Res 71(1):1–27
Deming WE (1986) Out of the crisis. The MIT Press
De Waal AA (2004) Stimulating performance-driven behaviour to obtain better results. Int J Product Perform Manag 53(4):301–316
Downs A (1967) Inside Bureaucracy. Little, Brown, Boston
Elmore RF, Abelman CH, Fuhrman SH (1996) The new accountability in state education reform: from process to performance. In: Ladd HF (ed) Holding schools accountable: performance based reform in education. Brookings, Washington DC, pp 23–64
Evans JR (2004) An exploratory study of performance measurement systems and relationships with performance results. J Oper Manag 22(3):219–232
Figlio D, Rouse C (2006) Do accountability and voucher threats improve low-performing schools? J Public Econ 90(1):239–255
Frey BS, Benz M (2005) Can private learn from public governance? Econ J 115(507):F377–F396
Garcia J (1995) In Dallas: District ranks school by how much they improve on traditional tests. Catalyst November:1–8
Goddard M, Mannion R (2004) The role of horizontal and vertical approaches to performance measurement and improvement in the UK Public Sector. Public Perform Manag Rev 28(1): 75–95
Goldstein H, Spiegelhalter DJ (1996) League tables and their limitations: statistical issues in comparisons of institutional performance. J R Stat Soc Ser A Stat Soc 159:385–443
Greener I (2005) The potential of path dependence in political studies. Politics 25(1):62–72
Gubman EL (1998) The talent solution: aligning strategy and people to achieve extraordinary results. McGraw Hill, New York
Hamilton LS, Stecher BM, Klein SP, National Science Foundation (U.S.) (eds) (2002) Making sense of test-based accountability in education. Rand, Santa Monica, CA
Han Chun Y, Rainey H (2006) Consequences of goal ambiguity in public organizations. Public service performance: perspectives on measurement and management, pp 92–112
Hanushek EA, Benson CS, Freeman RB, Jamison DT, Levin HM, Maynard RA, Murnane RJ, Rivkin SG, Sabot RH, Solmon LC, Summers AA, Welch F, Wolfe BL (1994) Making schools work: improving performance and controlling costs. The Brookings Institution, Washington DC

Hanushek EA, Raymond ME (2005) Does school accountability lead to improved student performance? J Policy Anal Manag 24:297–327

Hatry HP (2006) Performance measurement: getting results. Urban Institute Press, Washington DC

Heinrich CJ (2002) Outcomes–based performance management in the public sector: implications for government accountability and effectiveness. Public Adm Rev 62(6):712–725

Heinrich CJ (2007) Evidence-based policy and performance management challenges and prospects in two parallel movements. Am Rev Public Adm 37(3):255–277

Heinrich CJ, Marschke G (2010) Incentives and their dynamics in public sector performance management systems. J Policy Anal Manag 29:183–208

Heneman HG (1999) Assessment of the motivational reactions of teachers to a school-based performance award program. J Pers Eval Educ 12:143–159

Her Majesty's Inspectorate of Constabulary (HMIC) (2000) On the record: thematic inspection report on police crime recording. Home Office, London

Hodgson L, Farrell CM, Connolly M (2007) Improving UK public services: a review of the evidence. Public Adm 85(2):355–382

Hood C (2006) Gaming in target world: the targets approach to managing British public services. Public Adm Rev 66(4):515–521

Ittner CD, Larcker DF, Randall T (2003) Performance implications of strategic performance measurement in financial services firms. Acc Organ Soc 28(7/8):715–741

Jacobs R, Martin S, Goddard M, Gravelle H, Smith P. Exploring the determinants of NHS performance ratings: lessons for performance assessment systems. 2006.

Jaworski BJ, Young SM (1992) Dysfunctional behavior and management control: an empirical study of marketing managers. Acc Organ Soc 17(1):17–35

Jensen MC (2003) Paying people to lie: the truth about budgeting process. Eur Financ Manag 9(3): 379–406

Jordan PC (1986) Effects of an extrinsic reward on intrinsic motivation: a field experiment. Acad Manag J 29(2):405–412

Kaplan RS, Norton DP (1992) The balanced scorecard measures that drive performance. Harv Bus Rev 70(1):9

Kaplan RS, Norton DP (1993) The balanced scorecard-measures that drive performance; putting the balance scorecard to work. Harv Bus Rev

Kaplan RS, Norton DP (1996) The balanced scorecard: translating strategy into action. Harvard Business School Press, Boston

Kelley C (1999) The motivational impact of school-based performance awards. J Pers Eval Educ 12:309–326

Kelley C, Protsik J (1997) Risk and reward: perspectives on the implementation of Kentucky's school-based performance award program. Educ Adm Q 33:474–505

Kelman S, Friedman JN (2009) Performance improvement and performance dysfunction: an empirical examination of distortionary impacts of the emergency room wait-time target in the English national health service. J Public Adm Res Theory 19(4):917–946

Kemmerer FN, Windham DM (eds) (1997) Incentives analysis and individual decision making in the planning of education. UNESCO, International Institute for Educational Planning, Paris, pp 94–107

Klerks M (2013) The effect of school inspections: a systematic review. Manuscript submitted for publication

Kluger AN, DeNisi A (1996) The effects of feedback interventions on performance: a historical review, a meta-analysis, and a preliminary feedback intervention theory. Psychol Bull 119(2): 254

Kohn A (1999) Punished by rewards: the trouble with gold stars, incentive plans, A's, praise and other bribes. Houghton Muffin, New York

Koretz D (1996) Using student assessments for educational accountability. In: Hanushek EA, Jorgenson DW (eds) Improving America's schools: the role of incentives. National Academy Press, Washington DC, pp 171–196

Kreps DM (1997) Intrinsic motivation and extrinsic incentives. Am Econ Rev 87(2):359–364
Kunz AH, Pfaff D (2002) Agency theory, performance evaluation, and the hypothetical construct of intrinsic motivation. Acc Organ Soc 27(3):275–296
Ladd HF (1999) The Dallas school accountability and incentive program: an evaluation of its impacts on student outcomes. Econ Educ Rev 18:1–16
Lan Z, Rainey HG (1992) Goals, rules, and effectiveness in public, private and hybrid organizations: more evidence on frequent assertions about differences. J Public Adm Res Theory 2:5–28
Leeuw FL (2003) Evaluation and new public management in the Netherlands. In: Wollmann H (ed) Evaluation in public sector reform: concepts and practice in international perspective. Elgar, Bodmin, pp 104–117
Levine FM, Broderick JE (1983) Attribution and contrast: two explanations for the effect of external rewards on intrinsic motivation. Br J Psychol 74:461–466
Levine CH, Peters BG, Thompson FJ (1990) Public administration: challenges, choices, consequences. Scott, Foresman/Little, Brown, Boston
Li A, Maani K (2011) Dynamic decision-making, learning and mental models
Lingle JH, Schiemann WA (1996) From balanced scorecard to strategic gauges: is measurement worth it? American Management Association
Lynn LE (1996) Public management as art, science, and profession. Chatham House, Chatham, NJ
Lynch RL, Cross KF (1991) Measure up: yardsticks for continuous improvement. Blackwell, Cambridge
Luginbuhl R, Webbink D, De Wolf I (2009) Do school inspections improve primary school performance? Educ Eval Policy Anal 31:221–237
Mars G (1982) Cheats at work – an anthropology of workplace crime. Allen & Unwin, London
McCartney S, Brown R (1999) Managing by numbers: using outcome measures in the NHS. Int J Health Care Qual Assurance 12(1):6–12
McGinnes S, Elandy K (2012) Unintended behavioural consequences of publishing performance data: is more always better? J Community Inform 8(2)
Meier KJ (1985) Regulation: politics, economics and bureaucracy. St. Martin's Press, New York
Meyer MW, O'Shaughnessy KC (1993) Organizational design and the performance paradox. In: Swedberg R (ed) Explorations in economic sociology. Russell Sage Foundation, New York, pp 249–278
Micheli P, Neely A (2010) Performance measurement in the public sector in England: searching for the golden thread. Public Adm Rev 70(4):591–600
Neely A, Adams C, Crowe P (2001) The performance prism in practice. Meas Bus Excell 5(2): 6–13. https://doi.org/10.1108/13683040110385142
Neely A, Kennerley M, Martínez V (2004) Does the balanced scorecard work: an empirical investigation. Performance measurement and management: public and private. Performance Measurement Association, Edinburgh
Odden A, Kelley C (1997) Paying teachers for what they know and do: new and smarter compensation strategies for teachers. Sage, London
Ouchi WG (1979) A conceptual framework for the design of organizational control mechanisms. Manag Sci 25:833–848
Oppel R, Shear M (2014) Severe report finds V.A. hid waiting lists at hospitals. New York Times, 28 May
O'Reilly CA, Roberts KH (1974) Information filtration in organizations: three experiments. Org Behav Human Perform:253–265
Perry JL, Porter LW (1982) Factors affecting the context for motivation in public organizations. Acad Manag Rev 7:89–98
Pidd M (2005) Understanding perverse effects of public sector performance measurement. Lancaster University Management School Working Paper
Platts KW, Sobotka M (2010) When the uncountable counts: an alternative to monitoring employee performance. Bus Horizon 14(1):28–42

Pollitt C (2005) Indicators are indicators, work is work and politics are politics: do we have theories to handle this? Erasmus University Rotterdam

Radnor Z (2005) Developing a typology of organisational gaming. Presentation at the EGPA conference, Bern

Rainey HG (1993) Toward a theory of goal ambiguity in public organizations. In: Perry JL (ed) Research in public administration, vol 2. JAI Press, Greenwich, CT

Read WH (1962) Upward communication in industrial hierarchies. Hum Relat:3–15

Ridgway VF (1956) Dysfunctional consequences of performance measurements. Adm Sci Q 1(2): 240–247

Ronen J, Sadan S (1981) Smoothing income numbers – objectives, means and implications

Rosenthal L (2004) Do school inspections improve school quality? Ofsted inspections and school examination results in the UK. Econ Educ Rev 23:143–151

Rothstein R (2008) Holding accountability to account: how scholarship and experience in other fields inform exploration of performance incentives in education. National Center on Performance Incentives Working paper 2008-04, Peabody College, Vanderbilt University

Santos SP, Belton V, Howick S (2002) Adding value to performance measurement by using system dynamics and multicriteria analysis. Int J Oper Prod Manag 22(11):1246

Shalley CE, Oldham GR, Porac JF (1987) Effects of goal difficulty and expected external evaluation on intrinsic motivation. Acad Manag J 30(3):553–563

Shaw I, Newton DP, Aitkin M, Darnell R (2003) Do Ofsted inspections of secondary schools make a difference to GCSE results? Br Educ Res J 29:63–75

Simon DR, Eitzen DS (1986) Elite deviance. Allyn & Bacon, Boston

Simons R (1995) Levers of control: how managers use innovative control systems to drive strategic renewal. Harvard Business School Press, Boston

Sipkoff M (2007) Go carefully when measuring quality. Managed Care 16(9/September 2007): 25–26, 28, 31

Skinner BF (1938) The behavior of organisms: an experimental analysis. D. Appleton-Century Company, New York

Skinner W (1969) Manufacturing-missing link in corporate strategy. Harv Bus Rev

Smith KB (1998) What price commonwealth: democracy, markets, and education. University of Nebraska, Mimeo

Smith P (1990) The use of performance indicators in the public sector. J R Stat Soc Ser A 153:53–72

Smith P (1993) Outcome-related performance indicators and organizational control in the public sector. Br J Manag 4:131–151

Smith P (1995) On the unintended consequences of publishing performance data in the public sector. Int J Public Adm 18(2/3):277–310

Soobaroyen T. Management control systems and dysfunctional behaviour: an empirical investigation. 2005.

Shadow Strategic Rail Authority (SSRA) (2000) On track: a new way of measuring performance. OPRAF, London

Strand S (1997) Key performance indicators for primary school improvement. Educ Manag Admin Leadersh 25(2):145–153

Terman JN, Yang K (2016) Reconsidering gaming in an accountability relationship: the case of minority purchasing in Florida. Public Perform Manag Rev 40(2):281–309

Todd DP (2000) A "Dynamic" balanced scorecard. Management Science and Information Systems, University of Auckland, Auckland

Tyack DB, Cuban L (1995) Tinkering toward Utopia: a century of public school reform. Harvard University Press, Cambridge

Van Dooren W, Halligan J, Bouckaert G (2010) Performance management in the public sector. Routledge, New York

Van Thiel S, Leeuw FL (2002) The performance paradox in the public sector. Public Perform Manag Rev:267–281

Vasquez Heilig J, Darling-Hammond L (2008) Accountability Texas-style: the progress and learning of urban minority students in a high-stakes testing context. Educ Eval Policy Anal 30(2):75–110

Vaughn D (1983) Controlling unlawful organizational behaviour – social structure and corporate misconduct. University of Chicago Press, Chicago

Visscher AJ, Coe R (2003) School performance feedback systems: conceptualisation, analysis, and reflection. Sch Eff Sch Improv 14(3):321–349

Weiner MJ (1980) The effect of incentive and control over outcomes upon intrinsic motivation and performance. J Soc Psychol 112(2):347–354

Werner RM, Asch DA (2005) The unintended consequences of publicly reporting quality information. JAMA 293(10):1239

Wilson JQ (1989) Bureaucracy: what government agencies do and why they do it. Basic Books, New York

Wouters M, Wilderom C (2008) Developing performance-measurement systems as enabling formalization: a longitudinal field study of a logistics department. Acc Organ Soc 33(4): 488–516

Youngblood AD (2003) The use of multi-attribute utility theory to address trade-offs for the balanced scorecard. Department of Industrial Engineering Fayetteville, University of Arkansas, p 222

Zambrano MF (2015) Alcances e inconsistencias del índice sintético de calidad educativa, diseñado y aplicado por el ministerio de educación de Colombia en el año 2015 [Scope and inconsistencies of the synthetic index of educational quality, designed and applied by the Ministry of Education of Colombia in 2015]. IQ Matrix Foundation, Bogota, Colombia

Chapter 2
Implementing Performance Measurement Systems to Hold Public Schools Accountable for Educational Quality

2.1 Defining the Concept of Educational Quality

The quality of education is a matter of debate due to its complex nature. Some experts in education have declared that educational quality is the resource that allows children and youth to gain skills and abilities to overcome the modern challenges (Marquès Graells 2001). Others assert (Mestres i Salud 2004; Schmelkes 1995) that educational quality results from a pedagogical process by which organizational interests are aligned with student needs. Therefore, education should (1) foster the development of students' skills, (2) provide schoolteachers with tools that support their teaching activities, and (3) encourage families to be involved in the learning process of their children. In the same vein, Escudero Muñoz (2003) argues that school education reaches a high level of quality if it is designed to satisfy the needs and demands of key societal actors and if the learning goals are measurable and realistic. He also denotes that the quality of education should be promoted by a sociocultural approach that maximizes the resources to improve it and guarantees a wide access to the service. Similarly, Schmelkes (1995) affirms that "the ideal education system" should provide students with an easy access to didactic resources. Likewise, it should be designed to enhance student achievement by fostering innovation and transformation at an individual and institutional level. Such transformation may be supported by the participation of families in school activities, which may stimulate an active learning environment.

In the last few years, a growing interest in using learning outcomes to gauge educational quality has been observed. More specifically, learning outcomes—measured by standardized tests such as PISA—not only make operational quality indicators but also make it easier to compare schools in terms of pedagogical strategies and educational attributes (Sayed 2011).

The United Nations Educational, Scientific and Cultural Organization (2005) defines educational quality as a final outcome of school activities. Therefore, these

R. Salazar Rua, *Improving Consistency in Performance Measurement System Design*, System Dynamics for Performance Management & Governance 7,
https://doi.org/10.1007/978-3-031-24699-9_2

activities should be designed and assessed carefully to enhance student learning. Likewise, the Organisation for Economic Co-operation and Development (2010) describes educational quality as an ex post concept to determine the potential contributions of students in the social system if they acquire an adequate level of higher order thinking skills. In the Colombian context, the Ministry of Education is responsible for designing competency standards that allow policymakers to determine whether schools have achieved the minimum performance targets. Such competency standards are aimed at comparing the current state of the system with a desired goal. Hence, performance is measured across different school levels (i.e., elementary and secondary school) in the subjects that students are expected to learn (i.e., language, mathematics, sciences, and civic competences).

Competency standards are conceived as a guide to determine the type and quality of education that students can access in Colombia. These standards support the adoption of instruments, techniques, and methods to assess individual and organizational capabilities and achieve learning goals. In this regard, public schools have partial autonomy to adapt the educational contents and teaching strategies according to their needs. This implies that competency standards represent a flexible framework to design curricula, school projects, and teaching methodologies. In addition, competency standards set common criteria for both assessments and evaluations. For instance, standardized tests—that are aligned with competency standards—can be used to (1) monitor student progress over time and (2) design teaching strategies focused on enhancing the educational quality. Moreover, test results can be complemented with inspection processes to improve performance by taking into account the socio-economic conditions where schools operate (Colombian Ministry of Education 2013; Cajiao 2008).

The Synthetic Education Quality Index is used in Colombia to measure educational quality through four dimensions of performance (Zambrano 2015). They are described below.

Progress: It determines how much a school has improved its performance compared to the previous year. The main component to be tracked in this dimension is the change in the percentage of students who are in the lowest performance level in the standardized tests. The weight of this dimension in the final score is a maximum of four points.

Performance: It discloses the average score in mathematics and language that has been gotten by the students in the standardized tests. The higher the average test score, the higher the score in this dimension. The weight of this dimension in the final score is four points as a maximum.

Efficiency: It denotes the proportion of students who are promoted to the next school level at the end of the year. The weight of this dimension in the final score is as a maximum of one point.

School environment: It is used to measure both classroom conditions and the relation between teacher and student. In order to quantify the variables of this dimension, a socio-economic questionnaire must be completed by the students after taking the standardized tests. The specific criteria to assess the school environment are not revealed by the Colombian Ministry of Education. The weight of this dimension in the final score is a maximum of one point.

According to the Colombian Ministry of Education, this performance measurement system is designed to foster a balanced improvement in the four dimensions intended to gauge the quality of education. Therefore, if a given school devotes its organizational efforts to the enhancement of a specific dimension at the expense of the others, then its Synthetic Education Quality Index will not reflect an optimal result.

After introducing the concept of educational quality and its relationship with competency standards and performance measurement, the next section will cover relevant notions of school accountability. This will provide the reader with solid knowledge to understand the discussions about the Colombian case study in Chap. 4.

2.2 Measuring Educational Quality through Assessments and Evaluations

In many countries around the world, school accountability is considered a central component for the development of government reforms (Levitt et al. 2008). Several scholars define school accountability as a system to hold schools and teachers accountable for educational quality (Ladd 2007; Hopmann 2008). In the education sector, the adequate acquisition of knowledge and skills is often operationalized by measuring student performance in standardized tests. The measurement of this output is highly related to reward schemes designed to improve the service delivery by fostering individual and institutional efforts (Booher-Jennings 2007).

Accountability in public education has a managerial and outcome-oriented nature grounded in regulatory mechanisms that allow the central government to exercise control on the service by means of autonomous providers (Scott 2000; Verger and Parcerisa 2018). Assessment instruments—such as large-scale standardized tests—are used by this accountability approach to "steer at a distance" (Osborne and Gaebler 1993) and measure the performance level of students, teachers, and schools. Through these instruments, the state has control over school-level actors and forces them to explain the outcomes and outputs of their decisions to formal authorities and society as a whole. This condition makes educators receptive to criticisms and strengthens the presence of external powers (Hamilton et al. 2002; Olmedo and Wilkins 2017).

An important distinction between two types of school accountability models has been denoted in the literature (Adams and Kirst 1999; Firestone 2002; O'Day 2002; Garmannslund et al. 2008; Levitt et al. 2008). On the one hand, *external accountability models*—also known as bureaucratic accountability—are used to provide information about (1) quality and efficiency in the service delivery, (2) the compliance of formal regulations set by the government, and (3) the use of financial resources. The main characteristic of these models is their top-down approach by which the education service is subjected to measurement and control by national, regional, and local agencies. In this category, formal authorities often use standardized tests to track student progress in public schools, and they allocate financial incentives based on scores. On the other hand, *internal accountability models* are grounded on the idea that the education service should not be ruled through hierarchical divisions. Instead, schoolteachers should be made accountable by how they perform their professional tasks according to the perceptions of other colleagues. This implies the use of a bottom-up approach by which the teaching practice relies on the criteria of experienced teachers who have already worked in similar contexts. Therefore, internal accountability models connect compliance of formal regulations with professional standards and codes of conduct, which are evaluated by professional associations and peer review mechanisms. International organizations—such as the World Bank—support the implementation of internal accountability models in public schools since these models are seen as feasible answers to deal with the deterioration of the education service in underdeveloped economies (World Bank 2006, 2017).

Assessments and evaluations are usually considered as synonymous terms; however, experts in the educational field emphasize on their differences. School assessment implies the collection of statistical data for making decisions on how to enhance the achievement of student learning goals. On the other hand, school evaluation implies the scrutiny of procedures, curricula, and materials aimed at supporting the delivery of a high-quality service (Harlen 2007). Hence, inspections belong to the evaluation category and standardized tests belong to the assessment category.

The inspections should be seen as a complement to the standardized tests. Their main difference is the extent and character of the learning measures. The standardized tests are often administered at a national and regional level, while school inspections are more localized. The former allows policymakers to have a "whole picture" of school performance in a broad context through the identification of large-scale patterns. The latter provides a "local picture" of how public schools deliver the education service by taking into account their contextual factors. Moreover, the standardized tests are grounded on objective and consistent elements, while the inspections have a subjective character. Therefore, inspection mechanisms may be disproved by teachers; however, such mechanisms can put in context the adopted indicators through comprehensive judgments that reflect the intrinsic nature of the system where the schools perform their activities (Altrichter and Kemethofer 2015).

In the category of school assessments, the scientific literature (Popham 2003) distinguishes between two types of tests. They are norm-referenced and criterion-referenced. The former is used to compare learning accomplishment levels among students, while the latter is used to contrast the achievement of student outcomes with educational standards set by the government. Both types of assessments can also be used to measure the effectiveness of internal school policies and teaching strategies through individual data that reflect the acquisition of knowledge and skills by the students (Vaishnav 2005). A further distinction between high-stake and low-stake assessments is found in the literature. High-stake assessments imply that the standardized tests are used to foster behavioral and instructional changes to improve school performance. To this end, rewards and sanctions are allocated among schools, teachers, and even students depending on the scores got in national and international tests (Ryan 2004; Haertel and Herman 2005). On the other hand, low-stake assessments lack compensation schemes aimed at addressing institutional efforts in a particular direction. They have an informative character and are administered on a regular basis, without directly affecting the actions of street-level bureaucrats (Klein et al. 2000; Carnoy et al. 2003; Jacob 2005).

Formative and summative assessments have also been discerned in the literature (EPPI 2002; Harlen 2007). Formative assessments are mainly used by schoolteachers to identify learning gaps and adjust teaching strategies. These assessments use a holistic approach to track student progress in particular topics that have been taught during the lessons. On the other hand, summative assessments involve the achievement of learning goals set by the government, which enables students to advance in the next stages of the education system (OECD 2005). Standardized tests belong to this category. The main characteristic of such tests is its homogeneity in the design and implementation. This implies that conditions for administration, questions, scoring rules, and interpretations must be objective and consistent to make reliable comparisons among schools and students (Zucker 2004; Popham 1991). Traditionally, the standardized tests have been used as a tool to track progress on academic achievement.

Cultural and social factors influence the choice of school accountability systems (Bracci 2009; Darling-Hammond 2004). This implies that policymakers may decide whether to (1) measure certain outputs and outcomes (e.g., family involvement, teacher–student ratio, the number of students per classroom, and graduation and dropout rates) and (2) use particular assessments or evaluation methods (e.g., holistic activities, standardized tests, and inspections) depending on the needs and goals of a community. For instance, the American and Colombian education systems strongly rely on standardized tests and external controls to hold schools accountable for student performance, while the Finnish education system relies on formative assessments to enhance both pedagogical processes and student learning (Kupianinen et al. 2009). Likewise, Finnish schools have discretionary power and independency to design and administer their own curricula and assessments—in accordance with national regulations—to enhance educational quality (Niemi et al. 2016). This implies that teaching practices are performed in a supportive and free environment, which fosters the development of instructional methods aimed at satisfying special needs of school students. This endeavor is supported by an extensive use of student portfolios, self-evaluations, and learning feedbacks (Finnish National Board of Education 2004; Rinne et al. 2002; Hendrickson 2012).

In the same line, Hamilton et al. (2002) claim that the design of school assessments must consider the following elements:

Goals: These are the final outcomes of the education service (e.g., educational quality, civic-mindedness, and social awareness). They are usually delineated as statements that express individual and organizational desires through competency standards.

Measures: These are performance indicators that quantify the goals (e.g., scores in the standardized tests and graduation and dropout rates). Therefore, it is indispensable that the measures are correctly aligned with the standards.

Targets: These are school benchmarks designed to compare the current state of the system with a desired level. They are usually set on an annual basis.

Incentives: These are positive or negative consequences that result from the ability of public schools to attain targets.

In particular, this book is focused on analyzing behavioral distortions from the use of school assessment systems. Therefore, the next section will cover the historical background of large-scale testing to give the reader a broader view on this topic.

2.3 History of Test-Based Accountability Systems

Education reforms promote test-based accountability systems as a mechanism to achieve educational quality and transparency. Over the last few decades, an accelerated increase in using these systems has been observed due to advances in data analysis and data visualization of test results. It has amplified the acceptance of large-scale assessments in the political sphere and their relevance for decision-making (Verger and Parcerisa 2018). Regardless of the geographical context, accountability systems often share similar principles such as the allocation of rewards and sanctions aimed at fostering individual and organizational change. However, their design and implementation may substantially differ depending on intrinsic factors (e.g., political agenda, institutional norms, and relationships among stakeholders). Therefore, student assessments may also reflect political and economic interests from a local community (Kellaghan et al. 2009).

The measurement of educational quality has been influenced by the development of sophisticated testing systems that may produce valid and reliable results (McDonnell 2005). According to Hamilton et al. (2002), test-based accountability systems are a set of methods designed to allocate benefits and punishments—to schools, teachers, and students—based on test scores. Such tests are usually administered in regional, national, and international domains. Moreover, in an organizational level, high-performing schools may receive financial benefits and public recognition, while low-performing schools may be intervened or closed. In an individual level, high-performing students may receive rewards such as scholarships, while low-performing students may be retained in a particular school grade. These exams are mandated by the government, and they are designed by external parties to the schools (e.g., private companies). These tests are also characterized by having a homogenous format in the questions, conditions for administration, and scoring rules. However, such homogeneity does not imply the use of a particular type of question (e.g., multiple choice and essay writing) or type of test (e.g., norm-referenced and criterion-referenced).

Using large-scale tests and incentives to improve school performance dates back to the mid-1800s (Tyack 1974). However, these tests became popular since the twentieth century (Haertel and Herman 2005). In the USA, the state of New York was a pioneer in the implementation of large-scale testing in public schools. This state used test-based accountability systems before their popularity grew up (Allington and McGill-Franzen 1992). One of the first standardized tests that was administered to high school students was the Stanford Achievement Test (SAT-10) in 1923. The purpose of this exam was to assess the effectiveness of educational programs and compare schools (Resnick 1982). In the mid-1920s and the 1930s, the use of these assessments increased rapidly because they were conceived as a cornerstone in the development of education reforms (Haney 1981).

Posteriorly, between the 1940s and the late 1950s, the standardized tests were used to judge school curricula and assess organizational performance (Goslin 1963; Goslin et al. 1965). However, they were not used yet to hold schools and teachers accountable for student achievement. Likewise, incentives were not allocated according to test scores. However, the function of these assessments changed over the years.

In the 1960s, the National Assessment of Education Progress (NAEP) was introduced. This was a test that measured academic achievement in a representative sample of American students. Parallelly, the Elementary and Secondary Education Act (ESEA) was released, and compensatory education programs were implemented. In this period, the standardized exams were used as a mechanism to test the effectiveness of government compensations allocated to students with socio-economic disadvantages. According to some scholars, this was a step forward in the use of standardized tests to support school accountability (Airasian 1987; Roeber 1988).

In the 1970s, the "minimum-competency movement" emerged as a model to determine whether a student had acquired the basic skills to be promoted to the next educational stage (Jaeger 1982). Therefore, both students and teachers were hold accountable for performance, and the notion of testing as a driver of changes in teaching practices appeared (Popham et al. 1985).

In the 1980s, after Reagan administration released a report called "Nation at risk" (National Commission on Excellence in Education 1983), test-based accountability systems became popular in education. This report described how the American public schools failed to achieve societal outcomes. As a consequence of this report, ideas from the field of business administration were implemented into the public sector, intending to increase efficiency and effectiveness in the services delivered by modern bureaucracies. This paradigm is named in the scientific literature as "New Public Management" (Vigoda 2003). According to this economic model, autonomy in public schools should increase, and their performance should be assessed by external agencies aimed at monitoring teaching strategies and at allocating incentives (Scott 2000; Gunter et al. 2016). In addition, during this decade, the "education reform movement" replaced the "minimum-competency movement." The former movement emerged due to an extended concern about students failing to acquire the most basic learning competencies. More specifically, research observed that performance in American schools was lower than other countries, despite higher investments in educational resources (National Governors' Association 1989). As a result, the format of the tests changed (it was mostly based on multiple-choice questions and few writing sections), and students' results in the exams were linked to rewards and sanctions.

In the 1990s, a second "education reform movement" came up to deal with the criticisms related to the dysfunctional effects that previous accountability systems had caused. In particular, the "score inflation effect" was perceived as a consequence of devoting a significant amount of school hours to exam preparations. Likewise, it was observed that the curricula were narrowed to focus the teaching practice only on the tested subjects. This provoked an improvement in test scores, but it was not linked to better outcomes in education. During this movement, the concept of rewards and sanctions for high-performing and low-performing schools remained the same. However, the design of tests was intended to help students achieve learning outcomes, despite teaching was inexorably focused on improving scores. Therefore, the tests—that emerged during this period—were used to assess higher order thinking skills (i.e., complex skills that are required to approach novel situations such as critical thinking and problem-solving). In addition, standardized tests were complemented with student portfolios, essays, and teamwork. In this period, performance standards were often modified to satisfy community needs, and consequently, assessments were adapted to support such changes. In later years, the main adjustments in test-based accountability systems were (1) the participation of students with special needs during the administration of standardized tests and (2) an increase in the contents and performance standards to be assessed (Hamilton et al. 2002).

In 2001, the act of congress No Child Left Behind (NCLB) was released in the USA. This provoked a substantial increase in school accountability. Every state was mandated to assess their students in reading, mathematics, and sciences. Likewise, school performance was officially determined by the AYP (i.e., Adequate Yearly Progress), which was calculated as a compound measure of students' test results. In addition, rewards and sanctions were allocated based on scores. For instance, high-performing schools received significant investments for classroom materials, and their teachers received performance bonuses. Conversely, low-performing schools had to face the possibility of closure or be subjected to organizational changes (Figlio 2005). This act of congress was controversial in the academic community because several studies supported the benefits associated with high-stake testing (Jacob 2002; Figlio and Rouse 2004; Deere and Strayer 2001), while other studies emphasized the side effects of measuring school performance through the standardized tests (Jacob 2005; Cullen and Reback 2006).

In 2015, NCLB was replaced by Every Student Succeeds Act (ESSA), whose main purpose was to develop a flexible method to support school accountability. Therefore, failures of the previous education reform were approached by substantial changes in the system of rules (Korte 2015; Darrow 2016).

Nowadays, test-based accountability systems are the main mechanisms to track progress in the achievement of learning outcomes by monitoring annual improvements in specific outputs. Likewise, an emphasis on aligning the competency standards with the tests has been observed in recent years (Mathis and Trujillo 2016).

The Systems Approach for Better Education Results (SABER) is a relatively new product developed by the World Bank to promote school accountability and empower students' families (Edwards 2012). This standardized exam aligns the school curriculum with the current market dynamics, intending to use education as a leverage mechanism for economic growth. The World Bank expects to improve the quality of education through both regulatory and participatory approaches (World Bank 2015). Therefore, families, businesses, and non-profit organizations are encouraged to provide their inputs and actively support all the stages in the routes to accountability.

Alike the World Bank, UNESCO does not support the allocation of benefits and sanctions for schools and teachers. Instead, this organization focuses on building trust and shared purpose among the educational actors to foster an effective accountability. It implies the use of a multi-dimensional and participatory approach to monitor quality in education by holding accountable each stakeholder in the system (UNESCO 2009).

Despite several international organizations agreeing on the use of test-based accountability systems (e.g., OECD, UNESCO, World Bank), there are profound differences in the implemented approaches to achieve educational outcomes. For instance, the World Bank is in favor of using performance measurement systems aligned with market mechanisms. UNESCO advocates for inter-institutional and collaborative mechanisms in education to measure quality in the service delivery (Verger and Parcerisa 2018). And the OECD prefers the integration of large-scale testing with professional self-evaluations by means of the PISA test.

2.4 *The "Programme of International Student Achievement."* A Cornerstone in the Development of Education Policies

Nowadays, economic competitiveness is considered a major concern for most governments. According to many studies on this subject, learning achievement—measured by test scores and the number of years of education—has a strong influence on GDP per capita (Hanushek and Woessmann 2008; De Mello and Padoan 2010). Therefore, as an attempt to increase economic productivity and the attraction of foreign investment, educational reforms promote the acquisition of new skills and competencies that are tuned up with the current market trends (Carnoy and Rhoten 2002). In this context, test-based accountability systems play a relevant role in supporting political and socio-economic interests by schooling the population (Koyama 2013).

The Programme of International Student Achievement (PISA) is one of the most popular standardized tests around the world. This exam was designed by the OECD (Organisation for Economic Co-operation and Development) to measure students' abilities for solving real-life problems (OECD 1999). While other standardized tests (e.g., IEA, PIRLS, and TIMSS) are mainly used to assess student performance in common subjects from the school curricula of the participating countries (Breakspear 2014), PISA is used to gauge academic achievement through an innovative approach.

PISA results are conceived as a cornerstone in the development of education policies since its first administration in 2000. PISA is a comparative assessment that is administered to every 3- to 15-year-old students in over 60 countries (e.g., the USA, Colombia, Finland, Italy, Russia, China, and Indonesia). Each PISA cycle lasts 6 years and three exams are performed during this period. The first cycle was from 2000 to 2006, and the second cycle was from 2009 to 2015. In addition, the exams of each cycle have a different focus on testing domains. For instance, in 2000, reading was the main subject of the assessment. Therefore, a greater number of questions were devoted to it. In 2003, the attention was concentrated on mathematics, and in 2006, the focus was on science. The same pattern was repeated in the second PISA cycle.

PISA has multiple-choice and open-ended questions that must be completed in 2 h. Contextual factors are also taken into account by means of a 30-min questionnaire about learning habits and students' socio-economic conditions. A similar questionnaire must be completed by school managers to provide further information on the learning environment, teaching practices, and educational resources. This information allows policymakers to put into context the measures. Hence, competencies—that are relevant for the integral development of citizens—are assessed by gauging literacy in mathematics, sciences, and reading in students from different geographical and social contexts. Posteriorly, results are published through league tables, which are often used to support education reforms. The education systems of the countries ranked in the first and last positions are regularly named as high-performing and low-performing school systems, respectively (OECD 2004; Schleicher 2007).

Several researchers have claimed that school performance may be improved through test-based accountability systems (Carnoy 2001; Figlio 2005). However, other scholars have associated the use of standardized tests with important consequences over street-level bureaucrats and students (Hershberg 2002; Linn 2000; Noble and Smith 1994; Smith and Fey 2000). Likewise, the beneficial effects of school assessments have also been rejected by some schoolteachers who argue that the standardized tests are irrelevant to improve learning outcomes (Shohamy 2001). As a result, these assessments have not been implemented in all geographical contexts. For instance, Finland is considered a role model in education. This country has repeatedly ranked first in international tests such as PISA. However, its main characteristic is an education system that does not depend on standardized tests and school rankings (Gavrielatos 2009). Similarly, in Northern Ireland, performance indicators are not used to assess public schools because of the possible side effects that such measures may provoke (Bird et al. 2005; McGinnes and Elandy 2012).

Conversely, in Colombia, the results from PISA, ICFES, and SABER have been used to make policy decisions in education during the last few years. The next section will introduce the reader to the Colombian school system and the most common assessments that are used in this country.

2.5 Description of the Public School System and the Types of Assessments in Colombia

The Colombian education system has a decentralized and bureaucratic profile, whose goal is to ensure the fundamental right of education to children and youth in the country. This system implies the existence of a sectoral organization where each government level has competences and complementary responsibilities to improve the service delivery. At the national level, the Colombian Ministry of Education develops policies and targets, establishes rules, regulates the education service, and tracks evaluations. It also provides public schools with technical and administrative help to strengthen their management capacity, and it allocates financial resources based on specific criteria. At the local level (i.e., departments and municipalities), the central offices manage the service by leading, planning, and supervising physical, human, and financial resources. They are also responsible for the results in terms of educational coverage and educational quality. Moreover, they provide public schools with technical help, carry out teacher trainings, and apply rewards and sanctions (Codesocial 2009).

The Colombian school system consists of the following levels.

Pre-school: It is made up of three sub-levels. Pre-kindergarten, kindergarten, and transition. Children usually start this school level when they are between 3 and 6 years old. The aim at this stage is to properly develop cognitive, social, and emotional skills in children (Codesocial 2009).

Elementary school: It is made up of five grades. Children are usually between 7 and 11 years old during this school level. The aim at this stage is to build basic knowledge in languages, mathematics, sciences, and civic competences (Codesocial 2009).

Secondary school: It is made up of six grades. Youth are usually between 12 and 17 years old during this school level. The main objectives at this stage are to (1) reinforce the notions that have been developed in the elementary school, (2) build higher order thinking skills to provide students with tools for overcoming novel challenges, and (3) prepare youth for their university careers or their inclusion in the labor market (Codesocial 2009).

According to the Colombian constitution, education is a fundamental right for children and youth, and its effective delivery is a shared responsibility among the primary social stakeholders (i.e., government, schools, families, and enterprises). The Colombian constitution states that every student in the country must receive a high-quality education, regardless of his or her age, gender, race, religion, or economic condition. Public education in Colombia is mandatory and tax free for children and youth between 5 and 15 years old. In addition, The Colombian Ministry of Education allocates financial resources to the central offices based on annual percentages that are defined in political agendas. These resources must be used to (1) pay the salaries of schoolteachers and administrative staff, (2) finance the purchase of educational materials and perform school maintenance, (3) support the development of school projects, and (4) provide low-income students with free meals (Torres and Duque 1994).

The four major laws that regulate the Colombian school system are described below.

Law 2277 of 1979: It outlines the payment schemes and the conditions for promoting teachers that have worked in public schools before 2002 (The Constitution of Colombia 1991).

Law 115 of 1994: It describes the pedagogical guidelines that public schools must follow to deliver the education service effectively (The Constitution of Colombia 1991).

Law 715 of 2001: In the first instance, it defines the tasks and responsibilities for which the national government and the local authorities are made accountable. In the second instance, it denotes the financial mechanisms that the government uses to support the service delivery (The Constitution of Colombia 1991).

Law 1278 of 2002: It outlines the payment schemes and the conditions for promoting teachers that have worked in public schools after 2002 (The Constitution of Colombia 1991).

The Colombian Education System has four key actors: families, schools, government, and enterprises (Herrera Santana 2007). An effective collaboration among such actors can generate behavioral changes in students to properly face the challenges of modern societies. Therefore, local agencies should perform their functions in a coordinated way to achieve important outcomes for society. Unfortunately, at the present time, the efforts of the Colombian educational actors are not addressed toward the same path. First, the families are absent in the education process. Second, the link that should connect families, society, government, and business does not exist in practice (Llinás 1995). These problems are caused by a fragmental and static approach in designing education policies in Colombia.

Concerning state evaluations, they are designed by "the Colombian Institute for the Promotion of Higher Education" and monitored by the Ministry of Education. The standardized tests that are used to assess the Colombian schools are called "SABER" and "ICFES." The former is aimed at monitoring the progress in mathematics, language, sciences, and civic competences of the students who are in the third, fifth, and ninth grade of the education system. The latter is aimed at assessing a comprehensive knowledge acquisition by 11th grade students before their entrance to the university system (ICFES 2010). The format used in these tests is based on multiple-choice questions with single answers. Students must read a passage, interpret the information, and select the most appropriate answer for the questions (ICFES 2017a, b). This format assumes that students carry out a logic process to reach the correct answer. In addition, the Colombian government has also administered international tests—such as PISA (Programme for International Student Assessment) and TIMSS (Trends in International Mathematics and Science Study)—in public schools. These tests have been used to analyze the state of the education system by taking into account the socio-economic background where the Colombian schools operate (Cajiao 2008).

The use of standardized tests is a recent practice in Colombia. Fernández Gómez (2005) states that ICFES has been administered in Colombia since 1968 in order to assess student achievement at the end of secondary school. Likewise, the scholar declares that SABER was first administered to a small group of Colombian schools in the 1990s. In order to ensure the reliability of the data collected by this standardized test, a representative sample was built. Schools from Bogota, Barranquilla, Cali, and Medellin were chosen, as they were located in the most important cities of the country. This pilot test was successful. In the first instance, the participating schools received feedback to improve their pedagogical methodologies. In the second instance, the government could get a whole picture of how schools were performing their tasks to set the basis for future education reforms. Posteriorly, SABER started to be administrated on a large scale, and it acquired a mandatory character for schools across the country since 2001. Language, mathematics, sciences, and civic competences are the main knowledge areas to be assessed in this exam, as they are assumed to develop students' skills and aptitudes to overcome real-life problems. Nowadays, standardized tests are a cornerstone for the development of public policies in the Colombian education sector (Fernández Gómez 2005).

Colombia's participation in TIMSS and PISA has strengthened institutional efforts to enhance the quality of education. A brief description of their administration in Colombian schools is provided as follows.

TIMSS was first administered to the Colombian schools in 1995. The main goal of this test is to determine the mastery level in mathematics and sciences that students from all over the world have acquired over the years. In addition, this test assesses the level of congruence between the school curricula set by the government and the contents that are really taught in the classrooms. This exam has successfully been administered to students who are in the third, fourth, seventh, and eighth grades of the Colombian education system. Relevant information for policymaking has resulted from the analysis of this test. For instance, it was revealed that only 60% of the Colombian public schools consistently taught the curricula that were set by the Ministry of Education. In addition, only one-third of the contents were mastered properly by the students. TIMSS results have also emphasized the need to improve the education service, as the Colombian students have repeatedly had low performance levels compared to students from the other participating countries (Fernández Gómez 2005).

PISA was first administered to the Colombian schools in 2006. The main goal of this test is to assess student performance in mathematics, reading, and sciences. At this time, over 70 countries around the world participate in this exam. Whereas TIMSS is focused on the assessment of the contents that are taught in the classrooms, PISA makes emphasis on the assessment of students' skills and abilities to solve novel challenges (e.g., assertive communication, critical thinking, and problem-solving). PISA results are published by means of league tables, which allow decision-makers to make comparisons among the participating countries and design education policies intended to enhance educational quality. Historically, Colombian students have had an unsatisfactory performance in this test, which has raised multiple concerns on how to solve this issue. Therefore, several assessment methods and education reforms have been designed in Colombia to increase student achievement. In particular, the competency standards have been introduced to address teaching efforts toward the attainment of both outputs and outcomes (Colombian Ministry of Education 2013). This implies that outcome-oriented methodologies (e.g., dynamic performance management) intended to outline school performance measurement systems may represent a central element to improve the quality of education.

In Colombia, two types of assessments are mostly used to track student progress. They are explained below.

In the first instance, summative assessments are administered to measure performance through a benchmarking process. These assessments are used to compare student achievement with predefined learning targets. A series of exams are periodically applied to determine whether students are promoted to the next school level based on their scores (Scriven 1996). In Colombia, summative assessments are usually administered through both internal and external approaches. From an internal approach, schools are free to measure student progress through exams that are designed by a board of teachers. These exams use specific performance criteria—which have been set by the schools—to assess the competencies of their students. From an external approach, the Colombian Ministry of Education and "the Colombian Institute for the Promotion of Higher Education" design standardized tests (i.e., SABER and ICFES) that are administered on a large scale. This type of assessment provides politicians and principals with crucial information to outline policies and strategies intended to positively affect school efforts toward the delivery of a high-quality service (Cajiao 2008).

In the second instance, formative assessments are used to help students in their learning process through assignments, projects, and other school activities. Whereas summative assessments have a direct implication on students' promotion and retention rates, formative assessments only have a pedagogical impact because they are used as diagnosis tools to examine students' knowledge acquisition from an individual perspective (Camperos 1984).

The next section will discuss different perceptions, within the academic community, of using test-based accountability systems.

2.6 Opinions on the Use of Standardized Tests

Fernández Gómez (2005) states that standardized tests have become a cornerstone in Colombian education reforms since the last few years. Test-based accountability systems have been used to support the enhancement of the education service. In particular, compensation schemes—based on test scores—have been a key component in such accountability systems, as they have been implemented to foster individual and organizational efforts toward the attainment of school outcomes. This scholar is in favor of using standardized tests to measure and track student achievement. He argues that standardized tests allow principals and schoolteachers to (1) determine the competencies that students have acquired over the years and (2) develop school programs for reinforcing strengths and mitigating weaknesses in the learning process.

On the other hand, Fernández Gómez (2005) also recognizes that misinterpretations of test results and ignorance about test limitations may deteriorate school performance in the long term. For instance, comparisons among public schools—which have contrasting social and economic backgrounds—are a common problem of using league tables. Therefore, this researcher provides several recommendations to better use standardized tests. In the first instance, he declares that these exams should be designed just to compare schools that operate in similar contexts. In the second instance, he recommends that schools do not adjust the teaching contents to the standardized tests. Instead, he suggests that enhancing performance in the standardized tests be conceived as a mean, rather than an end, in the learning process. In the third instance, Fernández Gómez stresses the importance of improving the quality of formative assessments by holistic activities (e.g., inter-disciplinary projects and advisories). In the fourth instance, he advocates for a higher collaboration between principals and schoolteachers to identify student weaknesses and perform counteracting actions. Lastly, the scholar draws the attention to the government to design effective policies for enhancing equity and educational quality. To this end, he suggests that (1) school performance is not judged by the standardized tests, as the impact of students' socio-economic background in test results is difficult to quantify, (2) decisions based on league tables are avoided because school rankings encourage unfair competition, and (3) reward allocation does not depend on test scores since negative consequences might be expected.

Some scholars have also emphasized how test-based accountability systems may provoke relevant side effects, which have the potential to weaken the quality of education. For instance, the use of league tables and compensation schemes based on test scores may provoke irregular behaviors intended to reduce performance gaps and increase the allocation of individual rewards (Hamilton et al. 2002; Bosker and Scheerens 2000; Brown 2005). In a similar vein, Popham (2001) asserted that an improper use of the standardized tests may trigger consequences for the service delivery, such as investing a disproportionate time in exam preparations at the expense of other school activities. As a result, the score inflation effect could be expected. Therefore, scores might not reflect a real improvement in learning outcomes. Moreover, in the Colombian education system, bonus payments, public recognition, and prioritization in the allocation of financial resources are some rewards that teachers and schools receive based on test results (See Appendix A). In this regard, Linn (2000) declared that test-based accountability systems have lost reliability because their main components (i.e., standardized tests and incentives) have been associated with detrimental effects.

Hamilton et al. (2002) described the most common behavioral effects that may emerge in public schools due to government pressure for increasing performance in standardized tests. In the first instance, narrowing the school curricula to only teach the contents that might be assessed in the exams. In the second instance, increasing exam preparations to raise the chances of better results. And in the third instance, mimicking the format of the assessments to boost student confidence during the administration of national and international tests.

According to Martínez Rizo (2010), the measurement of school performance through standardized tests may represent a danger to the education system because these assessments rarely take into account the environment where the students develop their learning process. This scholar argues that tests do not reflect how results were achieved, as contextual factors and pedagogical methodologies are not gauged properly. He also declares that the extended use of standardized tests is strongly linked to political interests, and the enhancement of school outcomes—such as educational quality and civic-mindedness—should represent a priority in government plans. Therefore, transparency in accountability and an adequate management of the education service are crucial. Standardized tests are a mechanism to satisfy these social demands.

However, in the Colombian education system, the primary actors usually have a poor assessment culture that leads to an overestimation of the scope of the tests (Martínez Rizo 2010). These exams are used as an absolute method to determine educational quality. In particular, policymakers disproportionately rely on them to design education reforms. As a result, ineffective policies are implemented in the schools, as decisions are made by using an approach that mostly measures students' knowledge acquisition through exams. This implies a prioritization of summative over formative assessments, which narrows student achievement just to test scores. Instead, Martínez Rizo (2010) argues that standardized tests should be seen as a complement to the teaching strategies because learning goals should be measured only by competent teachers, who have tracked student progress over a relevant period.

Perhaps, two of the major limitations for implementing Martínez Rizo's position are (1) teachers with experience in designing high-quality formative assessments are required. Therefore, investments in human capital to enhance teachers' skills in this subject should be a priority, and (2) the use of formative assessments for external comparisons would imply the design of new strategies for data aggregation and quantification, as formative assessments are intended to measure students' knowledge acquisition only in the classroom.

To sum up, positive and negative effects can be experienced from the use of standardized tests. In order to reinforce the positive effects, tests should be seen as an opportunity to support student learning. Likewise, this type of assessment should be used to make an early diagnosis of students and schools in risk, which also requires a greater support from the educational authorities.

The next section will cover the most common criticisms associated with the use of test-based accountability systems.

2.7 Criticisms of Using School Performance Measurement Systems

The OECD is aware of the limitations of test-based accountability systems. Therefore, it emphasizes on the need for holistic approaches in education instead of focusing on learning outputs (OECD 2012). For instance, the OECD stresses on aligning evaluations in the classrooms with standardized tests. Likewise, this organization supports school comparisons by making the test results public. However, it also declares that such results should be interpreted through a careful analysis of additional socio-economic variables because "testing can be hard" for low-income countries with significant constraints in administrative resources (OECD 2013; World Bank 2017).

In the scientific literature, the use of test-based accountability systems and school inspections has been associated with the appearance of behavioral distortions in public schools (Jacob 2005; Cullen and Reback 2006; Figlio and Getzler 2006; Chapman 2001; Ehren 2006). De Wolf and Janssens (2007) have categorized such behavioral distortions into three categories.

The first category is related to intended behaviors such as window dressing, misrepresentation, fraud, and reshaping the test pool (De Wolf and Janssens 2007; Jones et al. 2017). *Window dressing* consists of planning atypical lessons to satisfy inspectors' criteria. Findings on the topic suggest that this behavior was widely spread in England during the late 1990s and the early 2000s (Brimblecombe et al. 1996; Case et al. 2000; Chapman 2001; Fitz-Gibbon and Stephenson-Forster 1999; Wilcox and Gray 1996). For example, in 1999, it was documented that 81% of British principals accepted that inspectors did not attend to conventional school lessons, as pedagogical methods were deliberately modified (Fitz-Gibbon and Stephenson-Forster 1999). This action was executed to improve the chances of a successful evaluation in the inspection process. *Misrepresentation* consists of attempting to produce a favorable school image—which is usually distant from reality—during the inspection. In the Netherlands, it has been found that some schools included outdoor activities during the classes, intending to fulfill the minimum number of hours per lesson (Ehren 2006). *Fraud* implies that schools and teachers adopt behaviors out of the system of rules, which has been set by an external institution. For instance, Jacob and Levitt (2003) detected alterations in test-paper answers in a small percentage of schools in Chicago. *Reshaping the test pool* is a deliberate misclassification of low-performing students in the standardized tests, or an encouragement to push out students who can negatively affect school accountabilities. For example, Schiller and Muller (2000) found that an inconsistent alignment between standardized tests and incentives may increase the chances of pushing low-performing students out of the testing pool. In this regard, research in the field suggests that greater school investments in human capital—such as the development of professional skills in the staff—may reduce the inclination to reshape the test pool, as teachers may feel more confident to satisfy students' needs (Rustique-Forrester 2005).

The second category refers to the emergence of unintended behaviors because of both assessments and evaluations. It includes (1) the use of teaching methodologies based on conservative approaches and (2) narrowing the curricula through teaching to the test or teaching to the inspection. These distortions may lead to isomorphism in the long term (De Wolf and Janssens 2007), as schools may be ruled by criteria and parameters that homogenize their teaching practices at the expense of innovation (i.e., ossification).

The third category covers the other side effects that do not belong to the previous classifications. On this subject, Perryman (2007) developed a theoretical framework that outlines how the pressure exercised by the inspection process may lead to unforeseen consequences in schools due to an increase in principals and teachers' stress levels. This side effect may deteriorate the achievement of learning outcomes in the long term.

In particular, an effect called *score inflation* has been highlighted in the literature of test-based accountability systems (Hamilton et al. 2002; Sturman 2003; Klein et al. 2000; Tymms 2004). This dysfunctional effect is the result of excessive government pressure to improve test scores. Such a phenomenon leads schools to adopt traditional (notion-based) teaching—rather than holistic education—because of its supposed suitability to improve performance in standardized tests. However, this approach usually implies allocating more time to exam preparations, narrowing the school curricula, and rote learning. On the other hand, holistic education or "project-based learning" supports the acquisition of higher order thinking skills (e.g., critical thinking, problem-solving, and teamwork aptitudes). Such skills allow students to link the concepts learnt by attending traditional lessons and to apply them in real life (Harel and Papert 1991; Strobel and Van Barneveld 2009; Walker and Leary 2009).

Pedulla et al. (2003) observed that some schoolteachers devoted more teaching time to the tested contents, despite knowing the limitations of notion-based learning. This was the consequence of high-pressure levels for improving performance in standardized tests. In addition, Tymms (2004) and Klein et al. (2000) detected an increase in test scores in British and American schools as a result of investing a substantial portion of school hours to train students for the exams. Similarly, Sturman (2003) found evidence on teaching to the test in British schools. Furthermore, Wiggins and Tymms (2002) carried out a comparative analysis between elementary schools in England and Scotland. They observed a greater incidence of dysfunctional behaviors from the use of performance measurement systems in

British schools. This outcome was associated with the use of league tables to rank schools in England, which differs from the Scottish context.

Test-based accountability systems have also made schools and teachers prone to game the system by excluding students from the exams (Vasquez Heilig and Darling-Hammond 2008; Darling-Hammond 1991; Smith 1986; Allington and McGill-Franzen 1992; Figlio and Getzer 2002). It has led to a substantial reduction in learning opportunities for minority groups—such as Hispanic and African-American students—whose test scores tend to be lower than white students. In particular, the scholars argue that score increases in the testing year may be associated with an increase in retention and dropout rates of the previous year (Allington and McGill-Franzen 1992; Wheelock 2003; Holmes 2006). Additional research supports this position by claiming that high-stake testing provides greater incentives for developing "gaming attitudes," which may lead to a higher number of students being retained or even dropping out of schools (Nichols et al. 2006; Clarke et al. 2000; Lilliard and DeCicca 2001; Wheelock 2003; Roderick et al. 1999). In the same line, findings on the topic reveal that students who are retained are more likely to drop out in comparison to students who are promoted (Heubert and Hauser 1999; Rumberger and Larson 1998). Schools may encourage low-performing students to abandon their studies by enforcing rigid attendance policies, implementing repetitive teaching formats, and neglecting special learning needs. Thus, an increase in test scores may be observed from reshaping the test pool, which may provoke long-term consequences in minority groups whose educational rights may be ignored (Owens and Ranick 1977; Vasquez Heilig and Darling-Hammond 2008).

In parallel, some researchers have detected that compensation schemes based on merit may induce individualistic behaviors in schoolteachers (Weibel et al. 2010; Ballou 2001; Firestone and Pennell 1993; Kohn 1986; Pfeffer and Sutton 2000), which may be detrimental for building professional networks and improving organizational performance in the long term. Findings on the topic suggest the existence of an inverse relationship between an increase in external incentives and the inclination for sharing information and collaborating with others (Heyman and Ariely 2004; Yang and Maxwell 2011). Therefore, social and professional structures in schools may be deteriorated unintentionally by test-based accountability systems.

Similarly, research in the field suggests that informal networks may affect school performance (Siciliano 2017). This implies that public schools with collaborative environments may display a higher level in teachers' performance compared to schools with less supportive environments (Jackson and Bruegmann 2009; Pil and Leana 2009). It represents a challenge for a fair allocation of external rewards from the use of test-based accountability systems, as equally talented workers may exhibit a lower performance due to contextual factors for which they should not be made accountable.

The next section will provide the reader with some suggestions, from experts in the field, on how to enhance test-based accountability systems.

2.8 Recommendations to Enhance School Assessment Systems

This book is mainly focused on the emergence of behavioral distortions associated with the use of school assessment systems. Therefore, this section portrays several recommendations—from the scientific literature—on how to enhance school performance measures and avoid their most common side effects. Such recommendations are described as follows.

Aggregation of cohorts: School performance is often assessed through annual test results. It may provoke that well-performing schools receive punishments due to unforeseen situations. It may also exacerbate the emergence of dysfunctional strategies in those schools whose performance is below the target. Hence, aggregating cohorts over time could be a solution to this issue. It may be conceived as a fair method to avoid sanctions in schools that are improving their performance consistently but are still under the benchmark (Figlio 2005).

Aggregation of learning outcomes: Figlio (2005) highlights that schools are often induced to narrow their curricula to teach only the contents that are assessed in the standardized tests. In order to overcome this unintended effect, it is proposed to merge a number of learning outcomes into a single indicator (Duncombe and Yinger 1998). This action may reduce the attention on specific subjects and increase the use of a holistic education in public schools.

Aggregation of student groups: Several test-based accountability systems are designed to monitor student performance by taking into account ethnicity, socio-economic conditions, and disabilities. This is done as a mechanism to promote social equity and track the performance of minority groups. However, findings suggest that errors in the measurements are related to highly heterogenous schools because some of them classify potential high-performing students as disabled, intending to meet the targets (Kane and Staiger 2002). Therefore, it is suggested to check whether a strict disaggregation of student groups is necessary in order to avoid this effect from the use of indicators (Figlio 2002).

Broadening the scope of measurement: A proper design of the tests is central to enhance school accountability. Figlio (2005) remarks the need of aligning benchmarks with standardized tests correctly, as it is expected that schools adapt their teaching strategies based on the exam structure (Figlio and Rouse 2004; Jacob 2002). As a result, schools may narrow the curriculum to teach specific contents and improve their outputs, with the purpose of receiving rewards and avoiding sanctions based on test scores. For instance, under the NCLB Act, American schools that were sanctioned could lose a significant portion of their Federal Title I aid (Figlio 2005). In order to cope with this side effect, the design of standardized tests should consider an extension of the scope of measurement by including a wide range of inter-disciplinary topics that foster the use of a holistic education. However, Figlio recognizes that an increase in the financial budget for the education service must be planned if this recommendation is taken into account.

In the same vein, the scholar suggests that non-tested measures (e.g., graduation rates, drop-out rates, and student mobility) are used to calculate school rankings.

Using value-added systems for school assessment: It consists of designing indicators that measure school performance based on test results over an extended period of time. This system highlights the need for assessing the impact of teaching strategies on the achievement of learning outcomes. In addition, the use of value-added measures may reduce the incidence of dysfunctional behaviors, as pressure for meeting annual targets is released. However, the interpretation of these results might be more difficult for the general audience, which is often used to analyze the effects of educational strategies in the short run (Figlio 2005).

In addition, some researchers assert that the compensation schemes should be coupled consistently with the school assessments (Stiefel et al. 2005). This implies that both incentives and performance measures should reflect individual, institutional, and inter-institutional interests. In this regard, Siciliano (2007) argues that the effect of external rewards on internal competition and the influence of collaborative settings on individual performance should be seen as central components in the design of school reward systems.

The emergence of dysfunctional behaviors in organizations is inherently associated with the link between accountability systems and compensation schemes. Under certain circumstances, meeting annual targets may induce schools to behave improperly. However, it may be mitigated partially by rewards and sanctions that discourage such behavior. For instance, the American government compensates public schools by the expected—rather than the current—number of disabled students. This implies that deliberately classifying regular students as disabled in order to meet targets in accountability systems may increase school expenses in the long term, as it must be demonstrated that students' special needs are covered (Figlio 2005). Thus, a separation between the standardized tests and the reward mechanisms in terms of the system of rules may contribute to a reduction of dysfunctional behaviors in public schools.

In parallel, a growing interest in enhancing higher order thinking skills (e.g., critical thinking, problem-solving, and teamwork aptitudes) in school students has been observed. In particular, a positive correlation has been found between the use of school practices oriented to develop inter-personal and intra-personal abilities and an increase in graduation and college entrance rates, compared to traditional teaching formats (Rickles et al. 2019). Likewise, students who have received a holistic education have been associated with higher proficiency levels in learning domains that require the application of basic notions in novel situations (National Research Council 2012). In the same line, previous research in the field suggests that students who have been enrolled in schools that foster deeper learning opportunities—through inter-disciplinary projects, student portfolios, advisories, internships, and technology integration—have improved their (1) grade point average (Collins et al. 2013), (2) test scores (Nichols-Barrer and Haimson 2013), (3) college readiness (Friedlaender et al. 2014), and (4) non-cognitive skills (Collins et al. 2013; Guha et al. 2014).

As a result, enhancing student achievement through a strategic alignment between traditional and holistic education may represent an alternative approach to build basic notions and higher order thinking skills. Such skills can be used to (1) overcome modern challenges, (2) succeed in civic life, and (3) reduce learning gaps among pupils with different socio-economic backgrounds (Rickles et al. 2019). In Chap. 4, dynamic performance management (DPM) is used to build a simulation model that integrates both types of education through an outcome-oriented approach. Before analyzing such model and its results, the background of outcome-oriented performance measurement systems will be explored in Chap. 3. In particular, the design of control systems and the theory of street-level bureaucracy will be discussed.

References

Adams JE, Kirst M (1999) New demands for educational accountability: striving for results in an era of excellence. In: Murphy J, Louis KS (eds) Handbook of research in educational administration. Jossey-Bass, San Francisco, CA

Airasian PW (1987) State mandated testing and educational re-form: context and consequences. Am J Educ 95:393–412

Allington RL, McGill-Franzen A (1992) Unintended effects of educational reform in New York. Educational Policy 6(4):397–414

Altrichter H, Kemethofer D (2015) Does accountability pressure through school inspections promote school improvement? School Effectiveness and School Improvement: An International Journal of Research, Policy and Practice 26(1):32–56. https://doi.org/10.1080/09243453.2014.927369

Ballou D (2001) Pay for performance in public and private schools. Econ Educ Rev 20:51–61. https://doi.org/10.1016/S0272-7757(99)00060-6

Bird SM, Cox D, Farewell VT, Goldstein H, Holt T, Smith PC (2005) Performance indicators: good, bad, and ugly. J R Stat Soc Ser A (Stat Soc) 168(1):1–27

Booher-Jennings J (2007) Michael Fullan, Turnaround leadership. Book Review. J Educ Chang 8(3):291–294

Bosker RJ, Scheerens J (2000) Publishing school performance data. Eur Educ 32(3):12–30

Bracci E (2009) Autonomy, responsibility and accountability in the Italian School System. Crit Perspect Account 20(3):293–312

Breakspear S (2014) How does PISA shape education policy making? How does PISA shape education policy making? Why how we measure learning determines what counts in education. Centre for Strategic Education

Brimblecombe N, Ormston M, Shaw M (1996) Teachers' perceptions of inspections. In: Ouston J, Earley P, Fidler B (eds) Ofsted inspections: the early experience. David Fulton Publishers, London, pp 126–134

Brown A (2005) Implementing performance management in England's primary schools. Int J Product Perform Manag 54(5/6):468–481

Cajiao F (2008) Diálogo nacional sobre la evaluación del aprendizaje en el aula [National dialogue about the learning assessment in the classroom]. Ministry of Education, Bogota, Colombia

Camperos M (1984) La evaluación formativa del aprendizaje [The formative evaluation of the learning]. Faculty of humanities. Mimeo, Caracas, Venezuela

Carnoy M (2001) El impacto de la mundialización en las estrategias de reforma educativa. [The impact of globalization on education reform strategies] en Revista de Educación. n.o extra, págs. 101–10

Carnoy M, Elmore R, Siskin L (eds) (2003) The new accountability: high schools and high stakes testing. Routledge Falmer, New York, NY

Carnoy M, Rhoten D (2002) What does globalization mean for educational change? A comparative approach. Comp Educ Rev 46(1):1–9

Case P, Case S, Catling S (2000) Please show you're working; A critical assessment of the impact of Ofsted inspection on primary teachers. Br J Sociol Educ 21:605–621

Chapman C (2001) Changing classrooms through inspections. School Leadership Manag 21:59–73

Clarke M, Haney W, Madaus G (2000) High stakes testing and high school completion. National Board on Educational Testing and Public Policy Statements

Codesocial (2009) Organización del sistema educativo: Conceptos generales de la educación preescolar, básica y media [Organization of the Education System: General concepts of the pre-school, basic and vocational education]. Ministry of Education, Bogota, Colombia

Collins S, Davis-Molin W, Conley D (2013) Journey toward deeper learning: an evaluation of the Roadtrip Nation Experience in the San Jose PLUS academies. Educational Policy Improvement Center, Eugene, OR

Colombian Ministry of Education (2013) Colombia en PISA 2012: Principales resultados [Colombia in PISA 2012: Main results]. Ministry of Education, Bogota, Colombia

Cullen JB, Reback R (2006) Tinkering toward accolades: school gaming under a performance accountability system. In: Gronberg T, Jansen D (eds) Advances in applied microeconomics (14). Improving school accountability: check-ups or choice. Elsevier Science, Amsterdam, pp 1–34

Darling-Hammond L (1991) The implications of testing policy for quality and equality. Phi Delta Kappan 73(3):220–225

Darling-Hammond L (2004) Standards, accountability, and school reform. Teach Coll Rec 106(6): 1047–1085

Darrow AA (2016) The Every Student Succeeds Act (ESSA): what it means for students with disabilities and music educators. General Music Today

De Mello L, Padoan PC (2010) Promoting potential growth: the role of structural reform, OECD Economics Department Working Papers No. 793. OECD Publishing, Paris, pp 1–29

De Wolf IF, Janssens FJG (2007) Effects and side effects of inspections and accountability in education: an overview of empirical studies. Oxf Rev Educ 33:379–396

Deere D, Strayer W (2001) Putting schools to the test: school accountability, incentives and behavior. Working paper, Department of Economics, Texas A&M University

Duncombe W, Yinger J (1998) School finance reform: aid formulas and equity objectives. Natl Tax J 51(2):239–262

Edwards DB (2012) The approach of the World Bank to participation in development and education governance: trajectories, frameworks, results. In: Education strategy in the developing world: revising the World Bank's education policy. Bingley, Emerald, pp 249–273

Ehren MCM (2006) Toezicht en schoolverbetering [School inspection and school improvement. PhD dissertation, University of Twente

EPPI (2002) A systematic review of the impact of summative assessment and tests on students' motivation for learning. EPPI-Centre, Institute of Education, University of London, London

Escudero Muñoz JM (2003) La calidad de la educación: Controversias y retos para la Educación Pública [Educational quality: Controversy and challenges for the public education]. University of Murcia, Murcia

Fernández Gómez HG (2005) Cómo interpretar la evaluación de las pruebas SABER. [How to interpret the SABER standardized exam]. Ministry of Education, Bogota, Colombia

Figlio D (2002) Aggregation and accountability. In: Finn C (ed) No child left behind: what will it take? Thomas B. Fordham Foundation, Washington, DC

Figlio D (2005) Measuring school performance: promise and pitfalls. In: Stiefel L, Schwartz AM, Rubenstein R, Zabel J (eds) Measuring school performance and efficiency: implications for practice and research, Yearbook of the American Education Finance Association. Eye on Education, Larchmont, NY

Figlio D, Getzer LS (2002, April) Accountability, ability, and disability: gaming the system? National Bureau of Economic Research, Cambridge, MA

Figlio D, Getzler LS (2006) Accountability, ability and disability: gaming the system. In: Gronberg T, Jansen D (eds) Advances in applied microeconomics (14). Improving school accountability: check-ups or choice. Elsevier Science, Amsterdam, pp 35–49

Figlio D, Rouse C (2004) Do accountability and voucher threats improve low-performing schools? (Working paper). National Bureau of Economic Research, Cambridge, MA

Finnish National Board of Education (2004) National core curriculum for basic education 2004. https://www.oph.fi/sites/default/files/documents/perusopetuksen-opetussuunnitelman-perusteet_2004.pdf

Firestone W (2002) Educational accountability. Encyclopedia of Education

Firestone WA, Pennell JR (1993) Teacher commitment, working conditions, and differential incentive policies. Rev Educ Res 63:489–525. https://doi.org/10.3102/00346543063004489

Fitz-Gibbon C, Stephenson-Forster N (1999) Is Ofsted helpful? An evaluation using social science criteria. In: Cullingford C (ed) An inspector calls: Ofsted and its effect on school standards. Kogan Page, London, pp 97–118

Friedlaender D, Burns D, Lewis-Charp H, Cook-Harvey CM, Darling-Hammond L (2014) Student-centered schools: closing the opportunity gap. Stanford Center for Opportunity Policy in Education, Palo Alto, CA

Garmannslund P, Elstad E, Langfeldt G (2008) Lærernes opplevelse av måling og rangering av kvalitetsaspekter ved undervisning og læringsprosesser. In: Langfeldt G, Elstad E, Hopmann ST (eds) Ansvarlighet i skolen: Politiske spørsmål og pedagogiske svar. Cappelen forlag, Oslo

Gavrielatos A (2009, November 23) League tables don't tell a school's whole story. Sydney Morning Herald

Goslin DA (1963) The search for ability: standardized testing in social perspective. Russell Sage, NY

Goslin DA, Epstein RR, Hallock BA (1965) The use of standardized tests in elementary schools. Russell Sage, New York

Guha R, Adelman N, Arshan N, Bland J, Caspary K, Padilla C, Biscocho F (2014) Taking stock of the California linked learning district initiative: fourth-year evaluation report. SRI International, Menlo Park, CA

Gunter HM, Grimaldi E, E, Hall D, Serpieri R. (2016) NPM and the dynamics of education policy and practice in Europe. In: Gunter HM, Grimaldi E, Hall D, Serpieri R (eds) New public management and the reform of education: European lessons for policy and practice. Routledge, London, pp 3–17

Haertel EH, Herman JL (2005) A historical perspective on validity arguments for accountability testing. Yearbook of the National Society for the Study of Education 104(2):1–34

Hamilton LS, Stecher BM, Klein SP, National Science Foundation (U.S.) (eds) (2002) Making sense of test-based accountability in education. Rand, Santa Monica, CA

Haney W (1981) Validity, vaudeville, and values: a short history of social concerns over standardized testing. Am Psychol 36(10):1021–1034

Hanushek EA, Woessmann L (2008) The role of cognitive skills in economic development. J Econ Lit 46(3):607–668

Harel I, Papert S (eds) (1991) Constructionism. Norwood, NJ

Harlen W (2007) Criteria for evaluating systems for student assessment. Stud Educ Eval 33(1): 15–28

Hendrickson KA (2012) Assessment in Finland: a scholarly reflection on one country's use of formative, summative, and evaluative practices. Mid-Western Educational Researcher 25(1/2): 33–43

Herrera Santana HJ (2007) Problemas críticos de Colombia [Critical problems of Colombia]. National university of Colombia, Bogota, Colombia

Hershberg T (2002) Comment. In: Ravitch D (ed) Brookings papers on education policy: 2002. Brookings Institution Press, Washington, DC, pp 324–333

Heubert J, Hauser R (eds) (1999) High stakes: testing for tracking, promotion, and graduation. National Academy Press, Washington, DC

Heyman J, Ariely D (2004) Effort for payment: a tale of two markets. Psychol Sci 15:787–793. https://doi.org/10.1111/j.0956-7976.2004.00757x

Holmes CT (2006) Low test scores + high retention rates = more dropouts. Kappa Delta Pi Record 42(2):56–58

Hopmann ST (2008) No child, no school, no state left behind: schooling in the age of accountability. J Curric Stud 40(4):417–456

ICFES (2010) Estructura general del examen SABER 3,5,9 [SABER test: general structure for 3rd, 5th and 9th grades]. Bogotá, Colombia, May 30. http://www2.icfes.gov.co/estudiantes-y-padres/pruebas-saber-3-5-y-9-estudiantes/estructura-general-del-Examen

ICFES (2017a) SABER 3: Guía de orientación [SABER test for 3rd grade: Guidelines]. ICFES, Bogotá, Colombia

ICFES (2017b) SABER 11: Guía de orientación [SABER test for 11th grade: Guidelines]. ICFES, Bogotá, Colombia

Jackson CK, Bruegmann E (2009) Teaching students and teaching each other: the importance of peer learning for teachers. Am Econ J Appl Econ 1(4):85–108. https://doi.org/10.1257/app.1.4.85

Jacob BA (2002) Accountability, incentives and behavior: the impact of high-stakes testing in the Chicago public schools. (Working paper 8968). National Bureau of Economic Research, Cambridge, MA

Jacob BA (2005) Accountability, incentives and behavior: the impact of high-stakes testing in the Chicago public schools. J Public Econ 89:761–796

Jacob BA, Levitt SD (2003) Rotten apples, an investigation of the prevalence and predictors of teacher cheating. Q J Econ 118:843–877

Jaeger RM (1982) The final hurdle: minimum competency achievement testing. In: Austin GR, Garber H (eds) The rise and fall of national test scores. Academic Press, New York

Jones K, Tymms P, Kemethofer D, O'Hara J, Mcnamara G, Huber S, Myrberg E, Skedsmo G, Greger D (2017) The unintended consequences of school inspection: the prevalence of inspection side-effects in Austria, the Czech Republic, England, Ireland, the Netherlands, Sweden, and Switzerland. Oxf Rev Educ:1–18

Kane T, Staiger D (2002) Improving school accountability measures (Working paper). University of California Los Angeles, Los Angeles, CA

Kellaghan T, Greaney V, Murray S (2009) Using the results of a National Assessment of Educational Achievement. The World Bank, Washington DC

Klein SP, Hamilton LS, McCaffrey DF, Stecher BM (2000) What do test scores in Texas tell us? Rand, Santa Monica

Kohn A (1986) No contest: the case against competition. Houghton Mifflin, Boston, MA

Korte G (2015) The Every Student Succeeds Act vs. No Child Left Behind: what's changed? USA Today news, published on 11 December

Koyama J (2013) Global scare tactics and the call for US schools to be held accountable. Am J Educ 120(1):77–99

Kupianinen S, Hautamaki J, Karjalainen T (2009) The Finnish education system and PISA. Ministry of Education Publications, Finland

Ladd H (2007) Holding schools accountable revisited. Spencer Foundation Lecture in Education Policy and Management. Presented at the 2007 APPAM Fall Research Conference, Washington DC, November 8

Levitt R, Janta B, Wegrich K (2008) Accountability for teachers: a literature review. Rand Cooperation, Santa Monica, CA

Lilliard D, DeCicca P (2001) Higher standards, more dropouts? Evidence within and across time. Econ Educ Rev 20(5):459–473

Linn RL (2000) Assessments and accountability. Educ Res 29(2):4–16

Llinás R (1995) Ciencia, educación y desarrollo: Colombia en el siglo XXI [Science, education and development: Colombia in the 21st century]. Magisterium of education, Bogota, Colombia, pp 24–49

Marquès Graells P (2001) La enseñanza. Buenas prácticas. La motivación [The teaching. Good practices. The motivation]. Autonomous University of Barcelona, Barcelona, Spain

Martínez Rizo F (2010) Uso de SABER: Llamados de atención y lecciones [Use of SABER: calls of attention and lessons]. Colombian Ministry of Education. Accessed 20 Oct 2018

Mathis WJ, Trujillo TM (2016) Lessons from NCLB for the Every Student Succeeds Act. National Education Policy Center

McDonnell L (2005) Assessment and accountability from the policymaker's perspective. In: Haertel EH, Herman JL (eds) Uses and misuses of data for educational accountability and improvement: the 104th Yearbook of the National Society for the Study of Education, Part II. Blackwell, Malden, MA

McGinnes S, Elandy K (2012) Unintended behavioural consequences of publishing performance data: is more always better? J Community Inform 8(2)

Mestres i Salud L (2004) Comunicación y pedagogía: Nuevas tecnologías y recursos didácticos [New technologies and teaching resources]. Center of communication and pedagogy, Madrid, Spain, pp 37–39

National Commission on Excellence in Education (1983) A nation at risk: the imperative for educational reform. A report to the nation and the secretary of education. Washington, DC

National Governors' Association (1989) Results in Education: 1989. Washington, DC

National Research Council (2012) Education for life and work: developing transferable knowledge and skills in the 21st century. The National Academies Press, Washington, DC

Nichols-Barrer I, Haimson J (2013) Impacts of five Expeditionary Learning middle schools on academic achievement. Mathematica Policy Research, Cambridge, MA

Nichols SL, Glass GV, Berliner DC (2006) High-stakes testing and student achievement: does accountability pressure increase student learning? Educational Policy Anal Arch 14(1) http://epaa.asu.edu/epaa/v14n1/

Niemi H, Toom A, Kallioniemi A (2016) Miracle of education: the principles and practices of teaching and learning in Finnish schools. 2nd Revised Education ed. Sense Publishers, Rotterdam

Noble AJ, Smith ML (1994) Old and new beliefs about measurement-driven reform: "The more things change, the more they stay the same" (CSE Technical Report No. 373). University of California, Los Angeles, CRESST, Los Angeles

O'Day J (2002) Complexity, accountability, and school improvement. Harv Educ Rev 72(3)

OECD (1999) Measuring student knowledge and skills. A new framework for assessment. OECD Publications, Paris

OECD (2004) Learning for tomorrow's world: first results from PISA 2003. OECD Publications, Paris

OECD (2005) Formative assessment: improving learning in secondary classrooms. OECD Publications, Paris

OECD (2010) DAC guidelines and reference series. Quality standards for development evaluation. OECD Publications, Paris

OECD (2012) Equity and quality in education. Supporting disadvantaged students and schools. OECD Publications, Paris

OECD (2013) Education at a Glance 2013: OECD indicators. OECD Publications, Paris

Olmedo A, Wilkins A (2017) Governing through parents: a genealogical enquiry of education policy and the construction of neoliberal subjectivities in England. Discourse: Studies in the Cultural Politics of Education 38(4):573–589

Osborne D, Gaebler T (1993) Reinventing government: the five strategies for reinventing government. Penguin, New York
Owens SA, Ranick DL (1977) The Greensville program: a commonsense approach to basics. Phi Delta Kappan 58(7):531–533
Pedulla J, Abrams L, Madaus G, Russell M, Ramos M, Miao J (2003) Perceived effects of state-mandated testing programs on teaching and learning: findings from a national questionnaire of teachers [Electronic version]. National Board on Educational. Testing and Public Policy, Lynch School of Education, Boston College, Boston
Perryman J (2007) Inspection and emotion. Camb J Educ 37:173–190
Pfeffer J, Sutton RI (2000) The knowing-doing gap: how smart companies turn knowledge into action. Harvard Business School Press, Boston, MA
Pil FK, Leana CR (2009) Applying organizational research to public school reform: the effects of teacher human and social capital on student performance. Acad Manag J 52:1101–1124
Popham WJ (1991) Why standardized tests don't measure educational quality. Educ Leadersh 56(6):8–15
Popham WJ (2001) The truth about testing: an educator's call to action. Association for Supervision and Curriculum Development, Hawaii
Popham WJ (2003) Test better, teach better: the instructional role of assessment. Association for Supervision and Curriculums Development, Alexandria, VA
Popham WJ, Cruse KL, Rankin SC, Sandifer PD, Williams PL (1985, May) Measurement-driven instruction: it's on the road. Phi Delta Kappan 66(9):628–634
Resnick D (1982) History of educational testing. In: Wigdor AK, Garner WR (eds) Ability testing: uses, consequences, and controversies, Part II. National Academy Press, Washington DC, pp 173–194
Rickles J, Zeiser KL, Yang R, O'Day J, Garet MS (2019) Promoting deeper learning in high school: evidence of opportunities and outcomes. Educ Eval Policy Anal 41(2):214–234. https://doi.org/10.3102/0162373719837949
Rinne R, Kivirauma J, Simola H (2002) Shoots of revisionist education policy or just slow readjustment? The Finnish case of educational reconstruction. J Educational Policy 17:643–658
Roderick M, Bryk A, Jacob B, Easton J, Allensworth E (1999) Ending social promotion: results from the first two years. Consortium on Chicago School Research, Chicago
Roeber E (1988) A history of large-scale testing activities at the state level. Paper presented at the Indiana Governor's Symposium on ISTEP, Madison, IN, February 10
Rumberger RW, Larson KA (1998) Student mobility and the increased risk of high school dropout. Am J Educ 107(1):1–35
Rustique-Forrester E (2005) Accountability and the pressures to exclude: a cautionary tale from Accountability Texas-Style England. Educ Policy Anal Arch
Ryan JE (2004) The perverse incentives of the No Child Left Behind Act. N Y Univ Law Rev 79: 932–989. https://doi.org/10.2139/ssrn.476463
Sayed Y (2011) After 2015: time for an education quality goal? Compare 41(1):129–130
Scriven M (1996) Types of evaluation and types of evaluators. Am J Eval 17:151–161
Schiller K, Muller C (2000) External examinations and accountability, educational expectations, and high school graduation. Am J Educ 108(2):73–102
Schleicher A (2007) Can competencies assessed by PISA be considered the fundamental school knowledge 15-year-olds should possess? J Educ Chang 8:349–357
Schmelkes S (1995) Hacia una mejor calidad de nuestras escuelas. Secretaría de Educación Pública. [Towars a better quality of our schools]. Secretary of Public Education, Mexico City, Mexico
Scott C (2000) Accountability in the regulatory state. J Law Soc 27(1):38–60
Shohamy E (2001) The power of tests: a critical perspective on the uses of language tests. Pearson Education, Harlow
Siciliano MD (2007) The effect of school choice on segregation and inequality in Charlotte Mecklenburg Public Schools: an analytic framework utilizing spatial and probability models. Association of Public Policy and Management Conference, Washington, DC

Siciliano MD (2017) Professional networks and street-level performance: how public school teachers' advice networks influence student performance. Am Rev Public Adm 47(1):79–101
Smith F (1986) High school admission and the improvement of schooling. New York City Board of Education, New York
Smith ML, Fey P (2000) Validity and accountability in high-stakes testing. J Teach Educ 51(5): 334–344
Stiefel L, Schwartz AE, Rubenstein R, Zabel J (eds) (2005) Measuring school performance and efficiency: implications for practice and research. American Education Finance Association 2005 Yearbook. Eye on Education, Larchmont, NY
Strobel J, Van Barneveld A (2009) When is PBL more effective? A meta-synthesis of meta-analyses comparing PBL to conventional classrooms. Interdiscipl J Problem-Based Learn
Sturman L (2003) Teaching to the test: science or intuition? Educ Res 45:261–273
The Constitution of Colombia (1991) Legis, Bogota, Colombia
Torres J, Duque H (1994) El proceso de descentralización educativa en Colombia [The decentralization process of the education in Colombia]. Colombian Review of the Education
Tyack DB (1974) The one best system: a history of American urban education. Harvard University Press, Cambridge, MA
Tymms P (2004) Are standards rising in English primary schools? Br Educ Res J 30:477–494
UNESCO (2005) Promover la seguridad humana: Marcos éticos, normativos y educacionales en América Latina y el Caribe [Promoting human security: ethical, normative and educational frameworks in Latin America and the Caribbean]. UNESCO, Paris
UNESCO (2009) EFA Global Monitoring Report 2009. Overcoming inequality: why governance matters. UNESCO, Paris
Vaishnav A (2005) Adding value to student assessment: does "value-added assessment" live up to its name? Harv Educ Lett 21(3):1–3
Vasquez Heilig J, Darling-Hammond L (2008) Accountability Texas-style: the progress and learning of urban minority students in a high-stakes testing context. Educ Eval Policy Anal 30(2):75–110
Verger A, Parcerisa L (2018) Test-based accountability and the rise of regulatory governance in education: a review of global drivers. In: Wilkins A, Olmedo A (eds) Education governance and social theory: interdisciplinary approaches to research. Bloomsbury, London
Vigoda E (2003) New public management. Encyclopedia of Public Administration and Public Policy 2:812–816
Walker A, Leary H (2009) A problem based learning meta-analysis: differences across problem types, implementation types, disciplines, and assessment levels. Interdiscipl J Problem-Based Learn
Weibel A, Rost K, Osterloh M (2010) Pay for performance in the public sector—Benefits and (hidden) costs. J Public Adm Res Theory 20:387–412. https://doi.org/10.1093/jopart/mup009
Wheelock A (2003) School awards programs and accountability in Massachusetts: misusing MCAS scores to assess school quality. Fair Test, Cambridge, MA
Wiggins A, Tymms P (2002) Dysfunctional effects of league tables: a comparison between English and Scottish primary schools. Public Money Manag 22:43–48
Wilcox B, Gray J (1996) Inspecting schools: holding schools to account and helping schools to improve. Open University Press, Buckingham
World Bank (2006) The World Bank Annual Report 2006. World Bank, Washington, DC
World Bank (2015) What matters most for school autonomy and accountability: a framework paper. SABER Working Paper Series, 9. World Bank, Washington DC

World Bank (2017) World Bank Annual Report 2017. World Bank, Washington, DC

Yang T-M, Maxwell TA (2011) Information-sharing in public organizations: a literature review of interpersonal, intra-organizational and inter-organizational success factors. Gov Inf Q 28:164–175. https://doi.org/10.1016/j.giq.2010.06.008

Zambrano MF (2015) Alcances e inconsistencias del índice sintético de calidad educativa, diseñado y aplicado por el ministerio de educación de Colombia en el año 2015 [Scope and inconsistencies of the synthetic index of educational quality, designed and applied by the Ministry of Education of Colombia in 2015]. IQ Matrix Foundation, Bogota, Colombia

Zucker S (2004) Administration practices for standardized assessments. Pearson Assessment Report

Chapter 3
Exploring the Conceptual Roots of Outcome-Oriented Performance Measurement Systems

3.1 Design of Control Systems in Organizations

The design of control systems must take into account the context where the organizations operate (Flamholtz 1983). This implies that a predefined set of targets and managerial procedures may not guarantee a successful policy implementation. In order to address efforts toward the achievement of community outcomes, a wider perspective in designing control systems is required. Policymakers often emphasize the need for monitoring, evaluating, and adjusting actions to meet targets. Likewise, they regularly conceive the allocation of incentives as a natural consequence of individual and organizational performance. In the scientific literature, this group of procedures and routines is either named as core *control system* (Flamholtz 1983) or *diagnostic control system* (Simons 1995).

According to Flamholtz (1983), the core control system is just one over four components that should be considered in the design of organizational control systems. Figure 3.1 provides a schematic representation of such components.

The core control system—represented by the innermost circle—is subdivided into the four components that are illustrated in Fig. 3.2.

The first component is the *planning system*. In this phase, organizational goals and strategic view are outlined to determine the steps to improve performance and achieve outputs and outcomes.

The second component are the *operations*. It refers to the tasks and activities that are crucial to achieve organizational goals.

The third component is the *measurement system*. In this phase, performance is gauged through operating results. Likewise, collected data are used to make short-term adjustments to the operations that are executed.

The fourth component is the *evaluation and reward system*. In this phase, incentives are allocated to reward organizational units based on their results. In

R. Salazar Rua, *Improving Consistency in Performance Measurement System Design*, System Dynamics for Performance Management & Governance 7,
https://doi.org/10.1007/978-3-031-24699-9_3

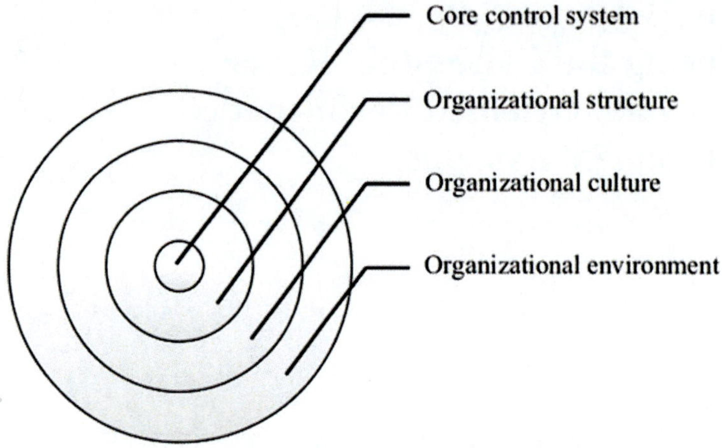

Fig. 3.1 Components of organizational control systems (Maciejczyk 2016, p.53 as in Flamholtz 1983)

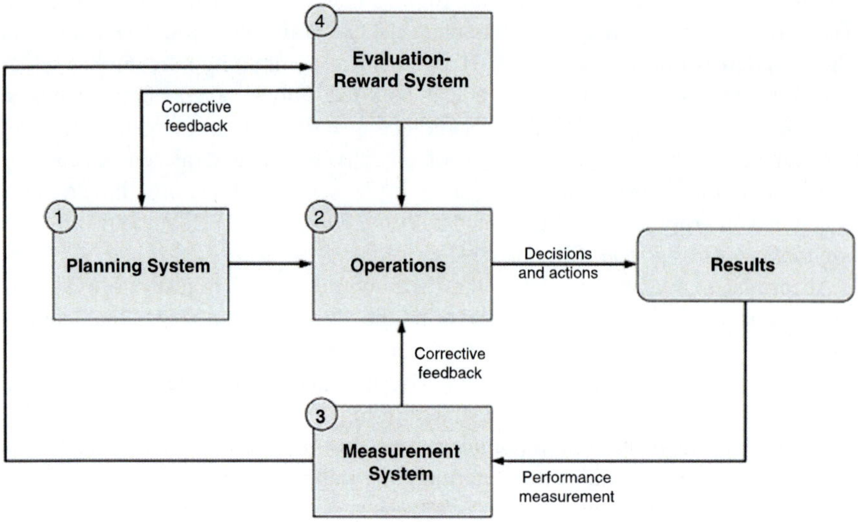

Fig. 3.2 Components of the core control system (Kuhlmann 2012, p.30 as in Flamholtz 1983)

addition, long-term adjustments may also be carried out to re-define both operations and goals.

The design of control systems should also take into account the organizational structure, the organizational culture, and the organizational environment.

Organizational structure refers to the set of rules that allow the core control system to work properly (e.g., the use of a centralized or decentralized structure and workforce specialization by means of organizational units).

Organizational culture describes the values and beliefs that govern the organization (e.g., the institutional vision and the assumptions associated with the managerial processes).

Organizational environment is the space where the above components converge. It also includes the internal and external forces that may affect performance (e.g., individual behaviors, management styles, government reforms, and stakeholder pressure).

Several years after Flamholtz's research was published, Kaplan and Norton (1992) proposed the "balanced scorecard" as a systematic framework to analyze critical performance factors in an organization. Such factors represented the key variables that should be monitored and assessed in diagnostic control systems, as they were the causes of either success or failure in policy implementation. Before the balanced scorecard was used in the public and private sector, performance was mainly conceived as the result of financial measures (e.g., operating income, return on equity, and cash flow). Therefore, this framework represented an important change in the organizational mindsets because the following non-financial measures were also taken into account. Customer measures (e.g., customer satisfaction, market share, and time delivery) were used to describe performance as a function of market conditions and customers' interests. Internal process measures (e.g., cycle time, unit cost, and capacity utilization rate) were used to enhance efficiency and effectiveness in organizational activities. Innovation and learning measures (e.g., training effectiveness, employee motivation, and time to launch a new product) were used to encourage best practices and attitudes among employees.

Later, Simons (1995) remarked the need for complementing the diagnostic control systems—also named as core control systems—with the interactive control systems to pursue strategic transformations in public and private organizations.

In the first instance, Simons (1995) was in favor of using the balanced scorecard as a framework to support the diagnostic control systems by gauging critical performance factors. In particular, the measurement of such factors should (1) be verifiable, (2) capture relevant behaviors in an organization, and (3) reflect individual efforts (Lawler and Rhode 1976). Unfortunately, control systems rarely meet such ideal conditions, and dysfunctional effects tend to emerge as a consequence of performance gaps (Simons 1995). Measuring the wrong variables and relying individual performance on external incentives (e.g., financial benefits) are frequently the causes of a gap between the current and the desired state of the system. In order to reduce this gap, Simons highlighted the importance of conceiving the diagnostic control systems as a set of principles and processes that use negative—also called corrective—feedbacks for enhancing the service delivery. A classic example that illustrates how a negative feedback works is the temperature regulation in a home by means of a thermostat. If the current temperature is too high compared to a desired level, then the thermostat will attempt to reduce it.

In the second instance, Argyris (1977) and Simons (1995) suggested to shift from a single loop learning to a double loop learning. The use of diagnostic control systems—based on a single loop learning—implies that *mental models* (Denzau and Roy 2005), goals, and performance targets are not questioned. Hence, the methods used in organizations to achieve performance targets are not altered. Once the organizational goals are set, predefined procedures are executed. It makes the organizations perform actions with a high level of autonomy and homogeneity, which eliminates the need for communication among stakeholders. However, today's world is characterized by fluctuating dynamics. It requires a high level of flexibility and innovation to approach novel events. In these circumstances, the use of interactive control systems—based on a double loop learning—that implies the participation of multiple stakeholders and fosters strategic dialogue is essential to review the underlying principles of organizational strategies. Therefore, interactive control systems should also be implemented in organizations.

The interactive control systems embody additional features that complement the diagnostic control systems. In uncertain environments, the design of adaptative strategies intended to align forecasts with organizational actions is a priority. In these circumstances, interactive control systems represent a powerful ally since a focus on processes—rather than predefined outputs and outcomes—becomes relevant to organizational performance. Likewise, these control systems stimulate dialogue and learning in organizations to enhance their operations. Moreover, the design of compensation schemes based on diagnostic control systems may be easier to manipulate than the design of rewards based on interactive control systems, as incentives in the former group are determined by fixed measures. Conversely, interactive control systems rely on subjective rewards that are allocated according to the superior's judgments about employees' performance. Therefore, innovative behaviors—which are difficult to quantify by using preset indicators—may be taken into account. It may also foster a positive—also called reinforcing—feedback based on information sharing among internal and external stakeholders (Simons 1995).

In a similar vein, Ouchi (1979) remarked that organizational profiles will never exhibit the characteristics of just one control mechanism, as they are usually made up of distinct kinds of controls. In particular, he argued that three types of control mechanisms can be identified and implemented in organizations depending on the clarity and congruence between their goals and the adopted measures. They are described below.

Market control is based on norms of reciprocity by which sanctions are executed if the parties do not behave properly during the financial transactions. In this control mechanism, people are rewarded by their precise contribution to the organization, which in turn implies that individual goals may be pursued if a reward reduction is accepted. The market control is appropriate in situations where outputs can be identified and measured accurately, such as in economic activities influenced by price dynamics (Peterson's 2010).

Bureaucratic control embodies not only norms of reciprocity but also a legitimate authority that leads the subordinates' actions toward the achievement of organizational goals through a set of rules. Both performance assessment and agreement on hierarchical structures are major characteristics of this type of control. In addition, rewards are allocated depending on the superiors' interpretations about individual contributions to the organization. This is the most extended control mechanism in public and private organizations.

Clan control uses social agreements based on traditions, shared values, and beliefs to foster knowledge exchange among people (Tighe 2019), and to encourage individual efforts through limited supervision (Peterson's 2010). This control mechanism is the most demanding to be implemented because (1) the employees' view should be congruent with the mission and vision of the organization beforehand and (2) the managers should periodically reinforce the internalization of values and beliefs among subordinates by rituals and ceremonies (e.g., professional training). The clan control is usually used if performance control (Mintzberg 1979) and behavior control (Govindarajan and Fisher 1990) are not sufficient for the attainment of outcomes. In these circumstances, strict selection criteria and high levels of socialization are used to promote cooperative mechanisms intended to reduce goal incongruence between organization and individuals (Brenner 2009). Thus, the clan control is appropriate in flexible settings where discretionary power is central, as different methods can be used to carry out the tasks.

In 1980, after the publication of Ouchi's research about types of control mechanisms (Ouchi 1979), Michael Lipsky contributed to the top-down *vs* bottom-up debate in policy implementation by coining the concept of street-level bureaucracy. His work stressed on the discretionary power and relative autonomy of front-line workers (e.g., schoolteachers, doctors, police officers, and judges), and he made organizational authorities re-think how such employees were held accountable.

The next section will expand the discussion on the theory of street-level bureaucracy, focusing on professional recognition, need for trust, and challenges in measuring performance.

3.2 The Theory of Street-Level Bureaucracy

Street-level bureaucrats directly interact with citizens and decide how to allocate benefits and sanctions. According to Lipsky (1980), these employees should be seen as "the ultimate policymakers" because public policy becomes effective through their routines to deal with pressure and uncertainty. This scholar also denotes that public service workers often act in ways that contradict policy statutes as their job makes it impossible to achieve managerial expectations completely. Similarly, other researchers assert that front-line workers operate in environments where goals are ambiguous and the need for action is imperative. Therefore, they are forced to make choices between dissimilar rules and routes of action (Davis 1970; Blau 1955; Merton 1957). It implies that rule application and discretion require a high level of trust in public service workers' abilities to achieve institutional outcomes and deal with complex situations in creative modes (Fox 1974).

Classical theories aimed at explaining changes in policy outcomes assume a straightforward link between policy design and policy implementation. That is, strategic plans are often outlined by top-down approaches that have little concern about how the operational process is performed at the bottom of the hierarchical structure (Dodds 2013; Weber 2009). However, reality is complex and limitations on the underlying principles become clear when service delivery takes place. Therefore, overlooking the role of street-level workers is not only a mistake, but it is dangerous for the achievement of community goals (Lipsky 1980).

If outputs and outcomes matter, then the role of front-line workers should be taken into consideration as they shape public policies by interpreting rules and allocating resources among citizens (Meyers and Nielsen 2012). In particular, linear and vertical integration from policy design to policy implementation should not be assumed, because the conversion of higher goals into operational variables depends on heterogenous dynamics. This situation makes public workers' actions diverge from general and abstract procedures that attempt to achieve uniformity in the provided services (May and Winter 2009).

Street-level bureaucrats have direct responsibility for public service delivery, since they behave as mediators in the relationship between the state and citizens (Lipsky 1980). In addition, their actions exercise a powerful influence on the community as they make final decisions on how to socialize government plans and how to allocate human efforts during the phase of policy implementation (Lipsky 2010). Likewise, street-level bureaucrats' behaviors have a personal and immediate effect on the community, characterized by a focus on the individual. On the one hand, it implies that these employees may affect citizens' life opportunities through their attitudes, behaviors, and beliefs toward the system where they operate. On the other hand, it also implies that a high commitment to the use of human-centered approaches for dealing with clients' reactions is expected (Maynard-Moody and Leland 2000). In short, the objects of street-level bureaucrats' decisions are people. Initially, their actions should be judged only through lenses that favor personal attachment and subjectivity because these are the foundations of autonomy and discretionary power. Posteriorly, their actions should be taken in concert to strengthen agency behavior (Brock et al. 1999).

In contrast to lower-level workers, street-level bureaucrats (e.g., schoolteachers, doctors, police officers, and judges) have enough freedom to determine the quantity and quality of the services provided by their agencies. For instance, schoolteachers have the power to determine who will be suspended from classes and how educational content will be taught (Costenbader and Markson 1998; Lipsky 2010). This does not mean that organizational rules and regulations do not constrain street-level workers' behaviors. In fact, political and administrative officials shape the major dimensions in public policy and structure the policy choices for street-level bureaucrats. However, discretion and autonomy are fundamental characteristics in the jobs of public service workers as they have to carry out complex tasks for which the application of rigid rules and guidelines would seriously compromise the balance between compassion, flexibility, and impartiality (Tummers and Bekkers 2014). To illustrate this, teachers are not only expected to teach school contents, but they should also identify the unique potential in children and should adjust their pedagogical strategies to enhance student learning. It means that children require a specific educational response related to their context, which can be achieved only through an environment where discretion and autonomy are promoted (Maynard-Moody and Musheno 2003).

In summary, street-level bureaucrats should be perceived as public officials since (1) the nature of their job implies deciding on citizens' lives, (2) their role involves resource allocation of relevant goods and services for society, (3) the outputs from their work are a subject of public accountability, and (4) their professional identity is shaped to a large extent by social relations rather than rules (Maynard-Moody and Musheno 2003; Lipsky 2010).

3.2.1 Professional Recognition and Need for Trust

Street-level bureaucrats perceive themselves as professionals and they expect to be treated accordingly. In defining the concept of profession, DiMaggio and Powell (1983) connect the legitimacy for occupational autonomy with work conditions and methods. In particular, these authors affirm that profession is an activity exercised by people who share technical skills and who enjoy societal recognition (e.g., police officers, social workers, doctors, and schoolteachers). These individuals do not need formal routines, regular supervision from superiors, and other bureaucratic controls to work correctly. Instead, internalization and standardization of values and beliefs are the crucial factors to guarantee their commitment to clients (Lipsky 1980). Therefore, they are granted high discretionary power and autonomy in their decisions, but they are expected to achieve similar results to those coming from environments where constant audits are the norm (Mintzberg 1979). In addition, professionals often operate in decentralized settings, which makes it difficult to plan a broad organizational strategy. As a result, their performance is the product of individual initiatives that are characterized by an entrepreneurship spirit (Brock et al. 1999).

Front-line workers are professionals who have a culture consisting of instrumental practices, contents, and codes of conduct associated with their work context. These workers perceive themselves as "insiders" as they know the inner workings in service delivery (Lipsky 2010; Hupe and Hill 2007). Therefore, they are more inclined to listen to their peers, rather than politicians and managers, on how to implement policies and strategies. For example, Hill (2003) found that math teachers prefer to consult other educators on how to prepare pedagogical contents instead of following institutional guidelines and educational standards.

Monitoring daily practices of actors who work in human service organizations is difficult because of the nature of their job (Lettinga and Moulijn 2004; Freidson 2001; Hasenfeld 1983). Hence, managers attempt to gain control by limiting street-level workers' discretion through strict rules and rigorous procedures. However, when conflicts of interests between supervisors and their employees rise, the underlying assumptions that shape the achievement of agency objectives, trust, and the influence flows in the organizational structure must be questioned (Prottas 1978; Hupe and Hill 2007). It should entail a detailed analysis of street-level bureaucrats' priorities and working conditions, intending to change the reward system (Lipsky 1980).

Street-level bureaucrats have different job priorities than managers. These workers are prone to implement strategies for minimizing discomfort and for maximizing income and personal gratification. If their individual interests differ from the organizational goals and the rewards do not address their behaviors toward a preset direction, then one may expect a lack of compliance (Weatherley 1979; Resh and Pitts 2013). This is true when they do not have affinity with the organizational culture and beliefs. As a result, orders from the superiors might not be seen as legitimate. For instance, teachers who are asked to participate in school activities that they do not support will exhibit resistance to policy objectives defined by hierarchical authorities (Lipsky 2010).

Another source of conflict between street-level bureaucrats and managers is a dissimilar complexity in the interactions with clients. While managers follow strict principles of fairness and equity to gain social legitimacy, street-level workers exercise discretion to decide whether or not treating people alike (Hupe and Hill 2007). This comes from understanding that individuals are more complex than their bureaucratic characteristics (e.g., age, gender, income level, and location), and it could be perceived as unfair to deny such attributes and neglect personalized approaches (Lipsky 1980). For instance, most teachers recognize that certain groups of students have better students' learning results when receiving greater attention (Tummers and Bekkers 2014). Therefore, using managerial principles of fairness and equity in the classrooms—while constraining teachers' discretion and autonomy—might erode the achievement of educational outcomes in the long term.

Compared to other types of settings, front-line workers often operate in environments where the level of scrutiny is minimum, as discretion requires trust and freedom to be exercised (Tummers and Bekkers 2014; Rothstein and Dietlind 2001). For instance, school inspections are rare and educators are usually notified in advance when such assessments will take place. This allows them to prepare lessons according to specific criteria. On the one hand, this limited supervision reflects a substantial respect for professional standards in school bureaucracies. On the other hand, it contributes to exacerbate problems in measuring educational quality because of weak mechanisms for peer evaluation (Lipsky 2010).

Constraint in resources is an additional characteristic of front-line workers' environment. It happens when demand for services is greater than supply. This limitation is usually expressed by poor time allocations and high client–worker ratios (Lipsky 1980). The first constraint emphasizes on the need for processing high volumes of administrative tasks in an expedited mode, intending to bring the attention back to the essential commitments. For instance, teachers have to complete paperwork and write reports/proposals to school authorities before focusing on learning activities. The second constraint remarks a lack of organizational resources for providing personal treatment to clients. In the school context, the higher the student–teacher ratio is, the more time to keep order in the classroom will be. It may affect efficiency in service delivery due to a reduction in the time allocated to teaching (Lipsky 2010). Therefore, teachers must exercise discretion to decide the best pedagogical strategies to perform when there is resource scarcity; otherwise, educational outcomes might be eroded. Under these scenarios, educators are often forced to implement reductionist strategies for balancing output efficiency and client-centered practices (Meyers and Vorsanger 2003).

Public workers create shortcuts and routines to deal with uncertainty and make their tasks easier. This is an expected consequence of their work environment that is characterized by limited resources, high demands, and goal ambiguity. They adjust their patterns of practice to maximize the use of resources (e.g., time, information, and efforts) and constrain citizens' demand (Lipsky 1980; Portillo and Rudes 2014). For instance, teachers often adapt their teaching strategies and school contents to enhance the learning process of particular students (Tummers and Bekkers 2014). However, street-level bureaucrats may also stereotype and privilege some clients over others as an effort to make "manageable" the pool of requests. This might cause creaming strategies that affect overall performance because street workers might focus their efforts on "more achievable targets" to gain social credibility and/or financial benefits (Vedung 2015).

Street-level workers will display resistance to managerial rules when they feel their sense of autonomy is affected, since their job requires freedom in how the service is delivered. It implies that they may perceive administrative and political efforts aimed at dictating service norms and at constraining individual actions as illegitimate (May and Winter 2009). To deal with this issue, public service workers will engage in gaming behaviors to maximize their pay-off according to the adopted measures. However, this dysfunctional attitude will only increase the gap between individual and organizational interests (Resh and Pitts 2013). Therefore, the way educators experience the system strongly influences their professional practices. In detrimental circumstances, it could lead to relevant discrepancies between outputs and outcomes in both institutional and inter-institutional levels.

3.2.2 Challenges in Measuring Performance

Street-level bureaucrats mostly operate in two social dimensions to perform their tasks and achieve organizational goals. *The first dimension* refers to horizontal relationships with peers who share similar duties and colleagues who have common interests even if they work in different fields. Intra-measures that reflect the judgment of both professional practices and conducts within a vocational association hold them accountable. *The second dimension* refers to vertical relationships with citizens, public managers, and appointed superiors. Inter-measures are used to gauge the quantity and quality of the delivered services, the synergy displayed among different internal and external stakeholders, and citizens' opinions as an expression of democracy (Lipsky 1980).

In contrast to business organizations, bureaucracies cannot be measured by market transactions. The rationale behind this statement comes from the idea that outcomes go beyond from what we could describe through a simple balance sheet. To be more precise, in contexts where organizational goals include social measures, and discretionary power is central for an effective service delivery, the accomplishment of strict rules and procedures could be highly detrimental (Lipsky 2010).

Role conflict and goal ambiguity are associated with the complex structure of street-level bureaucrats' jobs. If citizens cannot agree on what to expect from bureaucrats, then the latter will be less responsive to community orientations and will be prone to act by internally defined objectives (Resh and Pitts 2013). In the school context, there are several examples that illustrate this situation. Some parents want their children to be taught basic learning skills in the classrooms, whereas others prefer an education oriented to raise civic awareness. Likewise, some schools promote the preservation of the racial heritage, whereas others attempt to homogenize students who have different backgrounds (Lipsky 1980). Similarly, many parents conceive discipline as a crucial factor in fostering academic attitudes for their children, whereas others are in favor of more liberal approaches (Lipsky 2010).

Organizational evaluation in street-level bureaucracies requires a certain clarity in terms of goals since performance measurement in social context implies an operationalization of heterogenous views. This is not an easy task to perform because public workers must constantly deal with goal ambiguity and conflicting rules. Hence, results are often gauged without having clarity of the behaviors that were used to achieve them (Brodkin 2008). This implies that quantitative measures rarely reflect the rationales by which people are judged, because such measures do not support the qualitative content of the bureaucrats' actions (Lipsky 2010). Therefore, a management-by-enabling, instead of a management-by-incentive, approach is suggested (Brodkin 2011, 2012; Elmore 1979).

Despite the above difficulty in gauging performance on street-level bureaucracies, indicators are widely used to quantify individual and agency behaviors. For instance, graduation rates, attendance rates, and test scores are common measures to evaluate teachers and schools (Swanson et al. 2003; Fuhrman 1999). Problems emerge when such measures do not reflect actual changes in overall performance since they are not correctly aligned with institutional outcomes and they substantially limit public service workers' autonomy. This situation makes bureaucrats adjust their behaviors to maximize outputs and related benefits as a mechanism to keep a "safe" distance from organizational controls (Bianchi and Salazar Rua 2020). Therefore, measures do not always affect behaviors in the desired direction, and sometimes they might even be detrimental to long-term objectives in social contexts.

In particular, numerous side effects—such as goal conflict and dysfunctional behaviors—rise due to the discretionary power and autonomy of street-level bureaucrats, which hinders organizational controls (Resh and Pitts 2013; Lipsky 1980). This condition would make one think that the removal of both attributes would lead to a final solution. However, the activities that these employees must perform in public and private organizations are complex, and therefore narrowing the job substantially or setting rigid rules is not an optimal response. Both approaches are not sustainable in the long term because they cannot predict all behavioral scenarios that may appear during the execution of the activities (Portillo and Rudes 2014; Lipsky 2010). For instance, in education, every child is unique and has distinct learning needs. It makes almost impossible to implement a detailed instructional approach in all circumstances (Lipsky 2010). This implies that schoolteachers should use their discretion to decide the most suitable strategies for helping each student develop his or her skills.

To sum up, the discretionary power of street-level bureaucrats has two faces. On the one hand, it concedes substantial flexibility to state agents to answer citizens' demands. Discretion allows them to have more accurate responses to manage particular situations and provide personalized services. For instance, in education, this characteristic is central to enhance service delivery because abstract uniformity cannot recognize special needs of a certain group of students (Lipsky 2010). On the other hand, freedom to implement policies may lead to arbitrariness, unfairness, and corrupted behaviors (Della Porta and Meny 1997; Rothstein and Dietlind 2001). For this reason, some scholars stress on the importance of well-designed performance measurement systems and accountability criteria (Bianchi 2010; Hupe and Hill 2007).

Using an outcome-oriented view in outlining school performance measurement systems through dynamic performance management (DPM) is congruent with the ideas discussed above, and it allows one to pursue sustainable outcomes in the long term. In this book, DPM is used to frame problems with the use of output-oriented performance measurement S=systems (PMS) in Colombian public schools and to support the design of a robust set of performance measures based on an outcome-oriented view. The next chapter will cover these topics in detail by briefly explaining the DPM framework and applying it to the Colombian case study. Initially, the DPM-based model with the simulation runs will be analyzed. And posteriorly, conclusions, contributions to the body of knowledge, and research limitations will be discussed.

References

Argyris C (1977) Double loop learning in organizations. Harv Bus Rev September–October:115–124

Bianchi C (2010) Improving performance and fostering accountability in the public sector through system dynamics modeling: from an "external" to an "internal" perspective. Syst Res Behav Sci 27:361–384

Bianchi C, Salazar Rua RS (2020) A feedback view of behavioural distortions from perceived public service gaps at 'street-level' policy implementation: the case of unintended outcomes in public schools. Syst Res Behav Sci:1–22

Blau PM (1955) The dynamics of bureaucracy. Am Polit Sci Rev 49(4):1212

Brenner B. Management control in Central and Eastern European subsidiaries. 2009.

Brock D, Hinings CR, Powell M (1999) Restructuring the professional organisation: accounting, healthcare, and law. Routledge, London

Brodkin EZ (2008) Accountability in street-level organizations. J Public Adm 31(3):317–336

Brodkin EZ (2011) Policy work: street-level organizations under new managerialism. J Public Adm Res Theory 21(2):i253–i277

Brodkin EZ (2012) Reflections on street-level bureaucracy: past, present, and future. Public Adm Rev 72(6):940–949

Costenbader V, Markson S (1998) School suspension: a study with secondary school students. J Sch Psychol:59–82

Davis KC (1970) Discretionary justice. J Leg Educ 23(1):56–62

Della Porta D, Mény Y (1997) Democracy and corruption in Europe. Social change in Western Europe. Pinter, Washington DC

Denzau AT, Roy RK (2005) Mental models and game theory: cognitive constructions of multiple Nash equilibria. Presented at Public Choice Society, March

DiMaggio PJ, Powell WW (1983) The iron cage revisited: institutional isomorphism and collective rationality in organizational fields. Am Sociol Rev 48(2):147–160

Dodds A (2013) Comparative public policy. Palgrave Macmillan, Houndmills, Basingstoke, Hampshire

Elmore RF (1979) Backward mapping: implementation research and policy decisions. Political Sci Quart 94(4):601–616

Flamholtz EG (1983) Accounting, budgeting and control systems in their organizational context: theoretical and empirical perspectives. Acc Organ Soc 8(2/3):153–169

Fox A (1974) Beyond contract: work, power and trust relations. Farber & Farber, London

Freidson E (2001) Professionalism, the third logic: on the practice of knowledge. University of Chicago Press, Chicago

Fuhrman SH (1999) The new accountability. CPRE Policy Brief No. RB-27. Consortium for Policy Research in Education, University of Pennsylvania, Philadelphia, pp 1–14

Govindarajan V, Fisher J (1990) Strategy, control systems, and resource sharing: effects on business-unit performance. Acad Manag J 33(2):259–285

Hasenfeld Y (1983) Human service organizations. Prentice-Hall, Englewood Cliffs, NJ

Hill HC (2003) Understanding implementation: street-level bureaucrats' resources for reform. J Public Adm Res Theory 13(3):265–282

Hupe P, Hill M (2007) Street-level bureaucracy and public accountability. Public Adm

Kaplan RS, Norton DP (1992) The Balanced scorecard measures that drive performance. Harv Bus Rev 70(1):9

Kuhlmann D (2012) The interaction of contract, control, and relational norms as governance mechanisms in IS outsourcing relationships. Editor: Diplomarbeiten Agentur

Lawler EE, Rhode JG (1976) Information and control in organizations. Goodyear Publishing Company, Pacific Palisades

Lettinga B, Moulijn M (2004) Frontline professionals in the age of institutional rearrangement: the case of needs assessment for public services. Public Performance and Management Review. Routledge, London

Lipsky M (1980) Street-level bureaucracy: dilemmas of the individual in public services. Polit Soc 10(1):116

Lipsky M (2010) Street-level bureaucracy: dilemmas of the individual in public services. 30th Anniversary expanded edition. Russell Sage, New York

Maciejczyk A (2016) Challenges of control in functional organization structures: example of outsourcing sector. J Econ Manag 25:49–62

May PJ, Winter SC (2009) Politicians, managers, and street-level bureaucrats: influences on policy implementation. J Public Adm Res Theory 19(3):453–476

Maynard-Moody SW, Leland S (2000) Stories from the front lines of public management: street-level workers as responsible actors. In: Brudney A, O'toole J (eds) New developments in theory, methods, and practice, pp 109–123

Maynard-Moody SW, Musheno MC (2003) Cops, teachers, counselors: stories from the front lines of public service. University of Michigan Press

Merton RK (1957) The role-set: problems in sociological theory. Br J Sociol 8(2):106–120

Meyers M, Nielsen V (2012) Street-level bureaucrats and the implementation of public policy. In: The SAGE handbook of public administration. SAGE, UK, pp 305–318

Meyers MK, Vorsanger S (2003) Street-level bureaucrats and the implementation of public policy. In: The SAGE handbook of public administration. SAGE, UK, pp 245–256

Mintzberg H (1979) The structuring of organizations. A synthesis of the research. Prentice Hall, Englewood Cliffs, NJ

Ouchi WG (1979) A conceptual framework for the design of organizational control mechanisms. Manag Sci 25:833–848

Peterson's (2010) Peterson's Official Guide to Mastering Dsst Exams

Portillo S, Rudes DS (2014) Construction of justice at the street level. Annu Rev Law Soc Sci 10(1): 321–334

Prottas JM (1978) The power of the street-level bureaucrat in public service bureaucracies. Urban Aff Rev 13(3):285–312

Resh WG, Pitts DW (2013) No solutions, only trade-offs? Evidence about goal conflict in street-level bureaucracies. Public Adm Rev 73(1):132–142

Rothstein B, Dietlind S (2001) Social capital and street-level bureaucracy: an institutionnal theory of generalized trust. Center for Study of Democratic Politics, Princeton University, pp 1–49

Simons R (1995) Levers of control: how managers use innovative control systems to drive strategic renewal. Harvard Business School Press, Boston

Swanson DP, Cunningham M, Spencer MB (2003) Black males' structural conditions, achievement patterns, normative needs, and opportunities. Urban Educ 38(5):608–633

Tighe S (2019) Rethinking Strategy: how to anticipate the future, slow down change, and improve decision making, 1st edn

Tummers L, Bekkers V (2014) Policy implementation, street-level bureaucracy, and the importance of discretion. Public Manag Rev 16(4):527–547

Vedung E (2015) Autonomy and street-level bureaucrats' coping strategies. Nordic J Stud Educ Policy 2015(2)

Weber M (2009) From Max Weber: essays in sociology. In: Gerth HH, Wright Mills C (eds) Routledge classics in sociology. Routledge, Milton Park, Abingdon, Oxon

Weatherley R (1979) Reforming special education: policy implementation from state level to street level. MIT Press, Cambridge, MA, pp 1–21

Chapter 4
Using Dynamic Performance Management to Outline Outcome-Oriented Performance Measurement Systems in Public Schools

4.1 Dynamic Performance Management (DPM): An Instrumental View

DPM contributes to the enhancement of performance measurement systems (PMS) by providing decision-makers with effective tools to frame how indicators change over time for the implemented policies. In particular, DPM can enhance PMS by trade-off analysis in time and space (Bianchi and Williams 2015). Whereas trade-offs in time emphasize the effects of policies on different time horizons, trade-offs in space highlight the effects of policies on different subsystems. This distinction can help policymakers to (1) raise awareness about relevant delays in the system, (2) contextualize performance measures, and (3) broaden system boundaries through an inter-institutional perspective (Bianchi and Rivenbark 2014).

Figure 4.1 shows how DPM can support policymakers to outline sustainable policies by linking strategic resources with performance drivers, which in turn feed back into strategic resources through end results. This approach increases the understanding of complex dynamics in public organizations, improves the design of PMS, and fosters a cultural change toward an outcome-oriented view in performance by implementing modeling and simulation techniques from the field of system dynamics (SD). In particular, insight system dynamics models are used, and they provide the basis for more detailed modeling efforts as substantial data and time are available (Bianchi et al. 2008; Xavier and Bianchi 2019; Wolstenholme 1999; Ghaffarzadegan et al. 2011). Therefore, DPM could be seen as a bridge between two fields with differing tools and terminologies (i.e., system dynamics and performance management), which improves performance reports, communication among stakeholders, and the delivery of sustainable community outcomes (Bianchi 2021, 2022).

DPM charts illustrate strategic resources as tangible and intangible stocks (Morecroft 2007; Warren 2008) that decision-makers can use to develop their

R. Salazar Rua, *Improving Consistency in Performance Measurement System Design*, System Dynamics for Performance Management & Governance 7,
https://doi.org/10.1007/978-3-031-24699-9_4

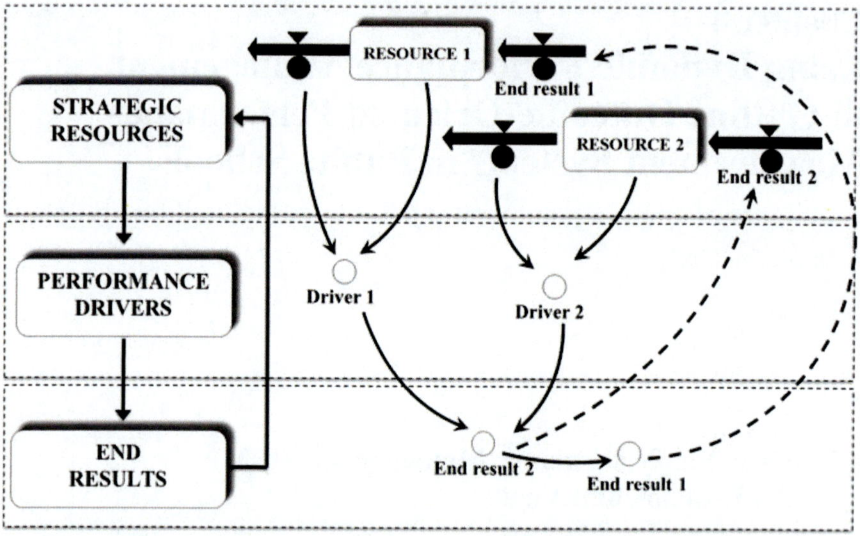

Fig. 4.1 An instrumental view of dynamic performance management (Bianchi 2016, p.73)

policies and meet performance targets. The availability of such resources varies depending on decision-makers' actions. On the one hand, deliberate decisions (e.g., staff hiring) will affect the change of strategic resources that can be obtained from the market. On the other hand, the outcomes of the implemented policies (i.e., end results) will affect the amount of such resources that cannot be acquired from the market (e.g., human capital, trust, image). These end results are symbolized as flows in Fig. 4.1. They accumulate in strategic resources, which are symbolized as stocks.

To some extent, strategic resources can be controlled individually. However, if an unbalanced growth of them becomes the rule, then organizations will not reach their maximum potential. Changes in the strategic resources provide a partial view about DPM. Therefore, critical success factors for achieving the end results (i.e., performance drivers) must also be identified to understand how the system works from a holistic perspective. Performance drivers are gauged as ratios between the current and the desired state of strategic resources (e.g., the student–teacher ratio divided by a benchmark). These drivers should be measured and tracked continuously to recognize symptoms of failure and enhance the attainment of end results (i.e., outputs and outcomes). Finally, end results feed back into strategic resources and the cycle of performance is repeated (Bianchi et al. 2017).

Learning in-and-about complex social systems (Sterman 1994) is possible through DPM. In particular, DPM describes organizational performance by making explicit the feedback structures, the mental models, and the decisions. This approach allows policymakers to deal with symptoms of crisis, such as the side effects of an inconsistent design of indicators. For instance, schoolteachers may narrow school curricula by focusing only on tested subjects to improve student performance in high-stake tests. However, this strategy may be detrimental for the achievement of school and community outcomes in the long term (e.g., educational quality, civic-mindedness, social awareness), as notion-based learning is prioritized over holistic education.

Traditional approaches in performance measurement often lead to a narrow and static view in data analysis and reporting, which often overlooks complexity associated with social issues, that is, how system's variables relate to each other from a causal dimension (Bianchi 2016; Bianchi et al. 2017). This implies that an output, rather than an outcome, view is mostly used to hold public schools accountable. DPM can support educational authorities in designing consistent PMS by using SD modeling and simulation. In addition, DPM charts can be used as a framework to foster policy discussion in education by outlining a shared view among multiple stakeholders through the identification of strategic resources, performance drivers, and end results.

Furthermore, DPM can enhance the comprehension of how dysfunctional behaviors affect school performance through an outcome-oriented view (Bianchi and Williams 2015). This view is crucial to avoid (1) the illusion of control as a result of a linear thinking (Langer and Roth 1975), (2) goal displacement due to an inversion between means and ends (De Lancer Julnes 2006; Sills 1957), and (3) performance adaptation as people learn how to game the measurement system, which causes that the indicators lose their capability to differentiate between low performers and high performers (Van Thiel and Leeuw 2002; Meyer and Gupta 1994).

In this book, DPM has been used to support the design of outcome-oriented performance measurement systems intended to prevent and counteract the emergence of behavioral distortions in Colombian public schools. To this end, a DPM simulation model—which takes into account trade-offs, non-linearities, and time delays—has been built.

In the Colombian case study, a broader perspective in school performance measurement systems is needed to (1) extend the boundaries of performance indicators from test scores to an assessment of holistic learning outcomes and (2) align external targets—set by the government—with measures that take into account the socio-economic background where public schools operate. An instrumental view of DPM is proposed to implement this perspective and pursue the achievement of sustainable outcomes. Thus, performance may be affected positively by focusing on end results (i.e., outputs and outcomes), performance drivers, and strategic resources.

The next section will explain the field research and the model building approach that were carried out in this study to be transparent on how the data were collected and how the DPM-based model was built.

4.2 Data Collection and Model Building Approach for the Colombian Case Study

A sequential exploratory design was used in this research. This is a type of mixed-methods strategy that consists of two phases to gather the primary data. The first phase involved the collection of qualitative data by means of semi-structured interviews and a questionnaire based on open-ended questions. The second phase involved the collection of quantitative data by further interviews, a second questionnaire based on numerical reasoning, and sessions for model validation.

In addition, the secondary data were collected and interpreted by document review. In the first instance, open government data, local newspapers, and school performance reports were examined. In the second instance, analysis of the scientific literature in the fields of performance management and education was carried out. As a result, several simulation parameters were estimated based on secondary data (Luna-Reyes and Andersen 2003).

In the first stage of this research, a preliminary field study—conducted through semi-structured interviews and a questionnaire—was carried out to determine how stakeholders in education impact on educational quality (Salazar Rua 2016). To guarantee the participation in this study of the main stakeholders in the Colombian education system (i.e., families, schools, government, and enterprises), a purposive sampling was used.

The sampling procedure is described below.

1. Colombian public schools—located in municipalities with different degrees of economic development—were chosen to carry out this research. For the highest category of economic development, schools from Bogota and Barranquilla were selected. For the intermediate category of economic development, schools from Monteria and Pereira were chosen. And for the lowest category of economic development, schools from Sabaneta and Sabanalarga were selected.
2. From the above-mentioned municipalities, several public schools were chosen. The selection criteria were (1) student–teacher ratios and test scores within the national average, (2) medium-sized schools (about 500 students), and (3) institutes with an educational offer for the school levels of interest (i.e., elementary and secondary school). A total of six public schools were involved in this stage.
3. Schoolteachers were asked to fill out a questionnaire based on open-ended questions (see Appendix B). In addition, several teachers agreed to be interviewed in order to share their experiences. A total of 20 schoolteachers completed the questionnaire.
4. Other stakeholders in the education system were also involved in this research. In particular, the sample included one person responsible for analyzing test results at "the Colombian Institute for the Promotion of Higher Education," two representatives from a local enterprise in Barranquilla, and four parents whose children study in the participating public schools.

In the second stage of this research, a questionnaire—based on open-ended questions and numerical reasoning—aimed at understanding the effects of standardized tests on the behavior of Colombian schoolteachers was administered (See Appendix C). In addition, facilitated sessions were performed in one of the public schools, intending to receive further help in the model validation phase. A total of 3 teachers attended the sessions, and 12 teachers filled out the questionnaire. In particular, the analysis of the collected data contributed to set the initial values of different variables in the DPM-based model, such as the teaching time allocated to traditional teaching and to holistic education and the time affecting the inflow and outflow of the stocks of higher and lower knowledge students. Thus, an experimental approach in building the simulation model was implemented.

This research design was chosen because it was appropriate to (1) answer the research questions and (2) triangulate the data in order to enhance the validity of the findings (Saunders and Lewis 2012). In particular, the information collected—through questionnaires, interviews, school reports, open government data, and model validation sessions—was used to estimate a range of values that performance variables and strategic resources would realistically portray in a medium-sized school for the investigated region. However, this was an exploratory research, which implies that the results should not be generalized since it used a small sample of individuals.

Concerning the model building approach, prior knowledge of the problem context was obtained by means of a preliminary field study carried out through semi-structured interviews with schoolteachers (Salazar Rua 2016). In particular, Colombian teachers perceive the assessment system as a potential disadvantage for schools in poor areas since it does not consider how the socio-economic conditions of the locations where schools operate affect institutional performance. In order to cope with this issue, schools have been inclined to increase their scores by allocating more teaching time to a narrow range of subjects, at the expense of others. As an unintended consequence of this unbalanced time allocation, the outcomes have been dropping despite slight increases in the scores. The risks of behavioral distortions, caused by such systems, require the use of proper methods that help the government design "robust" school performance measurement systems.

To start a dialogue with schoolteachers about the factors causing the described phenomena, two versions of a DPM-based model, which point out the interplay between strategic resources, performance drivers, outputs, and outcomes, were built. These versions of the model were focused on the unintended consequences from the use of inconsistent performance measurement systems and on the policy recommendations to counteract dysfunctional behaviors.

To this end, a focus on an empirical analysis—based on both primary and secondary data—was used. In particular, an exhaustive literature review, 12 questionnaires, 3 interviews with Colombian schoolteachers, document analysis, and sessions for model validation with the support of experienced teachers were carried out.

The model building process followed three steps.

In the first step, a problematic behavior was outlined by taking into account the literature review. In parallel, a causal loop diagram (CLD) was sketched to frame the feedback structure that describes the emergence of behavioral distortions in public schools due to the use of standardized tests.

In the second step, a DPM chart was developed to illustrate how the system structure—portrayed by the CLD—may support an understanding of the effects of myopic education policies on school performance. Two sessions of 3 h each were carried out to discuss both the CLD and the DPM chart with a group of three experienced teachers in a Colombian medium-sized school (about 500 students). The sessions raised several suggestions for improvement of both documents. The teachers, who attended the two facilitated sessions, were selected based on three main criteria: (1) qualification and experience, (2) involvement in school performance assessment, and (3) the status of the school, which in this case was of a public institution in a poor area of Barranquilla, Colombia.

In the third step, model structure and behaviors were validated—through group and individual meetings—to capture divergent ideas as suggested by Andersen et al. (2012). This analysis helped consider new insights, which led to iterative modifications in the initial model structure, parameter initialization, and simulation time. The validation of the model structure also contributed to test the hypotheses embedded in the CLD. In particular, building reliability in the DPM-based model was a gradual process that emerged as a result of the modeling experience during the facilitated sessions. Moreover, a qualitative, semi-formal, and non-technical process was used to assess the usefulness of the simulation model to foster policy discussion (Barlas 1996; Bianchi 2016).

Modeling strategy and simulation results were portrayed by following "the three C's" (Zhang and Shaw 2012). Therefore, completeness in data reporting, clarity in the process to attain research results, and credibility in such results by using a solid scientific reasoning were central to this investigation (See Appendix D).

The primary goal of the model was not to replicate a reference mode with actual data because such data were not available in detail due to the recent adoption of the test-based accountability system in the country. The model was rather intended to raise Colombian teachers' awareness about possible unintended consequences of policies focused on narrowing curricula to improve standardized test performance.

In the next section, the case of the Colombian public schools is used to frame problems triggered by the use of inconsistent performance measures.

4.3 Framing Behavioral Consequences from Colombian School Assessment Systems Through Dynamic Performance Management

According to the scientific literature, an excessive pressure to improve performance in standardized tests has caused the emergence of behavioral distortions in public schools (Hamilton et al. 2002). Such distortions are (1) narrowing the school curricula to focus the teaching practice only on tested subjects, (2) devoting a large portion of time in exam preparations, and (3) designing classroom assessments with a format similar to the standardized tests. These dysfunctional strategies may produce the "score inflation effect"—which implies a short-term increase in test scores but a long-term decrease in them—because higher-order thinking skills are not built appropriately (Fuller 2004 cited in Center on Education Policy 2009; Linn 1998; MassPartners for Public Schools 2005).

In Fig. 4.2, a causal loop diagram (CLD) outlines the effects associated with behavioral distortions in public schools due to perceived low performance ratios when static performance measurement systems are used.

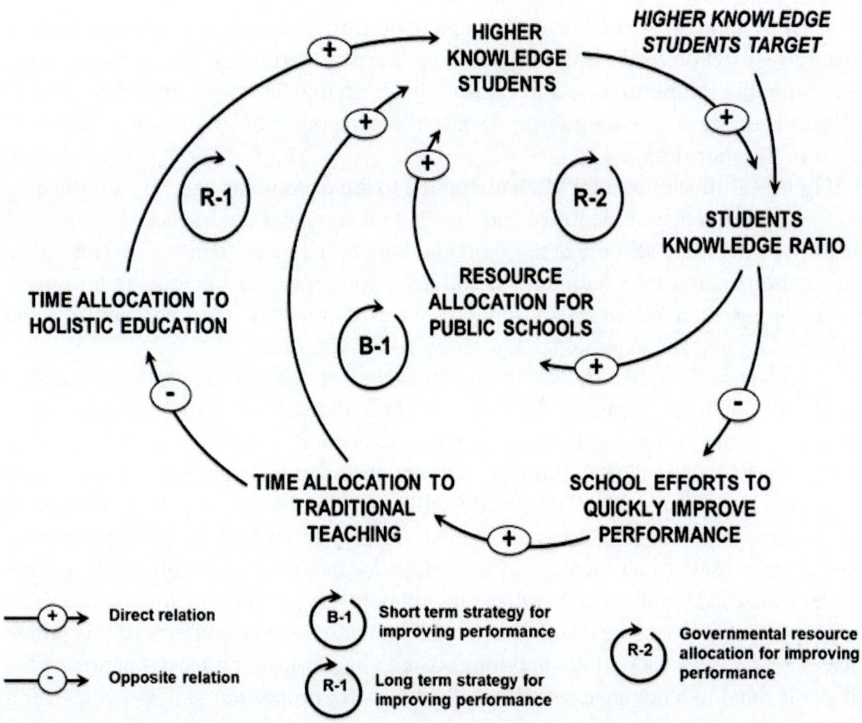

Fig. 4.2 Impact of government evaluation systems on school policies (Salazar Rua 2020, p.64)

A perceived low performance ratio may affect how teaching time is allocated in public schools. "Students' knowledge ratio" is the quotient between higher knowledge students (i.e., students who have acquired the knowledge and skills to approve the standardized tests) and a desired value. If such a ratio decreases, schoolteachers must face high pressure levels to improve student performance. Therefore, in the short term, an initial increase in the number of higher knowledge students (balancing loop "B1") is perceived because of the "score inflation effect" (Hamilton et al. 2002).

However, the above results are not sustainable in the long term since less time is allocated to pursue outcomes through a "holistic education." Therefore, a reduction in the number of higher knowledge students will be observed, which in turn will lead to lower perceived performance ratios (reinforcing loop "R1"). The reinforcing loop "R2" shows how a decrease in the student knowledge ratio generates a further reduction in the resources that the government allocates to such schools. This produces a lack of investments, which in turn causes lower performance levels.

In brief, if school activities are primarily based on traditional (i.e., notion-based) teaching, then a significant reduction in the time to build higher order thinking skills (e.g., critical thinking, problem-solving, and teamwork aptitudes) will be perceived. As a result, the number of higher knowledge students will decrease in the long term because the students will not be able to sustain high scores in the exams. Therefore, the acquisition of concepts in advanced stages of the student career will also be hindered due to a lack of accumulated knowledge.

A DPM approach has been used to model how government evaluation systems—mainly focused on test scores and the assessment of traditional teaching outputs—led Colombian public schools to narrow the curricula. This situation may weaken student learning in the long term, despite the perception of performance improvement in the short term.

Figure 4.3 illustrates a DPM chart applied to the Colombian case. By following a bottom-up sequence, it identifies end results to understand the impact of behavioral distortions in public schools in the short and long term. It also identifies performance drivers influencing the accumulation and depletion processes of strategic resources. By mapping such resources, it supports an assessment of the gaps between the current state and the external targets set by the government.

The change in higher knowledge students is affected over time by the dynamics of the following performance drivers: "% time allocated to tested subjects" (i.e., number of hours allocated to tested subjects over total school hours) and "balance and scope of learning" (i.e., number of hours allocated to tested subjects over time allocation benchmark). Such drivers describe a policy through which a reduction in the student knowledge ratio (i.e., number of higher knowledge students over a benchmark) leads to an increase in the time allocated to tested subjects. When the student knowledge ratio is lower than one, schools react by looking for an immediate improvement in test scores. However, in the long term, the side effects of this policy may negatively impact on the previous score gains. In fact, a greater fraction of the time allocated to a bounded set of tested subjects (in proportion to the total teaching time) would increase and worsen the balance and scope of learning.

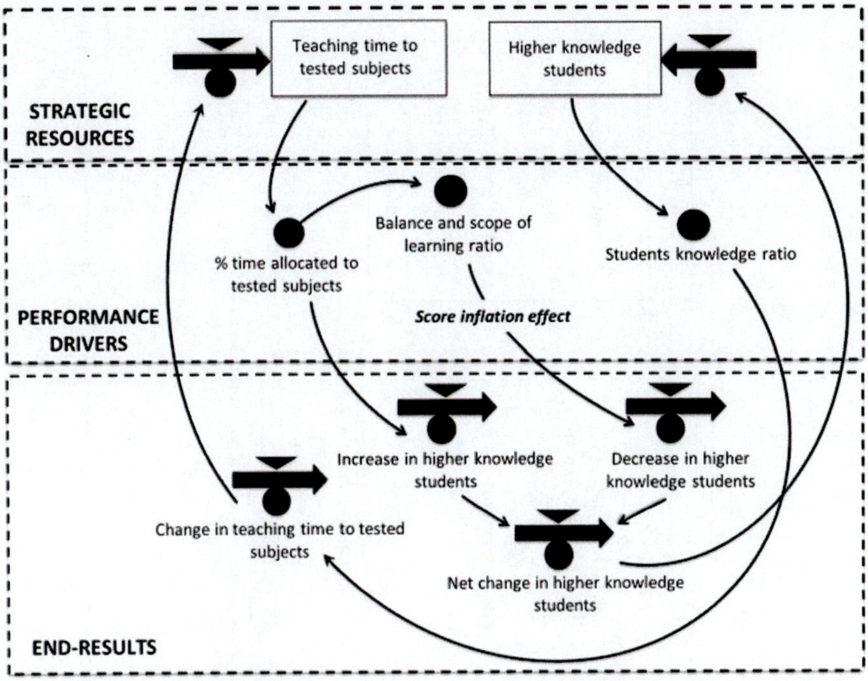

Fig. 4.3 Effects of myopic policies aimed at increasing the number of higher knowledge students in Colombian public schools (Bianchi and Salazar Rua 2020, p.12)

This would generate the "score inflation effect," rather than an improvement in student aptitudes and skills.

In the long term, also for those students who have initially been successful in their test scores, such "surface learning" (Marton and Saljö 1976a, b) could not sustain the achieved level in "higher knowledge students."

The next section will strengthen the previous analysis by discussing the DPM-based simulation model built for the Colombian case study. In the first place, problematic behaviors will be portrayed, and later, policy recommendations will be provided.

4.4 Using a DPM-Based Simulation Model to Analyze and Counteract Behavioral Distortions in Colombian School Assessment Systems

Figure 4.4 shows a DPM-based model designed to prevent and mitigate negative effects of school strategies based on an output-oriented approach. The main purpose of this model is to improve the understanding of complex dynamics rather than

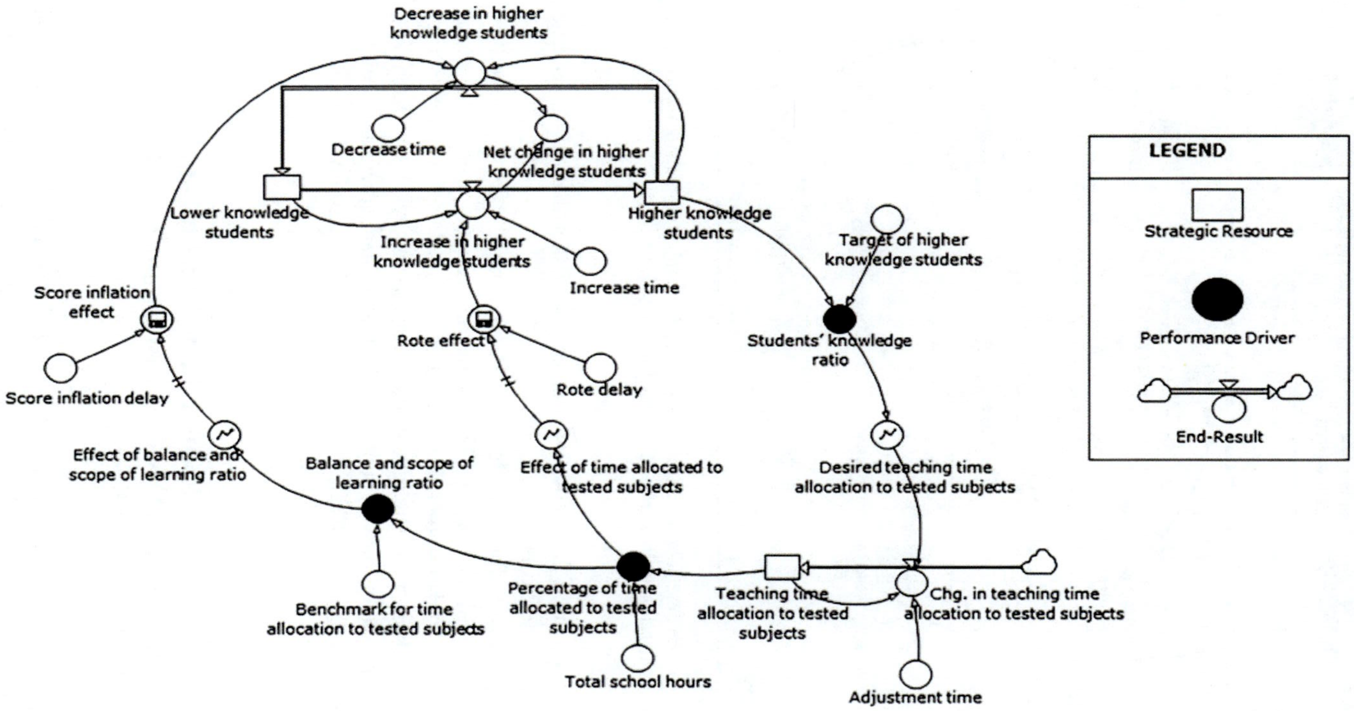

Fig. 4.4 DPM-based model illustrating the effects of myopic policies aimed at increasing the number of higher knowledge students in Colombian public schools (Salazar Rua 2020, p.69)

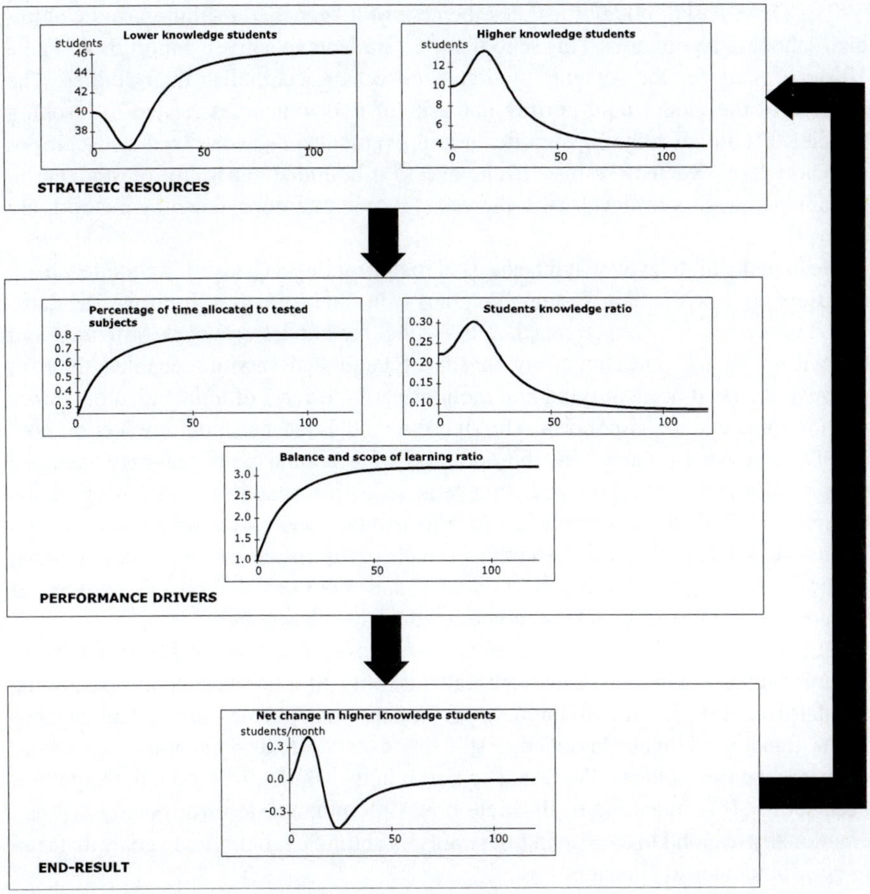

Fig. 4.5 Results from a simulation run illustrating the effects of myopic policies aimed at increasing the number of higher knowledge students in Colombian public schools (Salazar Rua 2020, p.70)

mimicking the behavior of historic time series (Bianchi and Winch 2006). Therefore, the parameters it embodies and the results it generates through simulations should not be taken as accurate estimates. Instead, the model can be used to develop an understanding of how feedback system structures affect system behavior.

Figure 4.5 portrays the results from a simulation run of a DPM-based model that was constructed by applying the previous myopic policies. The simulation has been run over a time horizon of 120 months (10 years). This time extension has been set to capture the unintended side effects of the current policies on learning outcomes.

The DPM-based model illustrates the case of a "generic" medium-sized Colombian school in a poor area. This school has 50 students in a given cohort. It will take 10 more years for the students of such a cohort to accomplish their studies. The purpose of the model is to portray the risks of poor outcomes related to teaching policies that aim at attaining significant improvements in standardized test scores in the short term. Such risks may occur due to a bounded capability of students to sustain high test scores levels over the years, if only traditional teaching methods are adopted.

Despite the high level of synthesis used in this model, a change in schoolteachers' mindsets on how to attain sustainable goals from their efforts to improve standardized test scores was experienced. Discussing the model feedback structure and behavior with the participants by means of facilitated sessions enabled them to perceive the need of identifying and monitoring the drivers of unintended outcomes from myopic education policies. Through the model, teachers could reflect on how silent phenomena—caused by the adoption of irrational policies—may generate poor results in the medium and long term (e.g., prioritizing rote learning to the detriment of holistic education in order to increase scores in standardized tests). Therefore, keeping the time behavior of such performance drivers under constant control may help schoolteachers detect and counteract early symptoms of unintended and unsustainable outcomes of the education policies.

The school described by the model has been unsuccessful in the past in achieving its targeted level of higher knowledge students: only 20% of total students passed the standardized tests. In order to improve performance and pursue a significant increase in the fraction of higher knowledge students over the total population, the school increases the time allocated to tested subjects. In the simulation model the target has been set to 90%, intending to illustrate how Colombian schools in poor areas may emotionally respond to the gap in test results by setting overambitious goals that lead to poor outcomes in education.

As an effect of this policy, in the initial 16 months the number of students who passed standardized tests increases, leading to a positive and rising net change in the number of higher knowledge students. It generates an improvement in the "students' knowledge ratio." However, since the level of such a ratio is still below the target, the school continues to divert the teaching time to the subjects on which standardized tests are done. This implies a prioritization of traditional teaching over holistic education. This phenomenon is captured by the increase in the "balance and scope of learning ratio": the higher this ratio is, the lower students' holistic learning will be, because of the diminishing residual teaching time allocated to inter-disciplinary projects, teamwork, and student portfolios. As a result, after the ninth simulation month, the net change in higher knowledge students progressively decreases and becomes negative after the 15th month. Such unintended outcomes are only latent until the 16th month. After that time, they become clear through a progressive decline in the number of higher knowledge students.

The difficulty to perceive such a phenomenon is associated with the feedback structure of the system, and namely, with the delays affecting the two flows that lead to the net change in higher knowledge students. It is also associated with the diminishing returns of additional teaching hours allocated to tested subjects. The understanding of such counterintuitive behavior portrayed by the dynamic complexity of the analyzed system is a benefit of using a DPM-based model to enhance performance management.

In Fig. 4.6, the model in Fig. 4.4 has been extended to show how an outcome-oriented view in performance management may positively affect the teaching time allocation in public schools. It has been proposed to assess outputs and outcomes from both traditional (i.e., notion-based) and holistic (i.e., project-based) education. While the former concerns the development of lower order thinking skills through rote learning and exam preparations, the latter refers to the development of higher order thinking skills (e.g., critical thinking, problem-solving, teamwork aptitudes) that will allow students to solve real-life problems.

Therefore, the suggested policy will allow public schools to complement the teaching of tested subjects (e.g., math, languages, social, and natural sciences) with the development of *deep learning* by the students, through activities such as interdisciplinary projects, group work, and student portfolios. Deep learning is the product from the implementation of a holistic education, which has been proved to be effective for enhancing performance in high-stake tests (Parker et al. 2011; Scogin et al. 2017; Ross et al. 2001; Levine 1994; Thomas 2000).

Figures 4.7 and 4.8 portray an explanatory causal loop diagram and the results from a simulation run of the extended model. According to the simulation results, the public school can reduce its performance gap by increasing the time allocated to holistic education, which leads to raise the percentage of time allocated to holistic learning. As a consequence, the change in the number of higher knowledge students is affected positively. In the extended model, the "students' knowledge ratio" is significantly lower than the benchmark. Initially, this condition makes the school increase the time allocated to traditional teaching to enhance such ratio (loop B-1 in Fig. 4.7).

As a result, in Fig. 4.8, a positive net change in the number of higher knowledge students is perceived during the first 16 months of the simulation. To keep the results of the above policy in the long term, the school must use overtime.[1] Therefore, the percentage of time allocated to holistic education (performance driver) must progressively be raised from 40% to 80% of the standard school hours.

After 16th month, a change in the loop dominance is perceived due to an increase in the performance driver (loop B-2 in Fig. 4.7). This condition leads to reach a substantial stability in the net change of higher knowledge students and to sustain the gains from the short-term policy illustrated in loop-B1.

[1] Colombian public schools often hire external contractors to train students for the standardized tests in out of school hours. In addition, regional governments and school district central offices invest financial resources to develop extracurricular activities for improving students' performance in standardized tests. The education policy, which is simulated by the extended model, implies the use of teacher overtime to foster a holistic education. This policy has been considered by most of the interviewees in this research as acceptable.

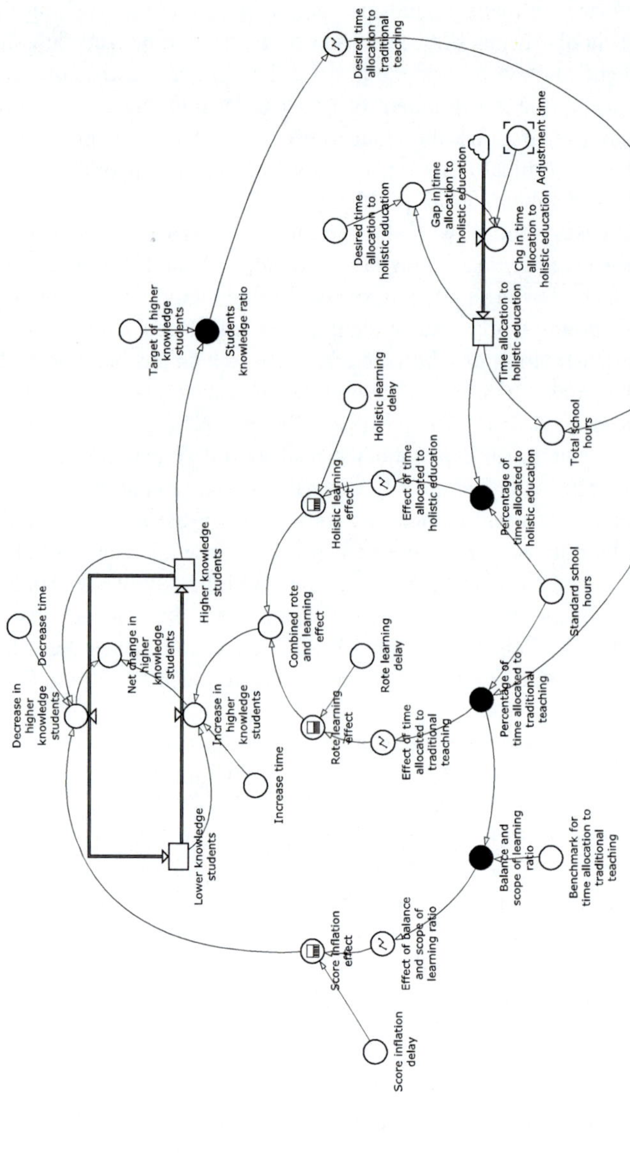

Fig. 4.6 DPM-based model illustrating a policy based on the joint use of traditional and holistic education in Colombian public schools (Bianchi and Salazar Rua 2020, p.14)

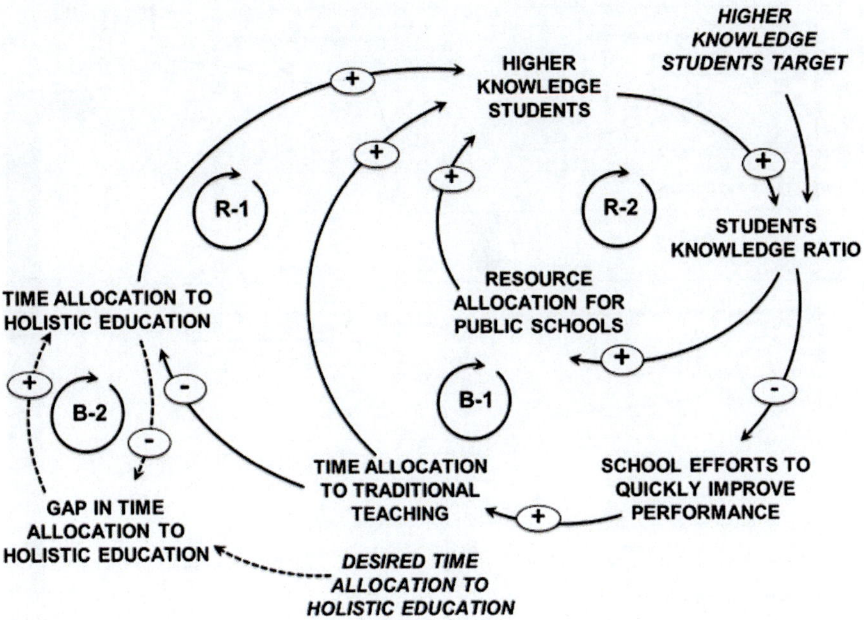

Fig. 4.7 Extending the scope of tests to holistic education in Colombian public schools (Salazar Rua 2020, p.75)

Posteriorly, from the 22nd month in Fig. 4.8, the school reduces the time allocation to traditional teaching from 60% to 25% due to an enhancement in its performance. It makes the "balance and scope of learning ratio" improve, which in turn slows down the loop "R1" in Fig. 4.7 since the increase in lower knowledge students is reduced. As a result, the net change of higher knowledge students is positively affected in the long term.

The above analysis shows how DPM can enhance the understanding of decision-makers in complex and counterintuitive phenomena by identifying feedback loops, strategic resources, and performance measures (i.e., performance drivers and end results). In particular, in the case study of the Colombian public schools, an instrumental view of DPM has been used to help system designers and decision-makers deal with behavioral distortions caused by the use of output-oriented performance measurement systems. Assessing school performance, through standardized tests, is a government mechanism to hold public schools accountable for student achievement in the short term.

However, dysfunctional behaviors may emerge because of (1) an excessive pressure for increasing test scores and (2) the adoption of inconsistent measurement systems. This condition may lead to a decline in school outcomes, such as educational quality and civic-mindedness, in the long term. To avoid the previous side effects, an extension in the boundaries of the adopted indicators has been proposed in this research by measuring the percentage of teaching time allocated to holistic

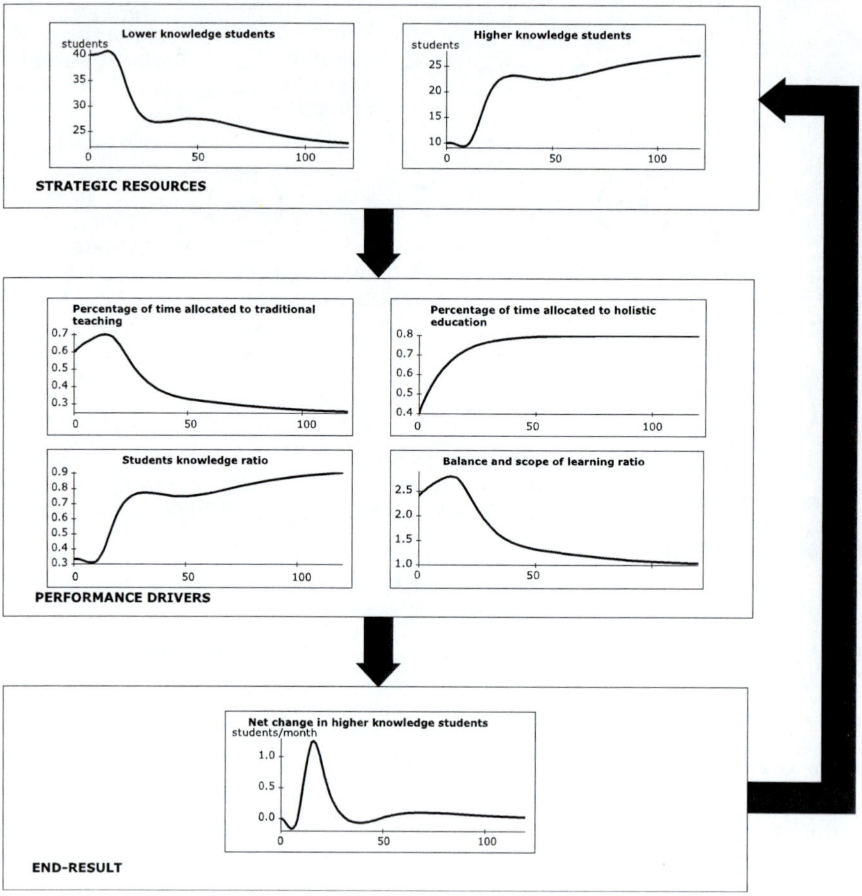

Fig. 4.8 Results from a simulation run illustrating the effects of combining traditional and holistic education in Colombian public schools (Bianchi and Salazar Rua 2020, p.16)

education and its effect on school performance. As a result of this policy, an improvement in outputs and outcomes may be expected in the long term.

In brief, the use of DPM to outline school assessment systems may foster a paradigm shift from an output to an outcome-based view in performance management. This view may support decision-makers to counteract the effects of static and simplistic indicators intended to measure student learning in public schools.

The next chapter will expand the boundaries of the school system with the purpose of improving institutional and community performance. To this end, boundary spanning initiatives in education and the adoption of an "outside-in" view of policymaking—through DPM—will be discussed.

References

Andersen DL, Luna-Reyes LF, Diker VG, Black L, Rich E, Andersen DF (2012) The disconfirmatory interview as a strategy for the assessment of system dynamics models. Syst Dyn Rev 28:255–275

Barlas Y (1996) Formal aspects of model validity and validation in system dynamics. Syst Dyn Rev 12:183–210

Bianchi C (2016) Dynamic performance management. Springer International, Zurich

Bianchi C (2021) Fostering sustainable community outcomes through policy networks: a dynamic performance governance approach. In: Meek J (ed) Handbook of collaborative public management, Chapter 22: 333–356. Edward Elgar

Bianchi C (2022) Enhancing policy design and sustainable community outcomes through collaborative platforms based on a dynamic performance management and governance approach. In: Peters BG, Fontaine G (eds) Handbook of policy design, Chapter 24: 411–433. Edward Elgar

Bianchi C, Bovaird T, Loeffler E (2017) Applying a dynamic performance management framework to wicked issues: how coproduction helps to transform young people's services in Surrey County Council, UK. Int J Public Adm 40(10):833–846. https://doi.org/10.1080/01900692.2017.1280822

Bianchi C, Rivenbark W (2014) Performance management in local government: the application of system dynamics to promote data use (with W. Rivenbark). Int J Public Adm 37

Bianchi C, Salazar Rua RS (2020) A feedback view of behavioural distortions from perceived public service gaps at 'street-level' policy implementation: the case of unintended outcomes in public schools. Syst Res Behav Sci:1–22

Bianchi C, Williams DW (2015) Applying System Dynamics Modeling to foster a cause and effect perspective in dealing with behavioral distortions associated with a city's performance measurement programs. Public Perform Manag:395–425

Bianchi C, Winch GW (2006) Unleashing growth potential in 'stunted' SMEs: insights from simulation experiments. Int J Entrep Small Bus 3(1):92–105

Bianchi C, Winch GW, Tomaselli S (2008) Management simulation as an instrument to aid turning "stunted growth" around in family businesses. Sinergie J 75:109–126

Center on Education Policy (2009) State test score trends through 2007-08, Part 2: Is there a plateau effect in test scores? Author, Washington, DC. https://files.eric.ed.gov/fulltext/ED506122.pdf

De Lancer Julnes P (2006) Performance measurement: an effective tool for government accountability? The debate goes on. Evaluation 12(2):219–235

Fuller B (2004) Accountability rises, scores fall. Los Angeles Times, 22 August. http://articles.latimes.com/2004/aug/22/opinion/oe-fuller22

Ghaffarzadegan N, Lyneis J, Richardson GP (2011) How small system dynamics models can help the public policy process. Syst Dyn Rev 27:22–44

Hamilton LS, Stecher BM, Klein SP, National Science Foundation (U.S.) (eds) (2002) Making sense of test-based accountability in education. Rand, Santa Monica, CA

Langer EJ, Roth J (1975) Heads I win, tails it's chance: the illusion of control as a function of the sequence of outcomes in a purely chance task. J Pers Soc Psychol 32:951–955

Levine D (1994) Instructional approaches and interventions that can improve the performance of African American students. J Negro Educ 63(1):46–58

Linn RL (1998) Assessments and accountability. CSE Technical Report 490. National Center for Research on Evaluation, Standards and Student Testing, Los Angeles, CA

Luna-Reyes LF, Andersen DL (2003) Collecting and analyzing qualitative data for system dynamics: methods and models. Syst Dyn Rev 19:271–296

Marton F, Säljö R (1976a) On the qualitative difference in learning I-Outcome and Process. Br J Educ Psychol 46:4–11

Marton F, Säljö R (1976b) On the qualitative difference in learning II-Outcome as a function of the Learner's conception of the task. Br J Educ Psychol 46:115–127

MassPartners for Public Schools (2005) Facing reality: what happens when good schools are labeled "failures"? Projecting adequate yearly progress in Massachusetts schools. Massachusetts Teachers Association

Meyer MW, Gupta V (1994) The performance paradox. Res Organ Behav 16:309–369

Morecroft J (2007) Strategic modeling and business dynamics. Wiley, Chichester

Parker W, Mosborg S, Bransford J, Vye N, Wilkerson J, Abbott R (2011) Rethinking advanced high school coursework: tackling the depth/breadth tension in the AP US Government and Politics course. J Curric Stud 43(4):533–559

Ross SM, Sanders WL, Wright SP, Stringfield S, Wang LW, Alberg M (2001) Two- and three-year achievement results from the Memphis restructuring initiative. Sch Eff Sch Improv 12(3): 323–346

Salazar Rua RS (2016) A Dynamic Performance Management approach to study the impact of education actors' decision making on the Colombian public schools' educational quality. European Master Thesis in System Dynamics

Salazar Rua RS (2020) Using Dynamic Performance Management to deal with behavioral distortions associated with school performance measurement systems. PhD dissertation in model based public planning, policy design and management at the University of Palermo. IRIS. http://hdl.handle.net/10447/394352

Saunders M, Lewis P (2012) Doing research in business and management. An essential guide to planning your project. Financial Times Prentice Hall, Harlow

Scogin SC, Kruger CJ, Jekkals RE, Steinfeldt C (2017) Learning by experience in a standardized testing culture: investigation of a middle school experiential learning program. J Exp Educ 40(1):39–57

Sills D (1957) The volunteers: means and ends in a national organization. Free Press, Glencoe, IL

Sterman JD (1994) Learning in and about complex systems. Syst Dyn Rev 10(2–3):291–330

Thomas JW (2000) A review of research on project based learning. The Autodesk Foundation. Project Based Learning (PBL), CA

Van Thiel S, Leeuw FL (2002) The performance paradox in the public sector. Public Perform Manag Rev:267–281

Warren K (2008) Strategic management dynamics. Wiley, Chichester

Wolstenholme E (1999) Qualitative vs quantitative modelling: the evolving balance. J Oper Res Soc. https://doi.org/10.1057/palgrave.jors.2600700

Xavier J, Bianchi C (2019) An outcome-based dynamic performance management approach to collaborative governance in crime control: insights from Malaysia. J Manag Gov 24:1089–1114

Zhang Y, Shaw JD (2012) Publishing in AMJ – Part 5: Crafting the methods and results. Acad Manag J 551:8–12

Chapter 5
Extending the System Boundaries to Enhance School Performance

5.1 The Need for Boundary Spanners in the Public Sector

Boundary setting is central to policymaking as it provides a sense of meaning and structure in complex social environments by delimiting fields of actions, reducing future uncertainties, and defining rules that prescribe specific behaviors in individuals.

In this respect, Hernes (2004) developed a comprehensive framework for setting and interpretating boundaries that distinguishes between mental, social, and physical limits. *Mental limits*, also called in the scientific literature as "mental models" (Doyle and Ford 1998; Sterman 2000), refer to the main ideas and concepts that describe how one perceives the world. These limits enhance the interactions among people who have a common background but hinder communication with outsiders that lack specific knowledge. For example, task efficiency can be improved by assigning roles and responsibilities to different departments in an organization; however, difficulties in coordination and collaboration will quickly become evident due to contrasting views and a lack of common professional jargon. *Social limits* emphasize group identity and normative behaviors that promote connection and belonging, for instance, feelings of loyalty and respect to certain individuals and to the entire organization. *Physical limits* are related to the set of rules and structures that regulate the relations with internal and external stakeholders. Some of them include provisions, access to information, institutional norms, communication channels, and governance modes.

Bureaucratic institutions organize their work through task specialization, hierarchical supervision, and a set of rules that guide operations toward the achievement of goals (Morgan 1997). In principle, one could think that this fragmented approach provides a high level of control by reducing uncertainty and increasing efficiency (Bianchi 2016; Van Meerkerk 2014; Dryzek 1990). In particular, splitting the goals into achievable objectives, making each organizational unit responsible for the

R. Salazar Rua, *Improving Consistency in Performance Measurement System Design*, System Dynamics for Performance Management & Governance 7,
https://doi.org/10.1007/978-3-031-24699-9_5

completion of a small set of functions, and having an executive figure that coordinates all the processes from the top of the pyramid seems practical and appealing. However, this approach is exclusively appropriate for well-defined problems enclosed in stable environments, for example, the optimization of resources in a manufactory setting.

Simon (1996) named the above-mentioned problems as *near-decomposable*, which means that each unit, department, agency, and/or institution can work in isolation and can concentrate their efforts on solving one specific set of tasks since the interaction among the parts is weak and, consequently, will not lead to relevant side effects. In short, the sum of the parts is equal to the whole. Conversely, when dealing with complex problems, implementing disaggregation strategies is counter-productive because it will displace issues across the institutional boundaries. As Dryzek (1990, p.62) stated "growing complexity can frustrate the decomposition strategy since the interactions across the boundaries of sets and subsets become too rich, irrespective of the quality and veracity of the theory informing decomposition."

Hence, despite boundaries being relevant to create a sense of order and structure in rapid changing environments, they also hinder communication and coordination across societal actors (Howes et al. 2015). In fact, boundary setting is perhaps the main cause of fragmentation in contemporary societies, but it is also the main driver for wealth. It implies that fragmentation is and will be an inherent organizational principle to any type of institution (Sørensen and Torfing 2009). Therefore, the dynamic nature of problems in the public sector requires a different set of strategies, by which boundaries are spanned without losing the benefits of task specialization.

In the last few years, the public sector has recognized the importance of activities aimed at closing internal gaps and fostering collaboration with other organizations from the same operational context. Such activities are referred in the scientific literature as *boundary spanning initiatives* (Atkinson 2019; Leifer and Delbecq 1978). Their implementation suggests a deliberate search for understanding among distinct stakeholders through adaptive leadership approaches, by which the agents bring diversity of opinions to the table, enrich discussions, build relations, and design action-oriented plans (Murphy et al. 2006). In particular, boundary spanning engages key actors in a strategic dialogue that leverages existing commonalities in order to reach consensus between competing perspectives and promotes cross-sectoral partnerships in highly fragmented environments (Atkinson 2019; Steadman 1992; Williams 2002). Moreover, boundary spanning has a dual nature since it could be perceived as both an activity and a role.

As an activity, boundary spanning focuses on the ability to gauge others' needs and find a common ground when dealing with dissimilar perspectives (Rhodes 2016; Atkinson 2019). This is done by facilitating the alignment of resources and the structural adaptation of the organizations belonging to the *governance networks* (Van Meerkerk 2014). Such networks are defined as patterns of social relations that emerge as a consequence of public issues and posteriorly evolve through the interactions among societal actors with mutual dependencies (Atkinson 2019; Koppenjan and Klijn 2004). Likewise, they develop around cross-sectoral issues that cannot be solved by fragmented approaches, which implies that organizations must join knowledge and resources to achieve community outcomes (Sørensen and Torfing 2009; Agranoff and McGuire 2001). However, completing this task is not easy as each actor has a different perception about how the system operates, the key problems to be addressed, and the solutions to be implemented. In addition, the involved actors are often made accountable for goals within their own institutional boundaries, which may slow down initiatives for cross-sectoral collaboration (Provan and Kenis 2008; Teisman 2000; Bianchi 2022). Under this scenario, an emphasis of connective capacities (Edelenbos et al. 2013) by means of linking roles is essential.

As a role, boundary spanning stresses on the work that boundary spanners execute at the periphery by performing tasks to link the organization they represent with its environment (Tushman and Scanlan 1981). These individuals connect processes and institutions at both sides of the boundary by analyzing, selecting, and exchanging information (Leifer and Delbecq 1978; Jemison 1984). It requires personal attitudes grounded on a genuine feeling for the social construction of other stakeholders in the governance network. As Tushman and Scanlan (1981, p.291) stated "boundaries can be spanned effectively only by individuals who understand the coding schemes and are attuned to the contextual information on both sides of the boundary, enabling them to search out relevant information on one side and disseminate it on the other."

Boundary spanners also translate information in a certain mode that is accessible to other members in the organization (Tushman 1977). In this context, translation means filtering data from external noise, drawing inferences from perceived facts, and making suggestions to design strategic plans aligned to relevant goals (Aldrich and Herker 1977). For instance, in the education context, central office administrators—who have boundary spanning roles—often collect data about the causes of poor school performance. These causes include variables such as school resources, neighborhood conditions, and teaching quality. However, delivering raw data to executive directors is highly unproductive. Therefore, information must be filtered by, first, selecting a data subset from the whole perceived system and, second, outlining specific recommendations (Leifer and Delbecq 1978).

According to Van Meerkerk and Edelenbos et al. (2013), the presence of boundary spanners leads to a higher level of trust in the governance networks, which mitigates the emergence of opportunistic behaviors, fosters collaborative actions, and improves performance. In this regard, Williams (2002) identified a number of traits that make boundary spanners competent for their role and that, in turn, boost their chances to build reliable relationships. Some of these traits are empathy, active listening, and assertiveness. Hence, the role of individuals who intentionally cross organizational borders and link organizations, by means of their connective capacities, is highly important. In a similar line, Klijn et al. (2010) highlighted the importance of trust in the social networks to achieve community outcomes. In particular, these scholars claimed that trust (1) creates a stable perception about others' intentions, which significantly reduces the risks associated with cooperative actions, and increases the probability to merge resources, (2) stimulates learning through information exchange and strengthens human capital, and (3) enhances innovation and makes vertical integration and coordination less necessary.

Furthermore, Scott (1981) argued that boundary spanners are responsible for managing the relationship between the organization and its environment by bridging and buffering. On the one hand, *bridging actions* are aimed at fostering mutual dependence among stakeholders, for instance, when boundary spanners assume the role of representatives and, consequently, perform tasks such as resource acquisition, maintenance of political legitimacy, and enhancement of institutional image. On the other hand, *buffering actions* attempt to preserve internal stability by mediating the relation between external disturbances and the organization itself. Information exchange is an example of this type of action since the boundary spanner is responsible for regulating the information from the environment to the organization, and the other way around. This is achieved by delaying and referring the information, in order to filter hostile situations that may put in risk regular operations (Murphy et al. 2006; Aldrich and Herker 1977; Leifer and Huber 1977).

Thus, public institutions must adapt their actions to satisfy demands from fluctuating environments and must develop relational capabilities to survive in circumstances out of the control of their members (Aldrich 1979). To this end, the organizational structure must be periodically adjusted as a strategic response to the identification of relevant threats and opportunities (Hannan and Freeman 1977). In particular, mechanisms to facilitate the relationship between the organization and its environment must be established (Murphy et al. 2006). Such mechanisms set and span boundaries within the organization by promoting the interaction between members and non-members (Scott 1981; Thompson 1967). In this regard, boundary spanners play an essential role as they transform the organizational practices by negotiating the interaction between internal and external stakeholders, which generates a better fit in shared interests and goals when dealing with wicked problems (Laegreid and Rykkja 2014; Lekvina and Vaast 2005; Steadman 1992).

The next section will focus on the rise of governance networks and the implementation of different kinds of boundary spanning initiatives and leadership styles in complex settings.

5.2 The Implementation of Boundary Spanning Initiatives and Adaptive Leadership Approaches Within the Governance Networks

In the scientific literature, the concept of *governance networks* has been broadly debated. For instance, Rhodes (2017) asserted that they are a set of formal and informal linkages between the government and other societal actors, who share common beliefs and interests in public policymaking and implementation. Similarly, Klijn (2008) indicated that governance networks refer to structured relationships between the government, local businesses, and civil society actors, and their goal is to enhance the design and implementation of community policies. Likewise, Hoppe (2011, p.167) stated that these networks could be understood as an "intellectual and practical reflection of a trend towards problem-specific, pragmatic arrangements for social and political decision-making." Correspondingly, in the preceding section, it was mentioned that governance networks develop around cross-sectoral issues that cannot be solved by fragmented approaches, which implies that different organizations must join knowledge and resources to achieve community outcomes (Sørensen and Torfing 2009; Agranoff and McGuire 2001).

The above definitions are a response to the socio-economic and cultural constraints imposed by modern institutions, as they challenge the status quo of the system by questioning the principles behind hierarchical structures and vertical accountability (Van Meerkerk 2014). In fact, it is difficult to determine who is responsible for the accomplishment of a particular task within the governance networks because they are grounded on flat structures and lack of clear relations of authority and subordination (Dryzek 2010). In other words, the boundaries among social actors (e.g., the government, local businesses, schools, universities, and families) are not easy to differentiate, which suggests that governing is the result of negotiated interactions rather than state mandates. For this reason, despite the governance networks often including politicians, these actors may not necessarily play a leadership role in decision-making.

Adding to the previous discussion, governance networks have a polycentric structure, which consists of different policy agents who attempt to find joint solutions to problems that demand collaboration (Van Meerkerk 2014). As a thumb rule, such networks are made up of private actors, public actors, and citizens. For instance, if the network is centered around issues concerning private and public actors, then collaboration between local businesses and political institutions might be expected. On the other hand, if the network is centered around issues concerning citizens, then the rise of civic initiatives—which will be discussed later—might be plausible (Van Meerkerk and Edelenbos 2017).

Therefore, governance networks offer an alternative way to make more participative the policymaking process by (1) mobilizing cross-sectoral resources, (2) enhancing the quality of decisions through integrated approaches, (3) improving coordination among mutually dependent actors, and (4) proposing innovative solutions based on synergistic efforts (Koppenjan and Klijn 2004). In short, they offer a greater sense of inclusiveness and engagement between the involved actors by emphasizing on cooperation and mutual adjustment rather than competition (Bekkers et al. 2007; Edelenbos et al. 2010; Hajer and Wagenaar 2003).

However, like collaborative governance (Bryson et al. 2006; Emerson and Nabatchi 2015; Xavier and Bianchi 2019), cooperation within the governance network is informal and highly implicit (Ansell and Gash 2007). Consequently, formal strategies to incorporate stakeholders into multi-lateral and consensus-oriented agreements are not a prerequisite to decision-making.

According to Emerson and Nabatchi (2015, p.18), *collaborative governance* refers to "the processes and structures of public policy decision-making and management that constructively engage people across the boundaries of public agencies, levels of government, and/or the public, private and civic spheres in order to carry out a public purpose that could not otherwise be accomplished." This concept emphasizes on how citizens and other actors from different policy fields and organizational levels join efforts to govern the society while keeping close proximity to political institutions (Ansell and Torfing 2016; Sørensen et al. 2020). It is also relevant to remark that collaborative governance has different strands of research, ranging from the study of how contrasting experiences lead to political innovation (Bommert 2010; Torfing 2016) to how cross-sectoral collaboration is used to deal with wicked problems, promote public deliberation, and empower communities (Fung and Wright 2003; Newman et al. 2004).

In order to promote cross-sectoral collaboration within the governance networks, the adoption of a facilitative management style (Van Meerkerk and Edelenbos 2017) and community entrepreneurship leadership (Selsky and Smith 1994) become central. Both concepts are explained below.

On the one hand, *facilitative management* relies on engaging activities that foster relationship building in the networks through the construction of social mechanisms for enhancing cross-boundary interactions and empowerment. Under this managerial style, workshops, team building events, and informal meetings are often preferred because they facilitate information exchange across the boundaries and, in particular, promote learning of tacit knowledge (Asheim et al. 2007). In addition, these events foster the development of a group identity by breaking down communication barriers and supporting a shared view of the problems (Gash 2008; Ostrom 1998). Therefore, facilitative management requires that executive directors focus their attention on creating the conditions to make stakeholders more receptive to cross-organizational boundaries and invest on boundary spanning initiatives.

For instance, in the public school context, implementing a facilitative management style implies that central office administrators must provide discretionary power to principals, who must in turn provide it to teachers. The rationale behind the previous idea is the fact that micro-managing daily operations will hinder project execution and will make the actors less responsive to the specific needs of their environment (Crawford et al. 2003).

On the other hand, *community entrepreneurship* is a leadership strategy based on a pluralist approach to social change. It leverages on the views and interests of stakeholders with distinct professional experience, organizational status, levels of power, and participation, and it aims at finding common understanding of complex public issues. This leadership strategy could be seen as both a method of inquiry and a method of participation. The former stresses on the analysis of active engagements to cope with external forces affecting the involved actors. For instance, understanding the motivations behind collective actions and gaining a deeper knowledge on how a broad range of efforts impact on the achievement of community outcomes. And the latter emphasizes on how resources are mobilized, how action plans are designed, and how collective capacities are built (Selsky and Smith 1994).

An unusual characteristic of community entrepreneurship is the fact that their leaders, also called as "community entrepreneurs" (Selsky and Smith 1994), do not exercise power relations directly. Instead, boundary spanning actions are used to influence on the governance networks through informal devices (Trist 1983). In other words, leaders in community settings mostly work in the background by innovating and mediating the claims and interests between the organization to which they belong and those of others (Burns 1978). This connective capacity along with an experimental attitude is critical for testing organizational limits and strengthening alliances in contexts whose outcomes are difficult to predict (Smircich and Stubbart 1985). Therefore, the willingness to learn from experimenting, failing, and reflecting is a major driver for developing a better understanding about the kinds of problems that community entrepreneurs must deal with (Selsky and Smith 1994).

In addition, community entrepreneurs' authority comes from non-traditional sources founded on the empowerment of disadvantaged groups through boundary spanning positions. Therefore, they are also responsible for raising the voice and serving the needs of citizens. To this end, they must look for win-win solutions based on collaborative advantages (Selsky and Smith 1994), because powerful actors within the governance networks will not easily quit their dominant status. In this context, seeking collaborative advantages implies questioning underlying assumptions about competition and, in turn, promoting partnerships, alliances, and mutual agreements that can mobilize collective resources (Smircich and Stubbart 1985). In particular, community entrepreneurs must establish cooperative arrangements by which the government and the citizens adapt their roles to better deal with deficiencies in service delivery (Van Meerkerk and Edelenbos 2016).

The above idea introduces the concept of *civic initiatives*, which refers to community actions whose goal is to address complex public issues through a lasting cooperation. It emphasizes on the use of a pragmatic view that constitutes a response to specific needs at the local level. Moreover, these bottom-up actions challenge the existing practices and power relations while encouraging coordination and collaboration for improving the socio-economic status (De Moor 2013; Van Meerkerk 2014). This is why the implementation of a hands-on approach is fundamental to promote autonomy, self-organization, and local ownership (Van de Wijdeven 2012). As a result, citizens exercise a direct influence on how certain public services are produced and managed, and community leaders become skilled in identifying strategic assets that are relevant for the achievement of societal outcomes (Wagenaar and Van der Heijden 2015).

Community leaders often emerge by implicit consensus within the governance networks (Lauman et al. 1978) and play an important role in developing collective capabilities among actors who share similar concerns regarding specific public issues (Selsky 1993; Selsky and Smith 1994). In this respect, Westley (1991, p.1015) stated that leaders in community settings must "exhibit the ability to both perceive and manipulate structure, to directly interact with groups and sentiments at the bottom of the organizational pyramid, and to have the ability to articulate and integrate, in powerfully symbolic forms, the various concerns of stakeholders." It implies that leaders must demonstrate relevant personal traits to successfully join cross-sectoral efforts. Some of them include (1) an entrepreneurial attitude to mobilize resources and manage events, (2) a reflective mindset to learn and adapt from changing conditions, and (3) the ability to integrate divergent perspectives in a common framework (Bolman and Deal 1991).

In order to reinforce the need for collaborative behaviors and reframe ambiguity in uncertain circumstances, community leaders often adopt a community-based approach that promotes change in institutional structures and norms (Kanter 1983; Gray 1989). However, the implementation of such an approach makes them face a complex dilemma. It consists of acting in total freedom without considering bureaucratic schemes, rigid rules, and government positions or modifying preferences to ensure political legitimacy. In the first scenario, civic initiatives have a high degree of autonomy but may lack institutional support to be sustained in the long term. In the second scenario, such initiatives may differ from the original proposals, but the chances of implementation are maximized (Van Meerkerk and Edelenbos 2016). Therefore, a middle point between both alternatives is preferred. It means that civic initiatives should partially consider public professionals' input to guarantee legitimacy, and the government should provide citizens with enough room to mature their proposals. This will enhance service delivery, trust in the institutions, and cross-sectoral collaboration.

In fact, the need for cross-sectoral collaboration is implicit in the goals outlined by the 2030 agenda for sustainable development (United Nations 2022). The achievement of such goals requires the implementation of initiatives intended to enhance communication, policymaking, facilitative leadership, and performance evaluation among stakeholders operating in the same context. However, civic initiatives cannot be always used to solve problems at the neighborhood level because, in certain circumstances, higher levels of governance are required. In this scenario, *collective impact initiatives* play a central role as they facilitate the interaction between local institutions and external actors (Turner 2015). This type of initiatives uses an innovative approach to deal with complexity in the public sector, which promotes collaboration among stakeholders that usually work in isolation through a centralized managerial structure.

Collective impact initiatives are also used to (1) deal with disturbances out of the neighborhood boundary by setting formal mechanisms of participation and leadership, (2) enable dialogue among actors belonging to the governance networks for building trust and creating common motivation, (3) collect data and measure results across all the participants for guaranteeing alignment and shared accountability, (4) coordinate mutually reinforcing activities that are agreed through common agendas, (5) encourage learning from success and failure for improving service delivery and finding solutions to specific public issues, and (6) promote citizen participation and organizational commitment toward the achievement of community outcomes (Kania and Kramer 2011).

However, implementing such initiatives in the education context is a complex endeavor because of the inherent difficulty to get reliable data and the lack of robust school performance management systems that provide educational authorities with effective tools to deal with wicked problems. Therefore, boundary spanning initiatives (i.e., civic initiatives and collective impact initiatives) must be coupled with collaborative platforms aimed at creating shared meaning and common understanding of the feedback structure behind the problems (Crosby et al. 2010; Van Meerkerk 2014; Bianchi 2016). Such platforms set the conditions for multi-level collaboration by strengthening accountability systems that track inputs, processes, and outcomes within the governance networks (Crosby et al. 2010). This implies that an "outside-in" view of policymaking—through dynamic performance management (DPM)—is required.

Before explaining how to adopt such a view in the context of the public schools, the next section will provide the reader with precise content about boundary spanning initiatives in education.

5.3 Dealing with Neighborhood Effects in Education Through Boundary Spanning Initiatives

The scientific literature on *neighborhood effects* mainly focuses on how the social environment impacts on individual outcomes. In particular, the composition and organization of the population at the neighborhood level are seen as the key drivers for such effects (Kauppinen 2007). However, if the concept of neighborhood is expanded by considering the interactions among physical environment, social structure, local institutions, and life opportunities, then the list of factors and pathways affecting the outcomes of interest would be extended equally (Galster 2012; Bernelius and Kauppinen 2012). In this respect, Friedrichs et al. (2003) built a framework that categorizes the main factors behind the emergence of neighborhood effects. They are neighborhood resources, model learning, socialization and collective efficacy, and resident perceptions of deviance. *Neighborhood resources* refer to services, installations, human capital, and educational culture in the schools. *Model learning* denotes the nature of peer groups, which is analyzed through behavioral assessments aimed at measuring team attitudes and time use at both schools and neighborhoods. *Socialization and collective efficacy* emphasize on the sense of social control in public places, and it is measured through students' opinions. Finally, *resident perceptions of deviance* refer to the perceived public safety in the local area.

When analyzing neighborhood effects on student performance, schools play a crucial role as they are part of the social and physical environment where pupils develop their daily life activities. In fact, in many geographical settings, students attend classes in institutions located in their own neighborhoods. This relationship between home and school location is relevant and, consequently, research that does not consider the importance of contextual variables could fail to depict the whole picture of institutional performance (Bianchi and Salazar Rua 2020; Bernelius and Kauppinen 2012). Hence, school effects should be partially understood through the same framework that describes neighborhood effects (Bramley and Karley 2007).

As a matter of fact, if the previous effects are analyzed in isolation, there is a high risk of overestimating their impact on student performance and educational attainment. The reason is that the demographic composition would not be entirely considered and, as a direct consequence, the understanding of the determinants of school outcomes would be limited (Nieuwenhuis and Xu 2021). In particular, both an increase in socio-spatial segregation and a lack of political interventions would lead to stronger neighborhood and school effects on education (Musterd and Ostendorf 2005; Andersson and Molina 2003; Kuusela et al. 2008).

For instance, depending on the neighborhood where students live, they could be exposed to contrasting adult role models, peer pressure, social controls, and educational opportunities (Galster 2012). These factors could impact on students' motivation to perform well at the schools and, in turn, such schools might have difficulties to provide a decent education service to the students if they are located in low-income areas. The rationale behind the previous idea is the fact that poor institutes often struggle at hiring competent teachers due to their deteriorated image and lack of financial resources (Wacquant 2008). Previous research has suggested that these issues reinforce the link between poverty and unsatisfactory individual outcomes (Nieuwenhuis and Hooimeijer 2016).

In the Colombian context, the main causes of neighborhood effects are perhaps the growing income differences between rich and poor, budget reductions in social projects, and low levels of education in adults living in deprived areas. In particular, the socio-economic background of the students is a relevant reason for educational success at both individual and school levels (Jakku-Sihvonen and Kuusela 2002; Kupari 2005). Such background could be seen as a structure consisting of mutually interconnected dimensions, whose configuration depends on the social composition of the local area. Therefore, contextual variables should be taken into account in order to design effective education policies that positively impact the school outcomes (Galster 2008; Bernelius and Kauppinen 2012).

The former education policies should also encourage a new type of relation between public actors, private actors, and citizens in order to enhance student learning, educational attainment, and the achievement of community outcomes (Honig et al. 2001; Crowson and Boyd 1993). As a general rule, these policies should foster (1) collaboration between schools, youth serving agencies, and citizens to define shared goals and implement community strategies and (2) partnerships between school district central offices and other public bureaucracies—such as healthcare agencies and social work institutes—to carry out initiatives that are sustainable over time (Honig 2006).

Moreover, school policies should be based on the assumption that local knowledge and empowerment are essential to improve student performance (Moll et al. 2005; Pittman and Cahill 1992). Therefore, educational authorities should support communication channels among teachers, families, universities, business owners, NGOs, and government by emphasizing on highly salient interests. To this end, public schools should have a flexible organizational structure that facilitates role innovation and recognizes the need for stakeholder collaboration to enhance service delivery. Perhaps, one of the biggest challenges for carrying out this proposition is the difficulty in understating the benefits of the boundary work, which may lead to a lack of cross-sectoral agreements that ensure policy continuity (Carey et al. 2017).

In order to mitigate the above-mentioned difficulty and promote legitimacy of boundary spanning initiatives, school district central offices must (1) partner with other community actors to learn about their system's perceptions, intending to develop integrative policies for meeting educational targets, and (2) respect the discretionary power of street-level bureaucrats (e.g., teachers and psychologists) and/or boundary spanners (e.g., principals) who work at the schools, because they have to deal with complex tasks that cannot be operationalized through rigid routines and bureaucratic schemes (Shulman 1983; Lipsky 1984).

These suggestions come from the fact that current education policies—aimed at improving student performance—have required that central office administrators coordinate efforts and mobilize resources between public schools and community agencies. It has led to the re-evaluation of traditional practices based on top-down approaches and to the emergence of new boundary spanning positions within the governance networks (Crowson and Boyd 1993; Jemison 1984). Such positions are often occupied by individuals who exhibit a high degree of empathy and interest for others' concerns, which support them in building trustworthy relationships (Sullivan and Skelcher 2002; Williams 2002; Edelenbos and Van Meerkerk 2015).

However, building trust in the social networks—through boundary spanning initiatives—is not an easy task for central office administrators, as they often lack local knowledge, have a limited authority to develop organizational strategies in spaces out of the school context, and do not receive enough institutional support to reinforce collaborative behaviors (Honig 2006). For instance, partnerships with community agencies may imply a certain degree of knowledge of human service systems, which goes beyond the traditional education that central office administrators receive at the university (Lawson 1998). Similarly, incentives for these public workers often emphasize top-down relations, instead of cooperative arrangements and flat structures (Malen et al. 1990; Walker 2002). Indeed, such incentives play an important role in test-based accountability systems because they intensify monitoring and control in the name of equity and efficiency, but they sacrifice the empowerment of hierarchical subordinates (Finnigan and O'Day 2003; Raywid and Schmerler 2003; Blau 1963).

In the context of public schools, such empowerment of hierarchical subordinates implies that (1) the government must progressively rely on citizens' self-organizing capabilities to develop policy options that mobilize support and validate decisions, (2) central office administrators must behave as communication bridges between the government and outsiders who desire to establish contact with political authorities (e.g., school principals, business owners, and researchers), and (3) the role of principals as boundary spanners is recognized, because they have to interact with parents who are external actors to the school system (Bidwell 1965; Warren 2009; Marien et al. 2010).

From a boundary spanning perspective, school principals' work is complex. They have to devote time on building the school–parent relation, but they also have to be careful enough not to produce an excessive interference from families in bureaucratic processes. To this end, principals must have a profound understanding of how the internal structure of the schools operates, and they must be aware of potential contingencies that parents may cause on teachers' work. This will allow them to strengthen the relation with the families while keeping internal stability in the organization (Goldring 1985). As an illustration, experienced principals know that both the administration of standardized tests and behavioral problems in the students will trigger parental reaction. Therefore, principals could anticipate such reactions by setting procedures for meeting special requests. Otherwise, schoolteachers would be forced to invest a significant portion of their time on dealing with heterogenous demands (Murphy et al. 2007).

Complementing the above idea, empirical research has found that principals spend a non-trivial amount of time on mediating the relation between schoolteachers and families. For example, Woolcott (1973) and Morris et al. (1984) reported that principals used about 8% of their time on regular interactions with parents. Likewise, after analyzing a broad number of administrative files to determine the effectiveness of school programs, Turner (1983) noted that principals behaved as bridges between parents and school staff. In a similar vein, Liphman and Hoeh (1974) suggested that parents preferred to interact with principals, rather than other educational authorities, because they are seen as the schools' representatives in the community.

In a nutshell, school principals acting as boundary spanners tend to focus on (1) buffering the relation between the central office, parents, and schoolteachers and (2) bridging the school with its environment to improve public relations and organizational image in the local area (Saxe 1975; Murphy et al. 2006). Performing these actions is beneficial to engage families in the school experience of the children because parents will be better informed about the dynamics inside the institution. It will make them more supportive during the implementation of novel approaches and educational strategies (Murphy et al. 2007; Hallinger and Murphy 1966). As a result, enhancing the relation between schools and families, through boundary spanning initiatives, promotes a stronger sense of commitment to the student learning process.

However, implementing the above-mentioned initiatives in the education context is a complex endeavor because of the inherent difficulty to get reliable data and the lack of robust school performance measurement systems that provide educational authorities with effective tools to deal with wicked problems. For instance, the identification of non-linear relations, trade-offs in time and space, material and perception delays, surrogates for intangible assets, and logic connections between strategic resources and performance measures at both school and community levels requires the use of hybrid approaches that combine performance management, system dynamics, and collaborative governance (Bianchi 2021). An "outside-in" view of policymaking—through DPM—could be seen as a response to this need.

The next section will emphasize on how to adopt such a view in order to design education policies that take into account the broader context where schools operate and promote the implementation of boundary spanning initiatives.

5.4 Supporting Boundary Spanning Initiatives in Education Through an "Outside-in" View of Policy Making

The implementation of boundary spanning initiatives is associated with high risks of failure, even if leaders within the governance networks exhibit well-intended behaviors. This phenomenon is explained by the following two reasons. On the one hand, policymakers have their own mental models about how the system operates, and their performance is usually measured through an "inside-out" view of policymaking that prioritizes the achievement of organizational objectives at the expense of community outcomes. It may lead to biased policy proposals and may hinder cross-sectoral collaboration (Klijn and Koppenjan 2000). On the other hand, a lack of robust frameworks that foster systems learning, shared accountability, consensus building, and facilitative leadership may result in poor results (Bianchi 2021). Therefore, in order to increase the chances of success when boundary spanning initiatives are carried out, a paradigm shift from an "inside-out" to an "outside-in" view of policymaking is required.

Performance measurement systems are often based on an "inside-out" view, which means that isolated efforts are used to achieve end results (i.e., outputs and outcomes). It implies that the input from the local area and the actions from other stakeholders in the operating system are not fully considered. Conversely, implementing an "outside-in" view, through the lenses of DPM, requires a connection between policies designed at the organizational level and policies designed at the community level. The adoption of this view allows stakeholders to formulate strategies that effectively impact on the endowment of shared strategic resources, which in turn strengthens the institutional performance (Bianchi 2021).

To promote a paradigm shift, actors within the governance networks must be supported by learning facilitators who help them notice how pursuing individual goals—without considering the operational context—may lead to diminishing returns in terms of performance (Bianchi and Vignieri 2020). Likewise, learning facilitators must emphasize on the need of boundary spanning initiatives, community entrepreneurship, and facilitative leadership to promote cross-sectoral collaboration and consensus building. As a result, not only the provision of public services will be improved, but also an increase in joint responsibility, shared accountability, information sharing, and citizen participation will be observed.

Applying DPM to outline an "outside-in" view of policymaking implies the compliance of laws and prescriptions mandated by political actors and the promotion of an active citizenship among the members of a community. This approach provides the basis for exercising civic commitment and developing a sense of belonging to the local area through the identification of shared goals and the proper management of collective assets (Cooper and Gulick 1984; Bianchi 2021). In particular, enhancing citizen participation is a relevant intermediate outcome for governance networks that adopt an "outside-in" view because a lack of active citizenship will make the collaboration efforts unsustainable in the long term. Moreover, low levels of citizen participation will negatively influence on community wellness, local area attractiveness, and quality of life since collective skills, integrated capabilities, and community experiences would not be synergically aligned to tackle wicked problems (Bianchi 2021; Ansell and Gash 2007).

As explained in Chap. 4, DPM is a robust framework aimed at enhancing the performance management practice in the public sector by embracing the dynamic complexity of wicked problems. In particular, DPM can support policymakers to outline sustainable policies by linking strategic resources with performance drivers, which in turn feed back into strategic resources through end results. This approach is useful for designing performance measures that take into account the operational context where organizations operate and for avoiding the effects of policy resistance (Bianchi 2016).

In particular, when the local area performance is the matter of assessment, implementing an "outside-in" view of policymaking is crucial. Before describing how to use DPM for that purpose, it is pertinent to explain in detail the notion of *local area performance*. This concept is the result of the need for building shared strategic resources in the community. These resources are called in the scientific literature as "common goods" (Hardin 1968), and they refer to tangible and intangible assets that define the socio-cultural identity of a group of people. It implies that they are not owned by a single institution, and, therefore, fragmented policies will not be effective to change their endowment over time. Some of them include active citizenship, civic-mindedness, social awareness, safety, financial stability, respect for the environment, quality of public services and infrastructures, human capital, employment rate, number of local businesses, and alliances between public and private institutions (Bianchi 2022).

Building up common goods is relevant to enhance the delivery of sustainable community outcomes, such as local area attractiveness and quality of life. These outcomes can be attained by means of an "outside-in" view of policymaking, which supports stakeholders in the design of group policies and complementary roles within the governance networks. These are fundamental conditions for managing shared resources, promoting performance sustainability, and pursuing societal outcomes (Bianchi 2021).

Figure 5.1 portrays how DPM can be used to implement an "outside-in" view of policymaking by connecting institutional interests with targeted community outcomes. To this end, DPM provides actors within the governance networks with a solid framework to cascade corporate policies at the departmental level and track implementation results through performance drivers, outputs, and outcomes. This control process is not only grounded on a feedback theory but also on a feedforward logic (Otley 1999). It implies that opportunities and problems at the departmental level have the potential to influence on policies designed at both the institutional and community levels. This approach—supported by learning facilitators—enables strategic dialogue and cross-sectoral collaboration. Likewise, it enhances the ability of outcome-oriented performance measurement systems to manage complex social issues (Bianchi 2021).

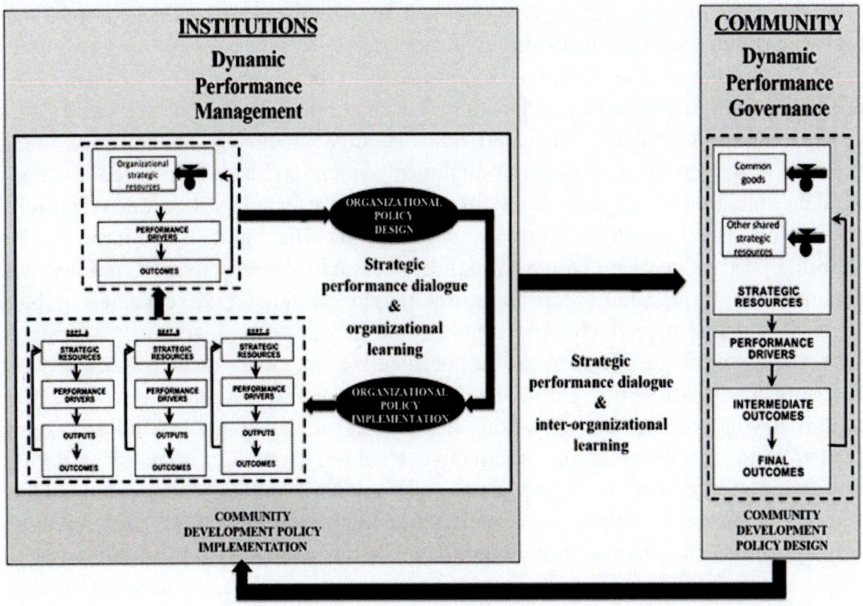

Fig. 5.1 Enhancing institutional and community performance through an "outside-in" view of policymaking (Bianchi 2021, p.342)

As an illustration of the previous approach, policies designed at the school level—such as the ones presented in Chap. 4—could be linked to community outcomes (e.g., quality of life and local area attractiveness). This would be done by extending the system's boundaries and considering—as a first step—how "school efforts to quicky improve performance" would distort teachers' efforts and would make them more inclined to focus on the best performing students in standardized tests. After a delay, this behavior would cause a higher student dropout rate because some pupils would feel neglected in the education process. As a result, the quality of education would be reduced despite possible score increases in state evaluations. In turn, a reduction of students attending the universities would be observed, which would lead to a workforce with lower technical skills to perform tasks that are demanded by society. It would affect collaborations and financial investments between public and private institutions in the local area, which would make neighborhoods poorer and the quality of future school students lower. All of this would end up in a reduction of the quality of life and local area attractiveness.

From the preceding example, the need for implementing educational policies that take into account the broader context where public schools operate and, in turn, promote collaboration within the social networks becomes evident. It requires the development of both an active citizen participation and a sense of belonging, intending to influence on community wellness. To support this endeavor, the adoption of an "outside-in" view of policymaking—through DPM—is proposed as follows.

Figure 5.2 illustrates the above-mentioned view by means of a DPM chart, which has been divided into the following two sections.

The left-hand section emphasizes on how to increase active citizenship by promoting a higher quality of education. To this end, change in active citizenship is modeled as a final outcome, and both "change in quality of education" and "change in citizen awareness about quality of education" are represented as intermediate outcomes. Change in quality of education is affected by the student learning ratio, which is a performance driver that determines the capability to improve the learning process at the classroom. Similarly, such a driver is affected by the dynamics of the strategic resources: municipal administrative capacity, public schools, financial investments (from public and private actors), and laws and regulations. Moreover, an increase in the intermediate outcome "change in quality of education" will strengthen the strategic resource "quality of education," which is central to (1) promote collaborations and events at the neighborhood level and (2) foster the implementation of effective teaching strategies in public schools. The compound effect of the previous variables will influence the intermediate outcome "change in citizen awareness about quality of education" that, in turn, will impact on the change in civic-mindedness by means of forums and projects aimed at enhancing the common goods (e.g., quality of public services, safety, and cultural integration). These cross-sectoral projects are carried out within the governance networks and can be perceived as a means of developing a broader sense of belonging to the local area (Bianchi 2021).

The right-hand section extends the policy configuration of the left-hand section by portraying how to use active citizenship to improve community wellness and neighborhood attractiveness. To this end, the change in community wellness is modeled as a final outcome, and the changes in neighborhood attractiveness, collaboration projects, and collaboration pacts are represented as intermediate outcomes. In the Colombian context, collaboration pacts are contracts that specify the technical details of the partnership between the municipality and citizens, and they are an indispensable requirement for developing collaboration projects. The change in collaboration pacts depends on the performance driver "time and efforts to foster collaboration ratio," which is influenced by the intangible assets: active citizenship and public administration transparency. In turn, the latter asset modifies the dynamics of the outcomes "change in trust and social capital" and "change in active citizenship" through "the information sharing ratio." In addition, the change in collaboration projects is driven by "the joint initiatives in policy design ratio," which is an aggregated measure of the impact of active citizenship, regulations for a participatory governance of the commons, and collaboration pacts. In the same line, the implementation of collaboration projects is measured through the performance driver "joint initiatives in policy implementation ratio," which relies on the strategic resources: public bureaucracy promptness and public administration transparency. An improvement in the previous ratio will lead to a positive change in neighborhood attractiveness, which will increase the level of community wellness, trust and social capital, and active citizenship (Bianchi 2021).

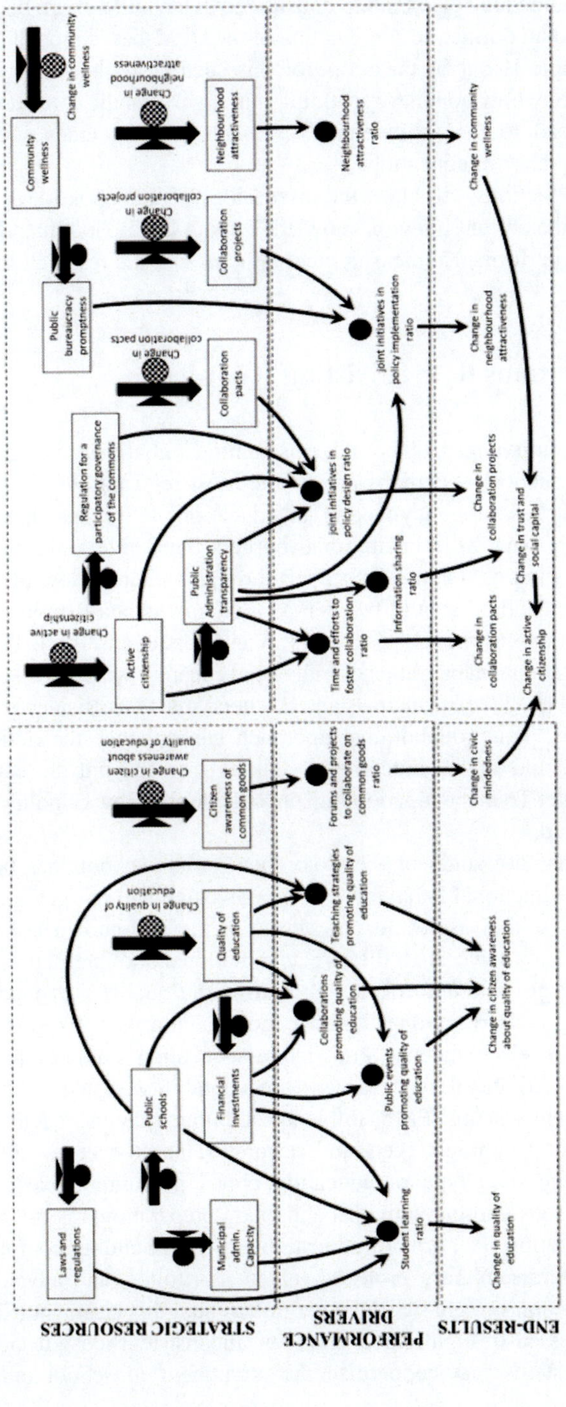

Fig. 5.2 Using dynamic performance management to support an "outside-in" view of policymaking in education (Adapted from Bianchi 2021, p.347)

For clarity, the effects of time delays associated with administrative procedures and capacity building are not explicitly considered in the DPM chart, but they should be taken into account during the construction of the DPM-based model. Nonetheless, this modeling topic is out of the scope of this section as the major goal was to qualitatively portray how boundary spanning initiatives—in the context of the public schools—may lead to the achievement of local area outcomes such as active citizenship and quality of education.

The next two sections will close the discussions of this book by explaining the contributions to the current body of knowledge about behavioral distortions associated with school performance measurement systems and the research limitations.

5.5 Contributions to the Existing Knowledge

Recognizing the importance of boundary spanning initiatives to improve performance in public schools is perhaps one of the most relevant contributions of this book for the practice of policymaking in education. It implies that the role of connecting people out of the institutional borders but inside the governance networks is crucial. Finding a common ground through a shared view of the problem and mutual understanding should be seen as a complementary strategy to internal re-organizations (Van Meerkerk et al. 2013; Van Hulst et al. 2012). Therefore, the investment on relational capabilities of individuals dealing with cross-sectoral problems at the school level (i.e., principals and teachers) must be considered a priority in government plans. This collaborative approach can enhance the delivery of the education service and school outcomes by taking into account the context where institutions operate. To this end, adopting an "outside-in" view of policymaking has been recommended.

In addition, the case study of the Colombian public schools has been used to illustrate how dysfunctional behaviors can be prevented and mitigated by outlining outcome-oriented school assessment systems. In the Colombian context, school performance has been lower than the targets in national and international standardized tests despite government efforts to improve the quality of education. It has caused a growing concern among the educational authorities. Experts in the field assert that a low level in students' knowledge acquisition is related to unintended behavioral effects triggered by test-based accountability systems. These performance measurement systems (PMS) influence on public schools' teaching strategies for the purpose of attaining targets and rewards (Hamilton et al. 2002). Hence, traditional teaching (i.e., type of education aimed at building basic notions in students through rote learning activities and exam preparations) is often prioritized over holistic education (i.e., type of education aimed at building complex learning skills through interdisciplinary projects, student portfolios, and advisories). As a result, school outcomes might be in danger due to goal displacement (Sills 1957).

This book has also highlighted how an inconsistent design of test-based accountability systems may jeopardize the attainment of school outcomes. An outcome-based view in performance management has been proposed to challenge the consistency of the adopted indicators and deal with behavioral distortions. Concerning this, the case study has shown how both an excessive government pressure for improving performance in high-stake tests and the adoption of output-oriented reward systems have provoked the emergence of gaming behaviors in the Colombian public schools (e.g., narrowing the curricula). In particular, an increase in the time allocated to traditional teaching to the detriment of a holistic education has been observed. This phenomenon is known as the *attribution problem*, and it hinders the implementation of an outcome-based view in performance management due to inconsistencies on framing the perceived causality between the adopted policies and their impacts on the final outcomes (Mayne 2007; Fasiello et al. 2021; Bovaird 2014).

Therefore, in order to mitigate the attribution problem and avoid the risk of data manipulation, an extension of the school boundaries, the involvement of different stakeholders, and the support of a transparent performance dialogue become central (Moynihan 2005; Bianchi and Williams 2015). Dynamic performance management (DPM) has contributed to the previous challenge by framing how policy levers and performance drivers impact on outputs and outcomes, and how the latter measures feedback into strategic resources. Likewise, this approach has been used to set the basis for policy discussion by filtering the insights from modeling and simulation through the lenses of decision-makers in public schools.

More specifically, concepts and experiences from the fields of performance management and education have been integrated by using *the instrumental view of DPM* (Bianchi 2016). This view enables the identification of strategic resources and performance measures (i.e., performance drivers and end results) to outline feedback mechanisms, which describe unintended behavioral effects. Likewise, efforts have been devoted to clarifying how an inconsistent design of test-based accountability systems may induce schoolteachers to game the system. A DPM-based model has been used to explain the previous phenomenon and provide policy recommendations on how to enhance school assessment systems through an outcome-oriented perspective. These policy recommendations have been intended to (1) mitigate behavioral distortions that may emerge as a response to both government pressure and myopic compensation schemes and (2) enhance collaboration among distinct stakeholders in the education sector. This implies that *common goods* should be identified and properly managed, with the purpose of increasing the effectiveness of public policies. Therefore, a consistent alignment between organizational outputs and community outcomes in the school context is crucial to improve the quality of education.

Moreover, the design of control systems (Flamholtz 1983; Simons 1995; Ouchi 1979) and the theory of street-level bureaucracy (Lipsky 1980; Mintzberg 1979) have been discussed to set the foundations on how to outline outcome-oriented PMS through DPM. On the one hand, *from an institutional perspective*, this implies that DPM-based models should subtly embody the notions of diagnostic and interactive control systems, the organizational structure, the organizational culture, and the organizational environment. Likewise, such models should be sketched by taking into account the characteristics of professional bureaucracies and clan controls. As a result, these simulation models could be used to foster feedback analysis, strategic dialogue, and information sharing in organizations. On the other hand, *from an inter-institutional perspective,* the attainment of outcomes requires that a multi-level and multi-actor approach is implemented. Such an approach can be supported by means of an alignment between organizational outputs and community outcomes. It suggests that pursuing outcomes in the local area goes beyond the design and implementation of performance measures and policies in single institutions. DPM can be used as an outcome-oriented approach to support decision-makers and system designers in outlining consistent PMS that foster a common shared view.

In particular, this book has proposed to combine traditional and holistic education for enhancing the achievement of school outcomes. The adoption of this approach implies that (1) a proper development of basic notions in different knowledge areas is a prerequisite for building higher order thinking skills in school students and (2) the use of standardized tests is an adequate practice because it can support pedagogical processes. However, decision-makers should know the limitations of the tests to maximize their benefits and minimize the emergence of behavioral distortions. These dysfunctional behaviors occur due to inconsistencies in the design of PMS and are reinforced by allocating rewards mechanistically if performance targets are achieved. It produces an increase in outputs in the short term and a decrease in outcomes in the long term (Bianchi 2016).

In the school context, there is a high risk that teachers adjust their actions to meet targets, regardless of the ability of such results to positively affect school outcomes (Bianchi and Salazar Rua 2017, 2020). Benchmarking is a common practice to hold public schools accountable for performance; however, PMS can be manipulated to maximize individual pay-offs (Radnor 2005; Smith 1995). As Campbell (1979, p.85) stated "The more any quantitative social indicator is used for social decision-making, the more subject it will be to corruption pressures, and the more apt it will be to distort and corrupt the social pressures it is intended to monitor."

Therefore, test results should not be conceived as an absolute truth for policymaking but as a tool to enhance teaching strategies. The findings from this research also suggest that an increase in test scores does not imply an increase in educational quality. However, it would be expected that an increase in educational quality would lead to an increase in test scores.

To sum up, performance measurement systems (PMS) affect actions toward a direction that contributes to achieve relevant outcomes. Such systems are never neutral to individuals and organizations. In particular, an inconsistent design of PMS—regarding the system of rules in an organization (Borgonovi 1996) and the outcomes of decision-makers' actions—may generate behavioral distortions (Smith 1993; Bianchi and Williams 2015). In this book, dynamic performance management (DPM) has been proposed as an outcome-oriented approach to counteract such distortions and extend the system boundaries with the purpose of fostering cross-sectoral collaboration in the school context.

5.6 Research Limitations

The results of this investigation should not be made generalizable since a small sample size of schoolteachers was used to collect the primary data. Instead, this research should be seen as an initial step to analyze behavioral distortions in public schools through an outcome-oriented perspective in performance management. To this end, dynamic performance management (DPM) has been applied to the case study of the Colombian public schools. As a result, a DPM simulation model has been built to deal with dysfunctional behaviors caused by the adoption of inconsistent performance measures. Such DPM-based model could be enhanced by (1) collecting future data from the implementation of the current standardized test system in Colombia and (2) working in a multi-stakeholder setting to identify shared strategic resources and outline sustainable school policies through a collaborative perspective.

This implies the need for broadening the focus of performance evaluation from test scores to the assessment of school and community outcomes (e.g., educational quality, quality of life, social capital, civic-mindedness, social awareness, attractiveness of the local area, and trust in government). In this respect, the last chapter was intended to introduce the reader on how to link institutional and community performance by adopting an "outside-in" view of policymaking through DPM. However, building a simulation model was out of the scope of that section.

Therefore, the development of further DPM-based models and case studies that take into account the impact of boundary spanning initiatives on performance at the institutional and community level will be needed to extend the boundaries of this research. This may lead to the analysis of different policy contexts and the incorporation of other important variables that may affect school performance (e.g., dropout rates, focus of teaching efforts on the best performing students, and family involvement).

References

Agranoff R, McGuire M (2001) Big questions in public network management research. J Public Adm Res Theory 11(3):295–326

Aldrich HE (1979) Organizations and environments. Prentice-Hall, Englewood Cliffs, NJ

Aldrich HE, Herker D (1977) Boundary spanning roles and organization structure. Acad Manag Rev 2:217–230

Andersson R, Molina I (2003) Racialization and migration in urban segregation processes – Key issues for critical geographers. In: Simonsen K, Öhman J (eds) Voices from the North. Ashgate, Aldershot, pp 261–282

Ansell C, Gash A (2007) Collaborative governance in theory and practice. J Public Adm Res Theory 18(4):543–571

Ansell C, Torfing J (2016) Handbook on theories of governance. Edward Elgar, Cheltenham

Asheim B, Coenen L, Moodysson J, Vang J (2007) Constructing knowledge-based regional advantage: implications for regional innovation policy. Int J Entrep Innov Manag 7(2–5): 140–155

Atkinson AB (2019) Measuring poverty around the world. Princeton University Press

Bekkers V, Dijkstra G, Fenger M (2007) Legitimacy and democracy: a conceptual framework for assessing governance practices. In: Governance and the democratic deficit: assessing the democratic legitimacy of governance practices. Routledge, London

Bernelius V, Kauppinen TM (2012) School outcomes and neighbourhood effects: a new approach using data from Finland. In: Neighbourhood effects research: new perspectives. Springer, Dordrecht, pp 225–247

Bianchi C (2016) Dynamic performance management. Springer, Zurich

Bianchi C (2021) Fostering sustainable community outcomes through policy networks: a dynamic performance governance approach. In: Meek J (ed) Handbook of collaborative public management. Chapter 22. Edward Elgar, pp 333–356

Bianchi C (2022) Enhancing policy design and sustainable community outcomes through collaborative platforms based on a dynamic performance management and governance approach. In: Peters BG, Fontaine G (eds) Handbook of policy design. Chapter 24. Edward Elgar, pp 411–433

Bianchi C, Salazar Rua RS (2017) Applying Dynamic Performance Management to detect behavioral distortions associated with the use of formal performance measurement systems in public schools: the case of Colombia. Paper presented at the APPAM Conference, Chicago

Bianchi C, Salazar Rua RS (2020) A feedback view of behavioural distortions from perceived public service gaps at 'street-level' policy implementation: the case of unintended outcomes in public schools. Syst Res Behav Sci:1–22

Bianchi C, Vignieri V (2020) Dealing with "abnormal" business growth by leveraging local area common goods: an outside-in stakeholder collaboration perspective. Int J Product Perform Manag 70(3):1–22

Bianchi C, Williams DW (2015) Applying System Dynamics Modeling to foster a cause and effect perspective in dealing with behavioral distortions associated with a city's performance measurement programs. Public Perform Manag:395–425

Bidwell C (1965) The school as a formal organization. In: March JG (ed) Handbook of organizations. Rand McNally, Chicago

Blau PM (1963) The dynamics of bureaucracy. Wiley, New York

Bolman L, Deal T (1991) Reframing organizations. Jossey-Bass, San Francisco

Bommert B (2010) Collaborative innovation in the public sector. Int Public Manag Rev 11(1): 15–33

Borgonovi E (1996) Principi e sistemi aziendali per le Amministrazioni Pubbliche [Management Principles and Systems for Public Administrations]. Egea, Milano

Bovaird T (2014) Attributing outcomes to social policy interventions – 'gold standard' or 'fool's gold' in public policy and management? Soc Policy Adm 48(1):1–23

Bramley G, Karley NK (2007) Homeownership, poverty and educational achievement: school effects as neighbourhood effects. Hous Stud 22(5):693–721

Bryson J, Crosby B, Stone M (2006) The design and implementation of cross-sector collaborations: propositions from the literature. Public Adm Rev 66(1):44–55

Burns JM (1978) Leadership. Harper, New York

Campbell DT (1979) Assessing the impact of planned social change. Evaluation Prog Plann 2(1): 67–90

Carey JC, Harris B, Lee SM, Aluede O, (eds). (2017) International handbook for policy research on school-based counseling. Springer International Publishing

Cooper TL, Gulick L (1984) Citizenship and professionalism in public administration/response. Public Adm Rev 44:143–151

Crawford BA, Bell RL, Blair LM, Lederman NG (2003) Just do it? Impact of a science apprenticeship program on high school students' understandings of the nature of science and scientific inquiry. J Res Sci Teach 40(5):487–509

Crosby B, Bryson J, Stone M (2010) Leading across frontiers: how visionary leaders integrate people, processes, structures, and resources. In: Osborne SP (ed) The new public governance? Emerging perspectives on the theory and practice of public governance. Routledge, New York

Crowson RL, Boyd WL (1993) Coordinated services for children: designing arks for storms and seas unknown. Am J Educ 101(2):140–179

De Moor T (2013) De Herontdekking van het Collectief [The rediscovery of the collective]. Samenleving en Politiek 4:29–38

Doyle JK, Ford DN (1998) Mental models concepts for system dynamics research. Syst Dyn Rev 14(1):3–29

Dryzek JS (1990) Discursive democracy: politics, policy, and political science. Cambridge University Press

Dryzek JS (2010) Rhetoric in democracy: a systemic appreciation. Political Theory 38(3):319–339

Edelenbos J, Bressers N, Scholten P (2013) Water governance as connective capacity. Ashgate, London

Edelenbos J, Van Meerkerk I (2015) Connective capacity in water governance practices: the meaning of trust and boundary spanning for integrated performance. Curr Opin Environ Sustain 12:25–29

Edelenbos J, Van Schie N, Gerrits L (2010) Organizing interfaces between government institutions and interactive governance. Policy Sci 43(1):73–94

Emerson K, Nabatchi T (2015) Collaborative governance regimes. Georgetown University Press, Washington DC

Fasiello R, Rivenbark WC, Bianchi C (2021) Exploring the shift from output measures to outcome measures of service delivery: insights from municipal research. Public Perform Manag Rev 45(2):428–447

Finnigan K, O'Day J (2003) External support to schools on probation: getting a leg up? Center for Policy Research in Education, Philadelphia

Flamholtz EG (1983) Accounting, budgeting and control systems in their organizational context: theoretical and empirical perspectives. Acc Organ Soc 8(2–3):153–169

Friedrichs J, Galster G, Musterd S (2003) Neighborhood effects on social opportunities: the European and American research and policy context. Hous Stud 18(6):797–806

Fung A, Wright EO (2003) Deepening democracy: institutional innovations in empowered participatory governance. Verso, London

Galster GC (2008) Quantifying the effect of neighbourhood on individuals: challenges, alternative approaches, and promising directions. Schmollers Jahr 128(1):1–42

Galster GC (2012) The mechanism(s) of neighbourhood effects: theory, evidence, and policy implications. In: van Ham M, Manley D, Bailey N, Simpson L, Maclennan D (eds) Neighbourhood effects research: new perspectives. Springer, Dordrecht, pp 23–56

Gash V (2008) Bridge or trap? Temporary workers' transitions to unemployment and to the standard employment contract. Eur Sociol Rev 24(5):651–668

Goldring E (1985) The relationship between organizations and their indirect clients: a study of principals' perceptions of parents. PhD dissertation, University of Chicago

Gray B (1989) Collaborating: finding common ground for multiparty problems. Jossey-Bass, San Francisco

Hajer M, Wagenaar H (2003) eds. Cambridge University Press, Deliberative policy analysis: Understanding governance in the network society

Hallinger P, Murphy JF (1966) The social context of effective schools. Am J Educ 94:328–355

Hamilton LS, Stecher BM, Klein SP, National Science Foundation (U.S.) (eds) (2002) Making sense of test-based accountability in education. Rand, Santa Monica

Hannan MT, Freeman J (1977) The population ecology of organizations. Am J Sociol 82:930

Hardin G (1968) The tragedy of the commons. Science 162:1243–1248

Hernes T (2004) Studying composite boundaries: a framework of analysis. Hum Relat 57(1):9–29

Honig MI (2006) Street-level bureaucracy revisited: frontline district central-office administrators as boundary spanners in education policy implementation. Educ Eval Policy Anal 28(4): 357–383
Honig MI, Kahne J, McLaughlin MW (2001) School-community connections: expanding opportunities to learn and opportunities to teach. In: Richardson V (ed) Handbook of research on teaching, 4th edn. American Educational Research Association, Washington DC, pp 998–1028
Hoppe R (2011) The governance of problems: puzzling, powering and participation. Policy Press
Howes M, Tangney P, Reis K, Grant-Smith D, Heazle M, Bosomworth K, Burton P (2015) Towards networked governance: improving interagency communication and collaboration for disaster risk management and climate change adaptation in Australia. J Environ Plan Manag 58(5):757–776
Jakku-Sihvonen R, Kuusela J (2002) Mahdollisuuksien koulutuspolitiikan tasa–arvo [Equality of opportunities in the educational policies]. Oppimistulosten arvointeja 7, 2nd edn. Finnish National Board of Education, Helsinki
Jemison DB (1984) The importance of boundary spanning roles in strategic decision-making. J Manag Stud 21(2):131–152
Kania J, Kramer M (2011) Collective impact. Stanf Soc Innov Rev 9(1):36–41
Kanter R (1983) The change musters. Simon & Schuster, New York
Kauppinen TM (2007) Neighborhood effects in a European city: secondary education of young people in Helsinki. Soc Sci Res 36(1):421–444
Klijn E (2008) Governance and governance networks in Europe. Public Manag Rev 10(4):505–525
Klijn E, Koppenjan J (2000) Public management and policy networks: the theoretical foundation of the network approach to governance. Public Management 2:135–158
Klijn EH, Steijn AJ, Edelenbos J (2010) The impact of network management strategies on the outcomes in governance networks. Public Adm 88(4):1063–1082
Koppenjan J, Klijn EH (2004) Managing uncertainties in networks. Routledge, London
Kupari P (2005) Kotitausta näkyy matematiikan oppimistuloksissa [Social background is reflected to PISA outcomes in maths]. In: Kupari P, Välijärvi J (eds) Osaaminen kestävällä pohjalla – PISA 2003 Suomessa. Jyväskylä University Press, Jyväskylä, pp 115–127
Kuusela J, Etelälahti A, Hagman Å, Hievanen R et al (2008) Maahanmuuttajaoppilaat ja koulutus – tutkimus oppimistuloksista, koulutusvalinnoista ja työllistämisestä [Immigrant pupils and education – A study of educational outcomes, school choice and employment]. National Board of Education, Helsinki
Laegreid P, Rykkja LH (2014) Governance for complexity: how to organize for the handling of "wicked issues"? Paper presented at the Policy and Politics Conference "The challenges of leadership and collaboration in the 21st Century", Bristol, September
Laumann E, Galaskiewicz J, Marsden P (1978) Community structure as interorganizational linkages. Annu Rev Sociol 4:455–484
Lawson HA (1998) Collaborative educational leadership for 21st century school communities. In: Veen Dv, Day C, Walraven G (eds) Multi-service schools: integrated services for children and youth at risk. Garant Publishers, Leuven/Apeldoorn, The Netherlands, pp 173–193
Leifer R, Delbecq A (1978) Organizational/environmental interchange: a model of boundary spanning activity. Acad Manag Rev 3(1):40–50
Leifer R, Huber GP (1977) Relations among perceived environmental uncertainty, organization structure, and boundary-spanning behavior. Adm Sci Q 22:235–247
Levina N, Vaast E (2005) The emergence of boundary spanning competence in practice: implications for implementation and use of information systems. MIS Q 29(2):335–363
Liphman JM, Hoeh JA (1974) The principalship: foundations and functions. Harper & Row, New York
Lipsky M (1980) Street-level bureaucracy: dilemmas of the individual in public services. Russell Sage Foundation, New York
Lipsky M (1984) Bureaucratic disentitlement in social welfare programs. Soc Serv Rev 58(1):3–27

Malen B, Ogawa R, Kranz K (1990) What do we know about school based management? A case study of the literature—A call for research. In: Clune W, Witte J (eds) Choice and control in American education. Decentralization and school restructuring. Falmer Press, Philadelphia

Marien S, Hooghe M, Quintelier E (2010) Inequalities in noninstitutionalized forms of political participation: a multi-level analysis of 25 countries. Political Stud 58(1):187–213

Mayne J (2007) Challenges and lessons in implementing results-based management. Evaluation 13(1):87–109

Mintzberg H (1979) The structuring of organizations. A synthesis of the research. Prentice Hall, Englewood Cliffs, NJ

Moll LC, Amanti C, Gonzalez N (2005) Funds of knowledge: theorizing practices in households and classrooms. Erlbaum, Mahwah

Morgan G (1997) Images of organisation. Sage, Thousand Oaks

Morris VC, Crowson R, Porter-Gehrie C, Hurwitz E (1984) Principals in action. Charles E. Merrill publishing, Columbus

Moynihan D (2005) Goal-based learning and the future of performance management. Public Adm Rev 65(2):203–216

Murphy J, Elliott SN, Goldring EB, Porter AC (2006) Learning-centered leadership: a conceptual foundation. The Wallace Foundation, New York

Murphy J, Elliott SN, Goldring EB, Porter AC (2007) Leadership for learning: a research-based model and taxonomy of behaviors. School Leadership Manag 27(2):179–201

Musterd S, Ostendorf W (2005) Social exclusion, segregation and neighborhood effects. In: Kazepov Y (ed) Cities of Europe: changing contexts, local arrangements, and the challenge to urban cohesion. Blackwell, Oxford

Newman J, Barnes M, Sullivan H, Knops A (2004) Public participation and collaborative governance. J Soc Policy 33(2):203–223

Nieuwenhuis J, Hooimeijer P (2016) The association between neighbourhoods and educational achievement, a systematic review and meta-analysis. J Housing Built Environ 31(2):321–347

Nieuwenhuis J, Xu J (2021) Residential segregation and unequal access to schools. Social Inclusion 9(2):142–153

Ostrom E (1998) A behavioral approach to the rational choice theory of collective action: presidential address. American Political Science Association, 1997. Am Polit Sci Rev 92(1):1–22

Otley D (1999) Performance management: a framework for management control systems research. Manag Account Res 10(10):363–382

Ouchi WG (1979) A conceptual framework for the design of organizational control mechanisms. Manag Sci 25:833–848

Pittman K, Cahill M (1992) Pushing the boundaries of education: the implications of a youth development approach to education policies, structures, and collaborations. In: Council of Chief State School Officers (ed) Ensuring student success through collaboration: summer institute papers and recommendations of the Council of Chief State School Officers. Washington DC

Provan KG, Kenis P (2008) Modes of network governance: structure, management, and effectiveness. J Public Adm Res Theory 18(2):229–252

Radnor Z (2005) Developing a typology of organisational gaming. Presentation at the EGPA conference, Bern

Raywid MA, Schmerler G (2003) Not so easy going: the policy environments of small urban schools and schools-within-schools. ERIC Clearinghouse on Rural Education and Small Schools, Charleston, WV

Rhodes CJ (2016) The 2015 Paris climate change conference: COP21. Sci Prog 99(1):97–104

Rhodes R (2017) Network governance and the differentiated polity: selected essays, vol I. Oxford University Press

Saxe RW (1975) School-community interactions. McCutchan, Berkeley

Scott WR (1981) Organizations: rational, natural and open systems. Prentice-Hall, Englewood Cliffs, NJ

Selsky J (1993) An exploration of developmental dynamics in maturing nonprofit federations. Working Paper 93/01. Management Department, University of Otago
Selsky JW, Smith AE (1994) Community entrepreneurship: a framework for social change leadership. Leadersh Q 5(3–4):277–296
Shulman LS (1983) Autonomy and obligation: the remote control of teaching. In: Shulman LS, Sykes G (eds) Handbook of teaching and policy. Longman, New York, pp 484–504
Sills D (1957) The volunteers: means and ends in a national organization. Free Press, Glencoe, IL
Simon HA (1996) The sciences of the artificial, 3rd edn. MIT Press, Cambridge
Simons R (1995) Levers of control: how managers use innovative control systems to drive strategic renewal. Harvard Business School Press, Boston
Smircich L, Stubbart C (1985) Strategy making in an enacted world. Acad Manag Rev 15(4): 724–736
Smith P (1993) Outcome-related performance indicators and organizational control in the public sector. Br J Manag 4:131–151
Smith P (1995) On the unintended consequences of publishing performance data in the public sector. Int J Public Adm 18(2–3):277–310
Sørensen E, Hendriks CM, Hertting N, Edelenbos J (2020) Political boundary spanning: politicians at the interface between collaborative governance and representative democracy. Polic Soc 39(4):530–569
Sørensen E, Torfing J (2009) Making governance networks effective and democratic through metagovernance. Public Adm 87(2):234–258
Steadman HJ (1992) Boundary spanners: a key component for the effective interactions of the justice and mental health systems. Law Hum Behav 16(1):75–87
Sterman J (2000) Business dynamics: systems thinking and modeling for a complex world. Irwin/ McGraw-Hill
Sullivan H, Skelcher C (2002) Working across boundaries: collaboration in public services. Palgrave/Macmillan, New York
Teisman GR (2000) Models for research into decision-making processes: on phases. Streams and decision-making rounds. Public Adm 78(4):937–956
Thompson JD (1967) Organizations in action. McGraw-Hill, New York
Torfing J (2016) Collaborative innovation in the public sector. Georgetown University Press, Washington DC
Trist E (1983) Referent organizations and the development of interorganizational domains. Hum Relat 36(3):269–284
Turner HE (1983) The department head: an untapped source of instructional leadership. NASSP Bull 67(464):25–28
Turner M (2015) A place-conscious approach can strengthen integrated strategies in poor neighborhoods. The Brookings Institution, Washington DC
Tushman ML (1977) Special boundary roles in the innovation process. Adm Sci Q 22(4):587–605
Tushman ML, Scanlan TJ (1981) Boundary spanning individuals: their role in information transfer and their antecedents. Acad Manag J 24(2):289–305
United Nations (2022) The sustainable development goals report. UN DESA. https://unstats.un.org/sdgs/report/2022/
Van de Wijdeven T (2012) Doe-democratie: Over Actief Burgerschap in Stadswijken [Do-Democracy: About Active Citizenship in Urban Districts]. PhD dissertation, Tilburg University
Van Hulst M, De Graaf L, Van den Brink G (2012) The work of exemplary practitioners in neighborhood governance. Crit Policy Stud 6(4):434–451
Van Meerkerk I (2014) Boundary spanning in governance networks: a study about the role of boundary spanners and their effects on democratic throughput legitimacy and performance of governance networks. Erasmus University Rotterdam, Rotterdam
Van Meerkerk I, Boonstra B, Edelenbos J (2013) Self-organization in urban regeneration: a two-case comparative research. Eur Plan Stud 21(10):1630–1652

Van Meerkerk I, Edelenbos J (2016) Complementary boundary-spanning leadership: making civic-induced interactive governance work. In: Edelenbos J (ed) Critical reflections on interactive governance. Edward Elgar

Van Meerkerk I, Edelenbos J (2017) Facilitating conditions for boundary-spanning behaviour in governance networks. Public Manag Rev 20(4):503–524

Wacquant L (2008) Urban outcasts. A comparative sociology of advanced marginality. Polity Press, Cambridge

Wagenaar H, Van der Heijden J (2015) The promise of democracy? Civic enterprise, localism and the transformation of democratic capitalism. In: Davoudi S, Madanipour A (eds) Reconsidering localism. Routledge, London

Walker EM (2002) The politics of school-based management: understanding the process of devolving authority in urban school districts. Educ Policy Anal Archiv 10(33)

Warren ME (2009) Governance-driven democratization. Crit Policy Stud 3:3–13

Westley F (1991) Bob Geldof and live aid: the affective side of global social innovation. Hum Relat 44(10):1011–1036

Williams P (2002) The competent boundary spanner. Public Adm 80(1):103–124

Woolcott HF (1973) The man in the Principal's office. Holt, Rinehart and Winston, New York

Xavier J, Bianchi C (2019) An outcome-based dynamic performance management approach to collaborative governance in crime control: insights from Malaysia. J Manag Gov 24:1089–1114

Appendices

R. Salazar Rua, *Improving Consistency in Performance Measurement System Design*, System Dynamics for Performance Management & Governance 7,
https://doi.org/10.1007/978-3-031-24699-9

Appendix A: Colombian Reward System: Incentives for Teachers and Public Schools

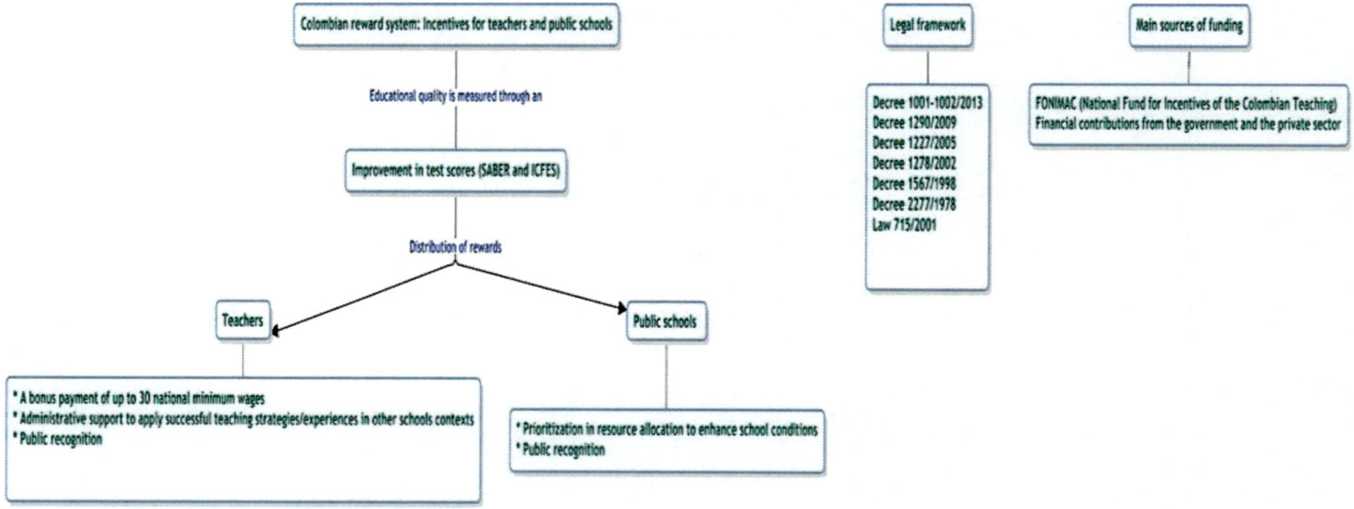

Appendix B: First Questionnaire

The purpose of this questionnaire is to determine how stakeholders in education impact on educational quality. To this end, the case of the Colombian public schools is analyzed through the lenses of Dynamic Performance Management (DPM).

In order to collect the primary data of this research, this questionnaire, based on open-ended questions, was designed. Participants are requested to analyze each of the questions carefully.

The information provided by the participants will be used for research purposes in academic publications. Participants are free to write their names, phone number and e-mail to be interviewed. The interviews will be recorded if the participants agree with it.

Thanks for your participation.

Participant details

Name:
Current job position:
Years of work experience in the education sector:
Phone number/e-mail for the interview:
Date:

Questions

1. How do you define the concept *"Educational quality"?*
2. What are your duties and responsibilities to improve student achievement in public schools?
3. What should the role of stakeholders (i.e., schools, government, families, enterprises) be to improve educational quality? Do you think a gap between role expectations and role performance exists in the Colombian education system? Why?
4. Do you think schoolteachers' role is properly valued in the Colombian society? If the answer is negative, do you think an undervalued perception of teaching activities may impact on educational quality? Why?
5. What is your opinion about using standardized tests to measure educational quality? Do you think it is an appropriate approach? If the answer is negative, what do you suggest for tracking student progress and school performance?
6. Which strategies have you used to improve student performance in the standardized tests? Do you think these strategies contribute to enhance the achievement of school outcomes? Have you ever observed unintended results of implementing such strategies in the long-term?
7. In the Colombian context, what are the main outcomes that might be attained if school performance is enhanced through a consistent design of education policies?

Appendix C: Second Questionnaire

This questionnaire has a dual purpose. In the first instance, it is aimed at understanding how standardized tests impact on the behavior of Colombian schoolteachers. In the second instance, it is aimed at validating the structure of a Dynamic Performance Management (DPM)-based model, which has been built to explain and counteract the negative effects of school practices aimed at increasing test scores.

In order to collect the primary data of this research, this questionnaire, based on open-ended questions and numerical reasoning, was designed. Participants are requested to analyze each of the questions carefully.

The information provided by the participants will be used for research purposes in academic publications. Participants are free to write their names, phone number and e-mail to be interviewed. The interviews will be recorded if the participants agree with it.

Thanks for your participation.

Participant details

Name:
Current job position:
Years of work experience in the education sector:
Phone number/e-mail for the interview:
Date:

Questions

1. What kind of education do you consider the most appropriate to build higher order thinking skills (e.g., critical thinking, problem-solving, and teamwork aptitudes)? Why?
2. Based on your experience, how long does it take a student to gain an adequate level of higher order thinking skills?
3. In an "ideal" school, how much time of the school year should be devoted to activities that look for developing higher order thinking skills? Why?
4. Do you think the current standardized tests are aligned with an education that fosters the development of higher order thinking skills? Why?
5. How much time of the school year does your school allocate to exam preparations? Is it possible to increase/decrease such time allocation under certain circumstances? If the answer is positive, could you quantify it?
6. Do you think school strategies intended to develop higher order thinking skills may impact on test scores? How?
7. Do you think resource investments to improve performance in standardized tests can also enhance the development of higher order thinking skills? How?
8. Based on your experience, how much time does a public school take to improve its performance in standardized tests significantly?
9. How often does your school adjust its policies and strategies?

10. What are the effects of prioritizing traditional education (i.e., teaching aimed at building basic notions in students through rote learning activities and exam preparations) over holistic education (i.e., teaching aimed at building higher order thinking skills)?
11. Which of the following options do you consider the closest to reality? Why? (Note: Read the statement carefully, and underline the most appropriate option in italics)
    - The time to increase the number of lower knowledge students to higher knowledge students is *greater than/equal to/less than* the time to decrease the number of higher knowledge students to lower knowledge students.
12. Could you quantify the previous statement?
    - Example: I think the time to increase the number of lower knowledge students to higher knowledge students is five times *greater than/less than* the time to decrease the number of higher knowledge students to lower knowledge students.

Graphs

You are requested to plot X vs. Y graphs in this section, intending to determine the relationship between the variables (Note: In all the statements, the first variable corresponds to the X-axis and the second variable corresponds to the Y-axis).

(a) Students' knowledge ratio (Higher knowledge students over Target of higher knowledge students) and Desired time allocation to traditional teaching
(b) Percentage of time allocated to traditional teaching and its effect on the increase in higher knowledge students
(c) Percentage of time allocated to traditional teaching and its effect on the increase in lower knowledge students
(d) Percentage of time allocated to holistic education and its effect on the increase in higher knowledge students

Appendix D: Model Documentation

D.1: DPM-Based Model (Initial Version)

Stocks and Flows

Higher knowledge students (t) = Higher knowledge students (t - dt) + (Increase in higher knowledge students − Decrease in higher knowledge students) * dt INIT = 10 (students)

Inflow:

Increase in higher knowledge students = ('Lower knowledge students' / 'Increase time') * 'Rote effect'

Outflow:

Decrease in higher knowledge students = ('Higher knowledge students' / 'Decrease time') * 'Score inflation effect'

Lower knowledge students (t) = Lower knowledge students (t − dt) + (Decrease in higher knowledge students − Increase in higher knowledge students) * dt INIT = 40 (students)

Inflow:

Decrease in higher knowledge students = ('Higher knowledge students' / 'Decrease time') * 'Score inflation effect'

Outflow:

Increase in higher knowledge students = ('Lower knowledge students' / 'Increase time') * 'Rote effect'

Teaching time allocation to tested subjects (t) = Teaching time allocation to tested subjects (t − dt) + (Chg in teaching time allocation to tested subjects) * dt INIT = 0.25 (dimensionless)

Inflow and Outflow:

Chg in teaching time allocation to tested subjects = ('Desired teaching time allocation to tested subjects' – 'Teaching time allocation to tested subjects') / 'Adjustment time'

Auxiliaries and parameters

Net change in higher knowledge students = 'Increase in higher knowledge students' – 'Decrease in higher knowledge students'

Students knowledge ratio = 'Higher knowledge students' / 'Target of higher knowledge students'

Percentage of time allocated to tested subjects = 'Teaching time allocation to tested subjects' / 'Total school hours'

Balance and scope of learning ratio = 'Percentage of time allocated to tested subjects' / 'Benchmark for time allocation to tested subjects'

Score inflation effect = DELAYINF('Effect of balance and scope of learning ratio','Score inflation delay',2,0)

Rote effect = DELAYINF('Effect of time allocated to tested subjects','Rote delay',2,0)

Effect of balance and scope of learning ratio = GRAPH('Balance and scope of learning ratio',1,0.2,{1,1.007,1.052,1.225,1.509,1.861,2.213,2.610,2.865,2.985,3//Min:1; Max:3//}

Effect of time allocated to tested subjects = GRAPH('Percentage of time allocated to tested subjects',0.25,0.055,{1,1.022,1.086,1.206,1.367,1.558,1.708,1.839,1.929, 1.978,1.993//Min:1;Max:2//})

Desired teaching time allocation to tested subjects = GRAPH('Students knowledge ratio',0,0.0833, {0.8,0.798,0.792,0.784,0.749,0.703,0.633,0.499,0.367,0.304,0.271,0.254,0.25-0//Min:0.25;Max:0.8//})

Adjustment time = 12 (months)

Score inflation delay = 36 (months)

Rote delay = 12 (months)

Increase time = 24 (months)

Decrease time = 3 (months)

Target of higher knowledge students = 45 (students)

Benchmark for time allocation to tested subjects = 0.25 (dimensionless)

Total school hours = 1 (dimensionless)

D.2: DPM-Based Model (Extended Version)

Stocks and Flows

Higher knowledge students (t) = Higher knowledge students (t − dt) + (Increase in higher knowledge students − Decrease in higher knowledge students) * dt INIT = 10 (students)

Inflow:

Increase in higher knowledge students = ('Lower knowledge students' / 'Increase time') * 'Combined rote and learning effect'

Outflow:

Decrease in higher knowledge students = ('Higher knowledge students' / 'Decrease time') * 'Score inflation effect'

Lower knowledge students (t) = Lower knowledge students (t − dt) + (Decrease in higher knowledge students − Increase in higher knowledge students) * dt INIT = 40 (students)

Inflow:

Decrease in higher knowledge students = ('Higher knowledge students' / 'Decrease time') * 'Score inflation effect'

Outflow:

Increase in higher knowledge students = ('Lower knowledge students' / 'Increase time') * 'Combined rote and learning effect'

Time allocation to holistic education (t) = Time allocation to holistic education (t − dt) + (Chg in time allocation to holistic education) * dt INIT = 0.4 (dimensionless)

Inflow and Outflow:

Chg in time allocation to holistic education = 'Gap in time allocation to holistic education' / 'Adjustment time'

Time allocation to traditional teaching(t) = Time allocation to traditional teaching (t − dt) + (Chg in time allocation to traditional teaching) * dt INIT = 0.6 (dimensionless)

Inflow and outflow:

Chg in time allocation to traditional teaching = ('Desired time allocation to traditional teaching' – 'Time allocation to traditional teaching') / 'Adjustment time'

Auxiliaries and Parameters

Total school hours = 'Time allocation to holistic education' + 'Time allocation to traditional teaching'

Percentage of time allocated to holistic education = 'Time allocation to holistic education' / 'Standard school hours'

Percentage of time allocated to traditional teaching = 'Time allocation to traditional teaching' / 'Standard school hours'

Balance and scope of learning ratio = 'Percentage of time allocated to traditional teaching' / 'Benchmark for time allocation to traditional teaching'

Combined rote and learning effect = 'Rote learning effect' * 'Holistic learning effect'

Net change in higher knowledge students = 'Increase in higher knowledge students' – 'Decrease in higher knowledge students'

Gap in time allocation to holistic education = 'Desired time allocation to holistic education' – 'Time allocation to holistic education'

Students knowledge ratio = 'Higher knowledge students' / 'Target of higher knowledge students'

Score inflation effect = DELAYINF('Effect of balance and scope of learning ratio','Score inflation delay',2,0)

Rote learning effect = DELAYINF('Effect of time allocated to traditional teaching','Rote learning delay',2,0)

Holistic learning effect = DELAYINF('Effect of time allocated to holistic education', 'Holistic learning delay',2,0)

Effect of balance and scope of learning ratio = GRAPH('Balance and scope of learning ratio',1,0.2,{1,1.007,1.052,1.225,1.509,1.861,2.213,2.610,2.865,2.985,3//Min:1;Max:3//})

Effect of time allocated to traditional teaching = GRAPH('Percentage of time allocated to traditionalteaching',0.25,0.055,{1,1.022,1.086,1.206,1.367,1.558,1.708,1.839, 1.929,1.978, 1.993//Min:1;Max:2//})

Effect of time allocated to holistic education = GRAPH('Percentage of time allocated to holisticeducation',0.25,0.055,{0.2,0.6,1.2,1.9,3.13,4.54,6.34,8.1,9.12,9.58,9.96//Min:0;Max: 10//})

Desired time allocation to traditional teaching = GRAPH('Students knowledge ratio',0,0.0833,{0.8,0.798,0.792,0.784,0.749,0.703,0.633,0.499,0.367,0.304,0.271, 0.254,0.250//Min:0.25;Max:0.8//})

Adjustment time = 12 (months)

Score inflation delay = 36 (months)

Rote learning delay = 12 (months)

Holistic learning delay = 24 (months)

Increase time = 24 (months)

Decrease time = 3 (months)

Target of higher knowledge students = 30 (students)

Desired time allocation to holistic education = 0.8 (dimensionless)

Standard school hours = 1 (dimensionless)

Benchmark for time allocation to traditional teaching = 0.25 (dimensionless)